576.15 M72h
Modeling the environmental
fate of microorganisms

WITHDRAWN

DATE DUE

#47-0108 Peel Off Pressure Sensitive

MODELING THE ENVIRONMENTAL FATE OF MICROORGANISMS

MODELING
THE ENVIRONMENTAL FATE OF
MICROORGANISMS

Edited by
CHRISTON J. HURST
U.S. Environmental Protection Agency
Cincinnati, Ohio

STAFFORD LIBRARY
COLUMBIA COLLEGE
1001 ROGERS STREET
COLUMBIA, MO 65216

AMERICAN SOCIETY FOR MICROBIOLOGY
Washington, DC

Copyright © 1991 American Society for Microbiology
1325 Massachusetts Avenue, N.W.
Washington, DC 20005

Library of Congress Cataloging-in-Publication Data

Modeling the environmental fate of microorganisms / edited by Christon J. Hurst.
    p.   cm.
Includes index.
ISBN 1-55581-031-4
    1. Microbial ecology—Mathematical models.   I. Hurst, Christon J.
QR100.M59   1991
576'.15'015118—dc20
                                                  90-20775
                                                    CIP

All Rights Reserved
Printed in the United States of America

To my children Rachel and Allen

Contents

Preface   ix

I. What, How, and Why?

1. Background and Practical Applications of Microbial Ecology
   David M. Updegraff   1

II. Fate in the Subsurface World

2. Problems with Using Existing Transport Models To Describe Microbial Transport in Porous Media
   Ruth A. Dickinson   21
3. Modeling Microbial Transport in the Subsurface: A Mathematical Discussion
   Marylynn V. Yates and S. R. Yates   48
4. Quantitation of Factors Controlling Viral and Bacterial Transport in the Subsurface
   Charles P. Gerba, Marylynn V. Yates, and S. R. Yates   77
5. Parameters Involved in Modeling Movement of Bacteria in Groundwater
   Ronald W. Harvey   89
6. Use of Models To Predict Bacterial Penetration and Movement within a Subsurface Matrix
   Michael J. McInerney   115

III. Fate in the Surface World

7. Using Linear and Polynomial Models To Examine the Environmental Stability of Viruses
   Christon J. Hurst   137
8. Development of Models To Explain the Survival of Viruses and Bacteria in Aerosols
   Alan Jeff Mohr   160
9. Models for the Survival of Bacteria Applied to the Foliage of Crop Plants
   Guy R. Knudsen   191

IV. Disinfection

10. Virus Inactivation by Disinfectants
    James M. Vaughn and James F. Novotny   217

11. Model of *Giardia lamblia* Inactivation by Free Chlorine
    Robert M. Clark   242

V. Biofilms

12. Background and Models for Bacterial Biofilm Formation and Function in Water Distribution Systems
    Betty H. Olson, Richard McCleary, and James Meeker   255

Index   287

Preface

Microorganisms are introduced into the environment through a variety of processes. These include the release of pathogens during disposal of septic wastes and sanitary sewage, following which the pathogens may complete cycles of environmentally related transmission to cause human illnesses associated with recreational water activities or the consumption of contaminated food and water. A wide variety of microorganisms may be released into the environment either accidentally or deliberately during the production, handling, and use of specialty microbiological products. Such products include genetically selected or genetically manipulated organisms that may be used in processes for treating hazardous or noxious wastes. Microorganisms may also be deployed as biological antagonists, intended to ecologically compete against or replace indigenous microflora in environmental settings. Many types of microorganisms, including viruses, are commonly released into the environment as biological pesticides.

This book addresses the application of statistical modeling to studies on the ability of microorganisms to persist in and potentially be transported through the environment following their deliberate or accidental release. The book additionally covers the use of chemical disinfectants to minimize waterborne disease hazards, and the subject of microbial biofilms within water distribution systems.

Accurate statistical models can serve both to help explain how a particular observed outcome occurred and to aid in predicting the results of related situations. The book includes a sampling of modeling applications from different topic areas and covers several classes of microorganisms. Many readers will find that their modeling interests and practical needs are already addressed in the book. Other readers whose modeling requirements for related organisms and topics are not specifically addressed may, nonetheless, find that the information provided comprises a firm basis from which they can move forward in their own endeavors.

I thank the many authors who generously contributed their time to preparing the chapters which appear in this book, and also Deanna K. Wild who served as the editor for my own chapter.

The United States Environmental Protection Agency was not involved with the editing of this book.

Christon J. Hurst

I. WHAT, HOW, AND WHY?

Chapter 1

Background and Practical Applications of Microbial Ecology

David M. Updegraff

The Place of Microorganisms in Nature 1
　Distribution of Microorganisms 2
　Evolution of Microorganisms 2
　Activities of Microorganisms 2
　Kinds of Microorganisms 3
　Origin of Life 4
Microbial Ecology 5
　Autecology and Synecology 5
　Methods in Microbial Ecology 5
　Interactions between Microorganisms and Other Organisms 8
Applied Microbial Ecology 11
　Human Pathogens 11
　Animal Pathogens 12
　Plant Pathogens 13
　Biological Pesticides 13
　Biogeochemical Cycling 13
　Biodeterioration 16
　Waste Treatment 17
　Drinking Water Treatment 18
　Microbial Leaching of Metals from Ores 18
　Microbial Enhanced Oil Recovery 18
　Pollution by Recalcitrant Chemicals 18
References 19

THE PLACE OF MICROORGANISMS IN NATURE

Why is it important to model the environmental fate of microorganisms? The answer to this question will lead us into a consideration of the relationship between an environment and its microbial inhabitants (microbial ecology) and the role of

---

David M. Updegraff • Department of Chemistry and Geochemistry, Colorado School of Mines, Golden, Colorado 80401.

these microscopic organisms in shaping the earth as we know it today (geomicrobiology).

Distribution of Microorganisms

Microorganisms, particularly bacteria, are found throughout the world in almost every conceivable environment, including some where no higher forms of life exist. Spores of bacteria are regularly recovered from the air high up in the atmosphere, and microorganisms are found throughout the oceans, freshwater lakes and rivers, and soils of the world. Bacteria are abundant and active in sediments at the greatest depths of the sea (>10,000 m) and in groundwater and petroleum reservoirs at depths of hundreds of meters or more. Bacteria are also found in extreme environments such as hot springs and geothermal vents beneath the sea (at temperatures up to 250°C) where no other forms of life exist (Baross and Deming, 1983). They grow vigorously in salt lakes, in lakes so acid (pH < 2) that no other organisms are found, or in highly alkaline lakes at pH values as high as 13 (Langworthy, 1978).

Aerobic bacteria grow well in environments containing oxygen, while anaerobic bacteria thrive in the complete absence of atmospheric oxygen. The total biomass of microorganisms on earth greatly exceeds the biomass contributed by all other forms of life. The principal producers of organic matter in the sea and in many freshwater lakes are microscopic algae, and the principal consumers are bacteria and fungi, which mineralize the organic matter in the great biogeochemical cycles of nature: the carbon, nitrogen, oxygen, phosphorus, sulfur, and iron cycles (Cloud, 1983).

Evolution of Microorganisms

Ancient stromatolite rocks tell us that microorganisms (bacteria) were active on earth more than 3.5 eons (3.5×10^9 years) ago (Schopf and Packer, 1987). The earth is only 4.6 eons old; thus bacteria have inhabited the earth for more than 76% of its life. By contrast, higher animals and plants have existed less than 400 million years.

Activities of Microorganisms

The biogeochemical activities of bacteria have profoundly modified the surface of the earth throughout geologic time and are continuing to do so today. At least one third of all chemical elements are recycled biologically. Living organisms are credited with an important role in the formation of rocks and mineral resources, limestone, oil shale, coal, petroleum, and natural gas. Sulfide ore deposits, from which we recover such valuable materials as gold, silver, copper, lead, zinc, molybdenum, and sulfur, owe their formation to the activities of sulfate-reducing bacteria in sediments (Trudinger, 1976). The oxygen in the air that we breathe was produced by the activities of photosynthetic bacteria and algae from about 2 eons ago to the present. On the negative side, many of our most serious diseases, as well as those of our domestic animals and plant crops, are caused by microorganisms.

Kinds of Microorganisms

Microorganisms are those forms of life which are so small that they are ordinarily visible only with the aid of a microscope. Modern biological taxonomy distinguishes four different kinds of microorganisms: bacteria, fungi, algae, and protozoa. A fifth entity, the viruses, are not considered living organisms at all, since they are not capable of self-replication, but live and multiply only within living cells of other organisms whose metabolic machinery is "fooled" into replicating the virus as well as their own nucleic acids and proteins. Viruses are noncellular (contain no cell membrane) and contain only a single type of nucleic acid (RNA or DNA) and a few proteins. All living organisms are cellular, having a cell membrane, and contain RNA, DNA, proteins, lipids, and carbohydrates.

Bacteria

Bacteria are the smallest and simplest of all living organisms. They have a very primitive cell structure devoid of an organized nucleus and are therefore called procaryotic organisms. They constitute one of the five kingdoms of living organisms, the *Monera* (Whittaker, 1969), also called Procaryotae (Krieg and Holt, 1984). All other organisms are eucaryotic, meaning that they have an organized nucleus containing DNA, RNA, and proteins, surrounded by a nuclear membrane. Eucaryotic organisms probably arose 1 to 2 eons ago by the entry of a small procaryotic organism into a larger procaryotic cell, with the smaller cell eventually evolving into a nucleus (Margulis, 1981). Bacteria are by far the most diverse metabolically of all living organisms. Some are anaerobic, growing only in the absence of oxygen, and some are aerobic. Some are photosynthetic autotrophs, deriving their energy from sunlight and their carbon from carbon dioxide; some are chemosynthetic autotrophs, deriving energy from the oxidation of an inorganic substance and carbon from carbon dioxide; still others are heterotrophs, deriving energy from the oxidation of an organic compound and carbon from the organic compound and carbon dioxide as well.

Fungi

The kingdom *Fungi*, another of the five kingdoms of living organisms, comprises heterotrophic organisms with a rigid cell wall, which assimilate their nutrients by diffusion through the cell wall and cell membrane. Some, such as the water molds, are microscopic, while others, such as mushrooms and puff balls, are quite large. Fungi are nearly all aerobic and are of great importance in mineralizing woody materials containing lignin in the soil.

Algae

The algae are photosynthetic autotrophs containing chlorophyll, the light-trapping pigment involved in photosynthesis. Again, some are very small, such as the diatoms, coccolithophores, and dinoflagellates, which are the principal producers of organic matter in the sea, while others, such as the brown algae (seaweeds), are very large. The microscopic algae are single-celled organisms belonging to the kingdom *Protista*, while the large, multicellular algae are members of the plant kingdom. The photosynthetic organisms, photosynthetic bacteria, algae, and plants, are the principal producers in nearly every ecosystem on earth, and all of the other organisms in these ecosystems depend on them for food.

Protozoa

The protozoa, single-celled organisms belonging to the kingdom *Protista*, clearly resemble primitive animals. They are heterotrophic, and most are active predators which can be seen swimming about capturing and swallowing prey such as bacteria, small algae, fungi, or other protozoa. Predation by protozoa may be an important factor in controlling bacterial populations in sewage sludge and other aquatic environments. Protozoa also cause some important diseases of humans and animals (malaria, amoebic dysentery, and giardiasis, for example).

Viruses

The viruses constitute a very different class of entities. They are noncellular (have no cell membrane), contain only a single nucleic acid (RNA or DNA) as their genetic material and protein, and are not capable of self-replication. Thus they may not qualify as living organisms at all. They may have been derived from portions of organism cells which had lost many of their characteristic biomolecules. Nevertheless, viruses behave somewhat like primitive bacteria and are considered to be microorganisms by many biologists. They are of extreme importance to humans since they cause many of our most serious infectious diseases as well as important diseases of domestic animals and plants which lead to enormous economic losses. Therefore it is vitally important to understand the factors involved in the transmission of viral diseases to the host organisms. Viruses also act as vectors to introduce foreign genetic material into other cells, and hence they are of great importance in studies of evolution and are of practical use in recombinant DNA technology and genetic engineering.

Origin of Life

Just how living organisms first developed has been the subject of intense speculation from the time of Charles Darwin to the present. Darwin seemed to favor heterotrophic origin from organic matter formed by chemical reactions in a primordial warm sea. Later, in the 1920s, Oparin and Haldane independently developed more detailed models for a heterotrophic origin. Recently, Wachtershauser (1988) developed a totally different model based on chemolithoautotrophy. In his model, the basic energy source for evolving life is proposed to be the exergonic reaction of hydrogen sulfide and ferrous iron to form pyrite (FeS_2). The positively charged pyrite crystals would then serve as a template upon which negatively charged biomolecules, such as phosphoric acid esters and carboxylic acids, could form. Wachtershauser presents compelling evidence from the structure of present-day biomolecules for his model. His suggested site for the origin of life is an environment of high temperature, hydrostatic pressure, and salinity where sulfide, ferrous iron, and dissolved carbon dioxide are present. Such an environment is found today in geothermal vents, located far beneath the sea in total darkness, where even today a unique food chain exists. The basic organic production is carried out by chemosynthetic autotrophic bacteria, which derive their energy from the oxidation of reduced sulfur compounds and their carbon from carbon dioxide (Karl et al., 1980).

MICROBIAL ECOLOGY

Autecology and Synecology

Microbial ecology is a very large subject, and only a brief summary can be presented here. For a thorough treatment, see the comprehensive treatise by Atlas and Bartha (1987). Investigations of microbial ecology may be pursued by two different approaches. Autecological studies are those involved with the behavior of individual populations, while synecological studies deal with interactions of populations of different kinds of organisms within an ecosystem. Most studies of infectious diseases of humans, animals, and crops are autecological studies involving the relationship between an individual pathogenic organism and its environment. Examples include survival and movement of viruses and bacteria in sewage systems, surface water, groundwater, and aerosols. Studies of indicator organisms, such as *Escherichia coli*, in sewage and receiving waters fall in this domain. Here the indicator organism may serve as a model organism, an indicator of fecal pollution. Epidemiological studies generally start with autecological studies of the habitats and life cycles of individual pathogenic organisms and the modes of transmission of these organisms to humans, domestic animals, or crops.

Habitat-oriented studies are generally synecological. Examples include studies of biogeochemical cycling (the carbon cycle, nitrogen cycle, sulfur cycle, etc.), microbial ecology of the rumen, microbial interactions in sewage, biodegradation of pollutants in soils and aquatic ecosystems, and the production of goods such as cheese, pickles, and fermented beverages. Studies on the killing of microorganisms by heat, radiation, or chemical treatment may be directed toward the elimination of all microorganisms present (synecological) or toward the control of a single species (autecological) such as in the treatment of a specific disease with an antibiotic.

Methods in Microbial Ecology

The study of microbial ecology has been driven by the development of suitable microbiological and biochemical methods.

Sampling

The sampling of a given environment for microbiological studies may be extremely simple, such as the collection of a blood sample by a syringe and needle, or very difficult and complex, such as the collection of water samples from geothermal vents deep below the surface of the sea. Indeed, the discovery of living bacteria in this environment could take place only after the development of deep-diving submarines capable of descending thousands of meters into the sea.

Once the sample is collected, there is a very large menu of microbiological and biochemical methods from which to choose in order to determine what kinds of microorganisms are present in the environment, how many of each kind, and what biological and chemical activities they are engaged in.

Isolation and identification of microorganisms

The isolation and identification of microorganisms usually require the separation of a single species from all others in the environment, followed by microscopic and biochemical studies. With large microorganisms such as certain algae and protozoa, this is often accomplished by direct selection of individual organisms under the microscope. With small organisms such as bacteria, the usual procedure is by plating on solid media, such as agar gels. The agar medium may be designed to permit the growth of as many kinds of organisms as possible (plate count agar), or it may be a selective medium designed to permit the growth of only a single kind of organism. Toxic or inhibitory chemicals may be included in the medium to prevent the growth of undesired organisms. Also, the medium may contain only a single source of carbon or energy for the selection of a single kind of microbe from the environment. For example, media containing only mineral salts, with bicarbonate as the only source of carbon, may be incubated under lights to permit the growth and isolation of algae, which derive their energy from light and their carbon from carbon dioxide. Likewise, media containing only inorganic salts with sulfur as the only energy source and bicarbonate as the carbon source may be used to cultivate *Thiobacillus thiooxidans*, which derives its energy from the oxidation of sulfur to sulfuric acid and its carbon from carbon dioxide. For the isolation of heterotrophic bacteria, organic substances must be added to the mineral salts medium. The addition of a single pure organic compound to the mineral salts medium permits the growth of only those organisms which can utilize the added organic carbon source as a source of energy. Carbon for growth is generally obtained from the organic compound, but some may also be obtained from carbon dioxide. This method constitutes the enrichment culture procedure, a powerful and important method for studies in microbial ecology.

Enumeration of microorganisms

Many ecological studies require enumeration of microorganisms. Procedures available include direct counting and viable counting methods. Direct counting is carried out under the microscope in a counting chamber or surface of known volume or area. Thus the numbers determined may be readily converted to the number of cells per unit volume of the original sample. A very useful procedure for direct counting involves staining the organisms with a dye such as acridine orange, which reacts with DNA and RNA. The dye fluoresces when exposed to UV light, and the cells thus become visible. The procedure, epifluorescent microscopy, distinguishes between microorganisms and particles of organic debris, but may not reliably distinguish living organisms from dead ones. Highly sophisticated methods have been developed using triphenyl-tetrazolium chloride dyes to enumerate microorganisms that are carrying out respiration; other methods, which enumerate bacteria that are carrying out DNA synthesis, involve adding tritiated thymidine to the cells, followed by autoradiography to detect radioactivity. Fluorescent-antibody techniques have been developed to enable the examination and enumeration of specific kinds of organisms in soils, cells, and tissues. The specimen is treated with an antibody which binds only to a specific organism, and the antibody is coupled to a fluorescent dye.

Two kinds of viable count procedures are available, the dilution count, or most-probable-number method, and the plate count. Each method has advantages and disadvantages, and both are very useful. There is no viable count procedure which will enumerate all kinds of microorganisms, or even all kinds of bacteria, and thus all are selective to some extent.

The dilution count is carried out by preparing serial dilutions (1/10, 1/100, 1/1,000, etc.) of the original sample in sterile water. Portions of each dilution are then transferred into 3 to 10 culture tubes containing the chosen culture medium. The medium may be one containing many nutrients, intended to cultivate as many kinds of bacteria as possible, or it may be so selective that it cultivates only a single type. The culture tubes are then incubated for a suitable period of time and examined for the presence of microorganisms. Those showing growth of microorganisms are then recorded as positive, and the most probable number of organisms is determined by looking in a statistical table based on the Poisson distribution.

The plate count procedure is carried out by transferring portions of each serial dilution to a petri dish of an agar gel medium. Two variations of the method are available, the spread plate method and the pour plate method. In the spread plate method a small amount, such as 0.1 ml of each dilution, is spread over the surface of the agar gel medium. In the pour plate method an appropriate volume of dilution is placed in a sterile petri dish and liquid agar medium is poured into the dish, where the medium then solidifies. Again, the plates are incubated for a suitable period of time, and the colonies of microbes are counted and reported as colony-forming units per milliliter of original sample. This may not be the same as the number of microbial cells, since a colony-forming unit may contain more than a single cell.

Biochemical methods

Since all living organisms utilize some of the same basic biochemical mechanisms, analyses for certain specific biochemicals may serve as a measure of microbial biomass in a system. Biomass is correlated with cell counts, provided that microbial size is considered. Thus ATP, which is a universal energy storage and energy transfer material, can be a good index of total microbial biomass. DNA, RNA, or total protein can also be used as a biomass indicator in some systems. Chlorophyll can be determined as an index of the biomass of photosynthetic organisms. Respiration rate can also be measured, by determining either oxygen consumption or carbon dioxide production, and is related to the biomass of aerobic microorganisms present. Anaerobic respiration or fermentation usually produces some carbon dioxide, and thus carbon dioxide formation in an anaerobic system may correlate with the biomass of anaerobic microorganisms. Both respiration and photosynthesis may be measured very sensitively and accurately by using radioactive ^{14}C compounds. Respiration may be measured by the formation of $^{14}CO_2$ from a ^{14}C-labeled substrate such as [U-^{14}C]glucose. Photosynthesis may be measured by the rate of consumption of $^{14}CO_2$. Stable isotopes such as deuterium (^2H), ^{15}N, or ^{18}O can also be used as tracers of biochemical reactions. For example, by the use of $H_2^{18}O$, it was shown that the initial attack on quinoline by both aerobic and anaerobic bacteria, in which the aromatic ring is hydroxylated, is carried out by a hydroxylase, not an oxygenase, since the ring substituent is an ^{18}OH rather than a ^{16}OH group (Pereira et al., 1988).

Respiration is a measure of overall energy metabolism, and photosynthesis is a measure of organic production. These general metabolic processes require the coordinated activity of dozens of enzymes, and there are many specific assays useful in microbial ecology that focus on these enzymes. These include evaluations of proteinases by measuring the hydrolysis of gelatin or some other protein; of cellulase by measuring the formation of cellobiose and glucose from cellulose; of nitrogenase by measuring the conversion of acetylene to ethylene (the enzyme carries out this reaction as well as its major role in converting nitrogen to ammonia); of phosphatase by measuring the hydrolysis of phosphoric acid esters to P_i; and of chitinase by measuring the conversion of chitin to water-soluble products. The rate of sulfate reduction in sediments, an important biogeochemical reaction, may be determined by measuring the rate of conversion of radioactive $^{35}SO_4^{2-}$ to $H_2^{35}S$. There are many other conversions of biogeochemical importance which can be measured by similar specific chemical reactions.

Interactions between Microorganisms and Other Organisms

Habitats containing only a single kind of microorganism are found only in the laboratory. Natural habitats contain many kinds of organisms which interact in complex ways. In mixed cultures, microorganisms in the population interact by both positive (increased growth rate) and negative (decreased growth rate) means. In very low populations there is little or no interaction. At a somewhat higher population, positive interactions may predominate, and the growth rate may increase. At maximum population, negative interactions (nutrient exhaustion, production of toxic metabolites) begin to predominate over positive interactions. These interactions are responsible for the normal bell-shaped growth curve of population versus time.

In natural ecosystems, populations of different organisms interact in many ways. Sometimes both populations are unaffected by the interaction (neutralism). In other cases one population may benefit, while the second population is unaffected (commensalism), or both populations may benefit (mutualism, symbiosis, synergism), or one population may benefit while the other is negatively affected (parasitism, predation). Finally, both populations may be negatively affected (competition).

Neutralism

Neutralism is the rule in environments where both populations are inactive (not multiplying), such as in the atmosphere or in a frozen environment. Environments with very low population densities, such as oligotrophic lakes or the ocean far from shore, favor neutralism because the low population densities allow for different spatial and temporal niches.

Commensalism

Commensalism, where one population benefits and the other is unaffected, is quite common in microbial ecosystems. An example often observed in eutrophic waters, soils, and sediments is the interrelationship between aerobic and anaerobic microbes. The aerobes grow and consume all of the oxygen, thus permitting the obligate anaerobes to grow. Thus the anaerobes are benefited by the interaction, but the aerobes are not affected. A population of algae growing in an anoxic habitat might produce oxygen, thus enabling aerobic bacteria to grow. The production of growth factors by one population may trigger the growth of a second population. In certain marine and freshwater habitats, the production of biotin or vitamin B_{12} by bacteria leads to blooms of certain algae. An important reaction in soils is the hydrolysis by fungi of cellulose to glucose which serves as a nutrient source for many kinds of bacteria. Two populations of bacteria may be in a competitive relationship because both need to utilize the glucose. Bacteria which grow on the surface of other organisms, such as epiphytic bacteria growing on algae and the nonpathogenic organisms which inhabit human skin, are commensal with the host organism.

Synergism

Synergism, sometimes called protocooperation, is a relationship between two microbial populations in which both populations are capable of surviving on their own, but their association offers benefits to both. In many cases synergism is due to syntrophism, the interaction of two populations of microorganisms which supply each other's nutritional needs. The nutrients supplied may be either organic or inorganic energy sources or growth factors such as vitamins or amino acids. For example, both sulfur-oxidizing bacteria (*Thiobacillus* species) and sulfate-reducing bacteria grow in the redox gradient zone in soils and sediments, where oxygen diffusing in from above enables the growth of the aerobic *Thiobacillus*, permitting this chemosynthetic autotroph to produce organic materials from carbon dioxide and sulfate from hydrogen sulfide diffusing up from below. Where the system is anaerobic because *Thiobacillus* and other aerobic organisms have removed the oxygen, sulfate-reducing bacteria grow, utilizing sulfate and organic matter produced by *Thiobacillus* and supplying the hydrogen sulfide required as an energy source for the *Thiobacillus*.

Many examples of syntrophism are cited in the literature on the microbial degradation of recalcitrant chemicals in soils and aqueous environments. These chemicals (including cyclohexane, parathion, propanil, and other pesticides) cannot be degraded by a pure culture, but instead require two different populations working together. One organism attacks the organic compound and converts it into a product which can be utilized by the second population. Sometimes this second compound can serve as an energy source for both organisms. In many cases the syntrophic consortium is then able to mineralize the organic compound.

Mutualism

Mutualism or symbiosis also benefits both populations, but is an obligate relationship which allows organisms to exist in habitats which neither could occupy alone. This relationship requires close and organized proximity of the two organisms. The resulting association may behave and appear entirely different from the separate species growing alone. Lichens are such associations, composed of a photosynthetic cyanobacterium or a green alga and a fungus. Lichens are capable of growing in habitats where other organisms cannot survive, such as rock surfaces. There they produce organic acids which liberate mineral nutrients from the rock and assist in the weathering of rock to form soil. A species of the protozoan genus *Paramecium* contains large numbers of the green alga *Chlorella* within its cells, thus deriving oxygen and nutrients from the photosynthetic organism while providing the alga with carbon dioxide and conveying it into locations where it derives the optimum amount of light for photosynthesis.

Methanogenic consortia contain populations of several different kinds of bacteria growing symbiotically in close contact. Tomei et al. (1985) have described such consortia, in which several kinds of methanogenic bacteria grew together in dense clumps with obligate proton-reducing, butyrate-degrading bacteria. The butyrate degraders produce acetate and hydrogen from butyrate, and the hydrogen is immediately consumed by the methanogens, resulting in an extremely low hydrogen concentration surrounding the cells of the butyrate degraders. The very low hydrogen concentration makes the free energy of reaction of butyrate to acetate and hydrogen favorable enough to permit the growth of the butyrate degraders, which could not grow at all without this effect. Also, the hydrogen serves as an important source of energy for the methanogens by the reaction: $CO_2 + 4H_2 \rightarrow CH_4 + 2H_2O$.

Another remarkable example of symbiosis is the association between autotrophic sulfur- and sulfide-oxidizing bacteria which grow within the tissues of the vestimentiferan tube worms (*Riftia pachyptila*) and produce sufficient organic nutrients for their animal host, which in turn supplies them with carbon dioxide and all other essential nutrients. Thus these unusual animals have no digestive tract (Jannasch and Taylor, 1984).

Competition

Competition is harmful to both populations. The two populations are competing for some resource in the environment which both require. The resource may be space, light, nitrogen, oxygen, carbon dioxide, or some organic nutrient. Where the two populations occupy the same habitat, it is likely that one population will eventually be eliminated. This has been shown by growing two species of *Paramecium* together in a culture flask with bacteria as a food source. If both species tend to occupy the entire flask, one will out-compete the other and the second species will be eliminated. However, if the two species tend to occupy different regions of the culture flask, both can survive. This is a good illustration of the principle of an ecological niche. An ecological niche is defined as a habitat in which only one of two competing species can survive. Thus, in the first case, where one of the species is eliminated, the flask contains a single ecological niche, but in the second example there are two.

In oligotrophic lakes and in the sea, bacteria compete with phytoplankton (algae) for available nitrogen. However, since the phytoplankton are primary producers and the bacteria are consumers, they occupy separate niches, and both survive.

Antagonism

Amensalism or antagonism is a relationship in which one population produces a substance that is inhibitory to others. Antibiotic production is an example of this effect, as the antibiotic producer gains a competitive advantage by excluding certain other microbes from its immediate environment. The production of lactic acid by lactose-fermenting bacteria in milk is another example. By this stratagem the lactose fermenters produce a pH value so low that other organisms cannot multiply. Proteolytic bacteria may produce high concentrations of ammonia, resulting in a strongly alkaline environment, which inhibits many competing bacteria.

Parasitism

Parasitism is a relationship in which the parasite benefits, but the host is harmed. Viruses are termed obligate intracellular parasites because they can live only within the cells of a host organism. Certain bacteria are also obligate intracellular parasites. Every kind of living organism is subjected to this kind of parasitism. *Bdellovibrio* sp. is an interesting small, highly motile bacterium which penetrates the cell wall of other larger bacteria and multiplies in the periplasmic space between the cell wall and the cell membrane, thus destroying the host cell. The myxobacteria constitute an important group of soil bacteria which secrete extracellular enzymes, digesting other bacteria as well as some algae, and assimilating the soluble organic products.

Predation

Predation, that relationship in which one organism attacks and ingests another, is common among microorganisms. Many species of protozoa are predators upon bacteria, fungi, algae, or other protozoa. Slime molds and a few algae also prey upon bacteria.

APPLIED MICROBIAL ECOLOGY

Applied microbial ecology examines those aspects of microbial ecology which are of importance to human health and welfare. Some of these aspects will be examined in later chapters of this book. Others, more difficult to model mathematically, must await future research.

Human Pathogens

The relationship of human pathogens to other microorganisms in the environment and to alternative host organisms is of obvious importance to human health and constitutes the discipline of epidemiology.

The human body is extensively colonized by microorganisms, some of which are neutralistic, some commensal, some synergistic, and some parasitic. The skin, oral cavity, intestinal tract, and genital organs are colonized largely by bacteria, while the bloodstream and internal organs are generally kept free of microorganisms by the immune system. Pathogenic organisms gain entry through the normal portals of entry, i.e., the gastrointestinal tract, the genital tract, or the respiratory tract, as well as through breaks in the skin caused by wounds or parasitic insects. Such insects, including mosquitos, fleas, ticks, and lice, are known as vectors, since they carry the infecting microorganism from some other environment, usually an infected animal, into the human bloodstream.

Pathogenic organisms may be viruses, bacteria, fungi, or protozoa. In some cases, the pathogen may invade the tissues of the host. In other cases the pathogen may be noninvasive. Food poisoning diseases, such as botulism and staphylococcal food poisoning, are caused by the microorganism growing in food and producing its toxin before the food is ingested.

Pathogenic organisms may be transmitted by direct contact (measles, mumps, influenza, venereal diseases), by airborne dispersal (whooping cough and other bacterial and viral respiratory infections), by waterborne dispersal (infectious hepatitis virus, cholera, typhoid, bacterial dysentery, amoebic dysentery, giardiasis), via food (the food poisonings, typhoid, salmonellosis), or by vectors (yellow fever, malaria, typhus, bubonic plague, tularemia). Some diseases of animals can be transmitted to humans. Such infections are known as zoonoses and include undulant fever, a bacterial disease of cows, where the bacteria are present in the milk; plague and tularemia, diseases normally associated with rodents and lagomorphs, respectively, which can be carried to humans by the bite of a flea; yellow fever, which is carried by monkeys although the virus can be carried to humans by the bite of an infected mosquito; African sleeping sickness, a protozoal infection of cattle which can be carried to humans by the bite of the tsetse fly; and Lyme disease, a bacterial infection of deer which can be conveyed to humans by the bite of a tick normally parasitic on the deer.

Our first line of defense against disease is the physicians of the country backed up by the U.S. and State Public Health services. Working together, these agencies can observe an epidemic in the earliest stages and can institute countermeasures. For water- and food-borne diseases the identification of the contaminated source must precede countermeasures, such as sterilization or destruction of the contaminated material. Waterborne infections frequently come from fecal contamination and may be controlled by chlorination or ozonation of the water supply. A special case is provided by Legionnaires disease, a waterborne bacterial pneumonia in which the organism grows in warm water and is dispersed as an aerosol. The source of the organism is often a contaminated air conditioning system, cooling tower, humidifier, or shower.

The control of infectious diseases has advanced dramatically since the discovery, during the latter part of the 19th century, that bacteria were responsible for some of the most serious epidemics. The greatest reduction in infectious disease was brought about by simple public health measures such as sewage treatment, treatment of drinking water supplies by chlorination or ozonation, and control of rats and insect vectors. Active immunization with vaccines brought about control of many serious diseases, including diphtheria, typhoid, tetanus, mumps, and measles, and the complete elimination from the entire world of one of our most serious epidemic diseases, smallpox. More recently the development of chemotherapeutic agents and antibiotics has provided successful treatment for nearly all bacterial, fungal, and protozoal diseases. This approach has been far less effective with viral infections, although a few antiviral agents are available and further research will doubtless discover more. However, as we succeed in eliminating pathogenic organisms and their resulting diseases, we may be opening up new ecological niches for other pathogens.

It is well to remember, as pointed out by Alexander (1978), that the great advances in disease control available to most people in the more advanced countries are not available to all in the less developed countries. Thus, for the majority of the world's population, microorganisms are still the major cause of human death, illness, and misery. Other major causes are starvation and chemical pollution of air and water. Unfortunately, the latter two problems are likely to worsen as population pressures increase and industrialization progresses.

Animal Pathogens

The diseases caused by microorganisms pathogenic to domestic animals are of great economic importance. The ecological principles governing such diseases are substantially the same as those described above for humans, and the same approaches are valid for preventing and treating these diseases. A controversial practice in animal husbandry is the administration of hormones and antibiotics to all of the animals, a very different approach from the use of antibiotics in medicine, where the antibiotic or hormone is administered only after diagnosis of a condition requiring such treatment in an individual patient. Indiscriminate administration of antibiotics and hormones to livestock may result in unwitting administration of these chemicals to humans who later consume the flesh or milk from such animals. Also, dosing of animals with antibiotics may encourage the selection of antibiotic-resistant bacteria in the animals. Antibiotic resistance is often controlled by genes on extrachromosomal organelles called plasmids, which are readily exchanged between different species of bacteria. Thus, human pathogens may acquire resistance to antibiotics, which may render their cure or control by antibiotic therapy difficult or impossible.

Plant Pathogens

Diseases of plants, particularly valuable crops, are also of great economic importance. Most of the recognized plant diseases are caused by fungi, but bacteria and viruses are involved in many cases. Plant pathogens may gain entry through the roots, stem, or leaves. Leaves contain pores called stomata, which are a common entry point for fungus spores and bacteria. Other pathogenic fungi, bacteria, and viruses enter through wounds, frequently caused by a vector organism. The vectors are frequently insects, but rootworms and other soil organisms may also be involved. The pathogenic organism multiples within the plant, causing symptoms such as root rot, wilt, or blight and sometimes leading to greatly reduced crop yields or death of the plant. The pathogens damage the plant by producing enzymes, toxins, and growth regulators (plant hormones). Enzymes such as polygalacturonase, cellulase, and hemicellulase destroy the structural material of the plant, resulting in soft rot. Plant growth regulators may induce dwarfism and other growth abnormalities. Toxins interfere with the metabolism of the plant. Blockage of water transport in the plant causes wilts. Bacteria, often specific to a single plant species, cause wilts in corn, cucumbers, and alfalfa. Bark beetles, which bore through the bark of trees into the phloem, may carry pathogenic fungi into the trunk of the tree where the fungi grow, prevent normal transpiration, and kill the tree; three such diseases are pine beetle disease, Dutch elm disease, and chestnut blight. Rusts and smuts are fungus diseases of plants which cause enormous losses in corn, wheat, oats, onions, and other crops. Rust fungi have a complex life cycle requiring two unrelated plant hosts.

Biological Pesticides

The idea of using microorganisms pathogenic to undesired pests such as insects, rats, rabbits, and weeds has received considerable attention and has resulted in some successes. A successful application requires that the pathogen be highly virulent and stable enough to be stored and applied under natural conditions. The pathogen must not cause disease in organisms other than the target species. There are many viral, bacterial, fungal, and protozoal pathogens which could conceivably be used in this manner, but only a few have been commercialized. *Bacillus thuringiensis*, a close relative of the common soil bacterium *Bacillus cereus*, produces endospores and a crystalline parasporal body containing toxins which are lethal to at least 140 species of insects. This organism is now available commercially in many countries as an alternative to chemical insecticides. Since *B. thuringiensis* attacks only insects and leaves no toxic residue, it is much safer to use than most synthetic chemical insecticides.

Biogeochemical Cycling

In the biosphere, that part of the earth inhabited by living organisms, many of the more abundant chemical elements are continuously recycled by living organisms. Since microorganisms, particularly bacteria, are ubiquitous and capable of metabolizing so much more rapidly than larger organisms, they are particularly important in these great biogeochemical cycles. The study of these organisms and their activities, geomicrobiology, is closely related to biogeochemistry, where the emphasis is on the chemical reactions of geological processes rather than the microorganisms involved. Of the 92 naturally occurring elements, only 27 are essential to living organisms, and no single organism requires all 27. For example, boron is required by plants but not by animals or bacteria. All living organisms are made up of the same kinds of major biomolecules—nucleic acids, proteins, carbohydrates, and lipids—and the bulk of these molecules consist of only six elements, which are therefore called the bulk elements: carbon, hydrogen, oxygen, nitrogen, phosphorus, and sulfur. Each of these elements is cycled through living organisms on a global scale, with profound effects on environmental conditions and on the geology of the earth.

The energy which drives all of these great cycles of nature ultimately comes from the sun. Cycles on the primitive earth, before the evolution of free oxygen, were very much simpler, as anaerobic reactions prevailed. It seems likely that the first organisms were chemosynthetic autotrophs, deriving energy from the reaction of hydrogen sulfide with carbon dioxide to form organic acids such as formic acid (Wachtershauser, 1988). Heterotrophic bacteria, which developed later, presumably consumed much of the primordial organic matter on the earth's crust. When photosynthesis evolved 3.4 eons ago, living organisms, at first the blue-green bacteria, followed by algae and higher plants, produced the earth's atmosphere by splitting water into its constituent elements, liberating the oxygen into the atmosphere and using the hydrogen to reduce carbon dioxide to form biomolecules. Thus the carbon cycle, the hydrogen cycle, and the oxygen cycle were established eons ago, but were modified by the later appearance of higher plants and animals. During this time carbon was actively cycled by living organisms through the pools of carbon from atmospheric carbon dioxide to biomolecules to carbon dioxide again by mineralization. In the oceans vast amounts of carbon were removed from circulation by precipitation and shell and coral formation as limestone and dolomite. A great deal of organic matter was also removed in the process of sedimentation, leading to the deposition of a vast amount of kerogen in shales and of petroleum and natural gas in porous sandstone and limestone reservoirs. On land, in swampy areas, algal and plant material accumulated to form coal deposits. The pool of carbon in the atmosphere as carbon dioxide is small, but has been growing steadily for at least the past 100 years from about 0.031% to about 0.034% of the earth's atmosphere. This rise of 20 to 30 ppm seems small, but it has created a great deal of concern because a continued rise in atmospheric carbon dioxide will result in trapping more of the sun's heat in the atmosphere, hence a rise in temperature of the earth with possible catastrophic changes in climate. This is the greenhouse effect, a topic of intense research and concern at present. The increasing carbon dioxide in the atmosphere is believed to be due to human intervention in the carbon cycle by the increasing burning of fossil fuels, i.e., coal, oil, and gas. Recent geological studies of ice in Greenland and Antarctica suggest that the average temperature of the earth has been closely correlated with the carbon dioxide content of the atmosphere over the past 160,000 years and that it has been higher at times in the past than at present, as well as dramatically lower during glacial periods (Schneider, 1989). The major sinks for carbon dioxide are the ocean and land plants. Carbon dioxide in the ocean is consumed by algae, and

Background and Practical Applications 15

some of this carbon dioxide is then removed by sedimentation. Algal growth in the sea is controlled by the availability of phosphate, nitrate, and iron, the limiting nutrients. Thus decreasing availability of any of these nutrients could account for a CO_2 increase. Land plants consume carbon dioxide, and the deforestation which has been proceeding at an increasing rate since the start of the industrial revolution may be a cause of a substantial part of the CO_2 increase. Under aerobic conditions carbon is transferred through food webs from the photosynthetic primary producers to grazers, predators, and finally to decomposers (bacteria and fungi) which mineralize the organic matter to carbon dioxide and water. Under anaerobic conditions the final products of decomposition are carbon dioxide and methane. This is the source of biogenic natural gas, which provides the larger portion of our gas reserves (Rice and Claypool, 1980).

The oxygen cycle also is closely connected to the carbon and hydrogen cycles. Rocks of the earth's crust constitute a very large but almost inert oxygen reservoir. Water and molecular oxygen in the atmosphere are much smaller reservoirs, but are actively cycled by photosynthesis and respiration. Geochemical evidence indicates that the oxygen concentration of the atmosphere was much higher in the carboniferous period 300 million years ago, probably due to very intense plant growth. Oxygen is essential as a terminal electron acceptor for aerobic organisms in the reactions which provide energy for both heterotrophic and chemolithotrophic organisms. Dissolved oxygen is the most important chemical agent for controlling the oxidation-reduction potential of surface waters and soils. Most obligate anaerobes grow only at very low oxidation-reduction potentials in the complete absence of oxygen.

The nitrogen cycle is more complex, involving many steps, each one mediated by living organisms. Nitrogen in the sea and in well-aerated soils is predominantly in the form of nitrate ion (NO_3^-). Nitrate is assimilated by photosynthetic organisms and converted into biomolecules, primarily proteins. This organic nitrogen, upon decay, is converted to ammonia (NH_3). In the presence of oxygen, chemolithotrophic nitrosifying bacteria oxidize the ammonia to nitrite. Nitrite does not accumulate, because another chemolithotrophic group of bacteria, the nitrifiers, oxidize it to nitrate (NO_3^-). Under anaerobic conditions nitrate may be reduced to nitrogen and lost to the atmosphere. To make up for this loss, other bacteria, the nitrogen fixers, convert nitrogen to ammonia. This reaction is absolutely essential in maintaining soil fertility in nature. The most important of the nitrogen-fixing bacteria in soils are the *Rhizobium* spp., which grow symbiotically on the roots of leguminous plants. Many species of blue-green bacteria are important nitrogen fixers in aquatic habitats.

The sulfur cycle is also a complicated one, largely controlled by microorganisms, and involving many oxidation and reduction reactions. The metastable form of sulfur in most soil and aquatic environments is sulfate (SO_4^{2-}). The ocean contains a large pool of sulfate. Sulfate is assimilated by photosynthetic organisms and converted to reduced sulfur compounds such as cysteine which contain sulfhydryl (—SH) groups. Many important enzymes and coenzymes contain sulfhydryl groups. The microbial degradation of these biomolecules leads to the formation of hydrogen sulfide (H_2S). Hydrogen sulfide is very toxic to animals and many other organisms. However, it is readily oxidized by molecular oxygen to sulfur (S^0). *Thiobacillus* species are chemolithotrophic organisms which oxidize H_2S, S^0, thiosulfate ($S_2O_3^{2-}$), sulfite (SO_3^{2-}), and other sulfur compounds to sulfate. The activities of the acidophilic species *Thiobacillus thiooxidans* and *Thiobacillus ferrooxi-*

dans are responsible for the serious water pollution problem of acid mine drainage. These aerobic bacteria oxidize H_2S and iron pyrite (FeS_2) to sulfuric acid and ferric sulfate and will grow and produce acid at pH 1. Another anaerobic group, the sulfate-reducing bacteria, reverse this reaction by reducing sulfate to hydrogen sulfide. These bacteria grow in sediments where sufficient sulfate and organic matter are available, and they reduce the sulfate to sulfide which then reacts with metal ions such as copper, zinc, lead, and iron to precipitate these toxic ions as insoluble sulfides. At the same time, bacterial action neutralizes the acid and raises pH. Thus, the bad effects of acid mine drainage can be reversed by passing the drainage through natural or constructed wetland ecosystems (Batal et al., 1989). On the other hand, the sulfate-reducing bacteria cause anaerobic corrosion of iron and steel, a problem of economic importance in oil fields.

The phosphorus cycle is also of great economic importance as the productivity of both sea and land depends on an adequate supply of soluble phosphate (PO_4^{3-}). A large pool of phosphate is locked up in the form of insoluble minerals in rocks, largely calcium and iron phosphates. Phosphate may be solubilized and made available to plants by the action of anaerobic bacteria in sediments and by the mycorrhiza fungi symbiotic with plant roots.

The iron cycle involves oxidation-reduction reactions which cycle iron back and forth between the ferrous (Fe^{2+}) and the ferric (Fe^{3+}) forms. Ferric iron is very insoluble at any pH value above 5. Thus iron is present at very low concentrations in many aquatic ecosystems and may constitute a limiting nutrient for algal growth. Iron is an essential part of many enzyme systems. In anaerobic sediments Fe^{3+} may serve as an electron acceptor for certain bacteria, just as oxygen, nitrate, and sulfate do for others (Lovley and Phillips, 1986). Insoluble iron oxides and hydroxides are thus converted to more soluble ferrous (Fe^{2+}) compounds. Sulfate-reducing bacteria precipitate iron as insoluble sulfide minerals (FeS) which later become converted to pyrite (FeS_2). Many of our most important ore deposits of gold, silver, copper, zinc, lead, and molybdenum occur in such depositional environments along with pyrites (Trudinger, 1976).

Other elements involved in biogeochemical cycling include manganese, calcium, and silicon. Manganous ion (Mn^{2+}) is oxidized by aerobic bacteria at pH 5 to 7 to insoluble MnO_2. Manganese dioxide nodules form at the bottom of the sea and freshwater habitats when soluble Mn^{2+} diffuses upward into an aerobic habitat. Some bacteria can also reduce MnO_2 to Mn^{2+}. Calcium is deposited as calcium carbonate by corals and shellfish in the sea and is precipitated as sedimentary limestone when the pH is raised. Silicon is essential for the marine planktonic diatoms, silicoflagellates, and radiolaria. These organisms take up silicon dioxide from sea water and precipitate it in their shells. Such dead shells accumulate in extensive deposits of diatomaceous earth on the ocean floor.

Biodeterioration

Virtually all naturally occurring organic substances are subject to attack and degradation by microorganisms, and the prevention of such deterioration involves many ecological problems. Food and beverages are protected by killing microorganisms by heat or radiation, by adding preservatives which inhibit microbial growth, by removing the microorganisms mechanically, as in the partial sterilization of beer by filtration, or by freezing or refrigeration.

Canning is an effective process for killing all microorganisms in the foodstuff so that it will keep almost indefinitely. This requires heating to temperatures of 110 to 120°C for several minutes to kill the spores of such resistant organisms as *Clostridium botulinum*, which can produce a fatal toxin in food not sufficiently heated. Acidic foodstuffs need not be heated to as high a temperature as neutral or alkaline materials, since the spores are more easily killed in an acidic environment. Chemical preservatives such as benzoate, propionate, esters of *p*-hydroxybenzoic acid, and acetic, lactic, and citric acid are used in this way. These agents retard the growth of common food spoilage organisms, but do not prevent the growth of all organisms. Refrigeration retards the growth rate of microorganisms, thus permitting foodstuffs to be kept for many days. Freezing completely stops microbial growth so that the food may keep almost indefinitely.

Foodstuffs prepared by microbial transformations involve deliberate treatment to encourage the growth of one kind of microorganism, which grows rapidly and produces conditions which exclude all other organisms. Examples are the manufacture of yogurt, cheese, pickles, and alcoholic beverages.

An ecologically interesting food-borne infection is viral hepatitis, acquired by eating raw oysters or clams. These animals are filter feeders and concentrate the virus in their digestive tracts when growing in polluted water. Immersion in clean seawater for a sufficient time (depuration) can remove the viruses (Power and Collins, 1990).

Wood, cotton, paper, and leather are readily degraded by fungi and occasionally by bacteria. Keeping the material dry is often enough to prevent deterioration, since all microorganisms require water to grow. Toxic chemicals such as copper arsenate and creosote are effective in preventing the deterioration of wood products in contact with damp soil.

The fouling of surfaces immersed in the sea is an economically important biodeterioration process which starts with microorganisms such as bacteria and algae and then goes on to large seaweeds, barnacles, mussels, and many other marine animals. The process can be prevented by the application of paints containing toxic agents which slowly leach out into the water, such as copper oxide or toxic organotin compounds.

Waste Treatment

Since microorganisms are the great destroyers of organic materials in nature, it is not surprising that they have long been used to treat waste products. The most common method of sewage treatment is the activated sludge process. The sewage is usually subjected to forced aeration for 8 h or more. The bacteria and fungi normally present in the sewage multiply and oxidize most of the organic matter in the sewage to carbon dioxide and water. The residual sludge remaining, mostly microbial cells, is removed by settling in a clarifier. Regulations established by the U.S. Environmental Protection Agency require that the treated effluent must have been reduced in organic content (as measured by biochemical oxygen demand) by 85% or more. The sludge is often treated by anaerobic digestion, which may convert at least 50% of the organic matter to methane and carbon dioxide. The liquid effluent is generally sterilized by chlorine before release to a river or lake. Chlorine may react with aromatic compounds in the effluent to form toxic or carcinogenic agents. Thus there is a growing interest in the use of ozone, which is more effective than chlorine and produces no toxic reaction products, although it is more expensive (Rice, 1988).

Drinking Water Treatment

In all of the more affluent highly developed nations, drinking water for city water supplies is treated to kill or remove all pathogenic organisms. Chlorination or ozonation is used to kill microorganisms. Larger pathogens, such as protozoal cysts, can be removed by filtration. Some viruses are more resistant than bacteria to chlorine and ozone. Those viruses which are of particular concern are ones that cause infectious hepatitis, polio, and viral gastroenteritis.

Microbial Leaching of Metals from Ores

An interesting and economically important process is the microbial leaching of metals from ores. Microbial leaching was practiced at Rio Tinto, Spain, more than 300 years ago to recover copper, but the role of bacteria in the process was not understood until recently. To carry out the process, heaps of ore are spinkled with water. The water trickles down through the ore and is collected in a pond at a lower level. The pond contains a high concentration of cupric ion, Cu^{2+}, and is very acid. The leaching is carried out by the same acidophilic *Thiobacillus* species (*T. ferrooxidans* and *T. thiooxidans*) described earlier as the cause of acid mine drainage. They produce sulfuric acid and ferric ion, which oxidizes copper-containing minerals and releases the soluble cupric ion. The same process has also been used to recover uranium and could be used for other minerals as well (Lundgren and Silver, 1980).

Microbial Enhanced Oil Recovery

Microbial activities have also been employed to increase the amount of oil recovered from oil fields. Normal methods of primary recovery, including flowing wells and pumping, ordinarily recover only about one-third of the oil in place. Secondary recovery by water and gas injection may recover another third. This still leaves one-third of the original oil in the ground. The objective of enhanced oil recovery is to recover some of this residual oil. Many methods are available, including steam injection, surfactant injection, and polymer injection. In most cases these methods are uneconomic today, but microbial enhanced oil recovery offers the possibility of a more economical method of enhanced recovery, since bacteria injected into the formation may be able to manufacture the necessary surfactants, polymers, acids, or gases to enhance recovery. Results on laboratory tests and a field test with injection of *Clostridium acetobutylicum* in a molasses medium were encouraging (Yarbrough and Coty, 1982), but much more research is needed to assess the reliability and cost effectiveness of this recovery method.

Pollution by Recalcitrant Chemicals

The development of synthetic chemicals provided many extremely useful materials such as plastics, pesticides, insulators, coolants, and flame retardants. Unfortunately there has been an unforeseen side effect which is proving very difficult and expensive to clear up. Many of these materials have never before existed in nature and hence are called xenobiotic compounds. Because they have not been found in nature, microorganisms do not possess enzymes for degrading them. Hence they have been described as nonbiodegradable or recalcitrant. The terms are not perfectly accurate, however, since many compounds previously thought to be nonbiodegradable have now been shown to be slowly degraded in soil or water environments. Examples of such toxic compounds which have caused ecological disasters include DDT (dichlorodiphenyltrichloroethane), polychlorinated biphenyls, and pentachlorophenol. Originally these three were thought to be nonbiodegradable, but now it is known that they are slowly degraded by bacteria in soils and river bottom sediments.

Recalcitrant compounds usually do not serve as sources of carbon or energy for microbial growth. If they are degraded at all, it is by the process of cometabolism, in which some other nutrient supplies carbon and energy for growth (the cometabolite) and the target chemical is also degraded. Many compounds are more readily degraded aerobically than anaerobically, but the reverse is true in some cases. Thus the insecticide methoxychlor showed no degradation under aerobic conditions, but was attacked under anaerobic conditions and converted to products which could be mineralized by further aerobic treatment, an example of sequential anaerobic/aerobic treatment which could accomplish what neither could alone (Fogel et al., 1982). Aerobic bacteria readily degrade aliphatic hydrocarbons but anaerobes do not, a fortunate situation for our petroleum companies, since oil deposits would not be here if hydrocarbons were readily degraded anaerobically (Singer and Finnerty, 1984). Many kinds of bacteria working together may accomplish what a single organism cannot. An amazing example was described by Zhang and Wiegel (1990): 2,4-dichlorophenol was completely transformed in a freshwater sediment under anaerobic conditions to methane and carbon dioxide. Five steps were required, each step carried out by a different organism.

An important question in microbial ecology is whether microorganisms will eventually develop to mineralize all of the xenobiotic compounds which have been buried in landfills and otherwise distributed in the environment. We should hope so, because some of these compounds are highly toxic or carcinogenic and will surely pose serious health problems to future generations, if not to our own.

REFERENCES

Alexander, M. 1978. Microbial ecology: what for science, what for society, p. 2–6. *In* M. W. Loutit and J. A. R. Miles (ed.), *Microbial Ecology*. Springer Verlag, New York.
Atlas, R. M., and R. Bartha. 1987. *Microbial Ecology: Fundamentals and Applications*, 2nd ed. Benjamin-Cummings, Menlo Park, Calif.
Baross, J. A., and J. W. Deming. 1983. Growth of "black smoker" bacteria at temperatures of at least 250°C. *Nature* (London) **303**:423–426.
Batal, W., L. S. Laudon, T. R. Wildeman, and N. Mohdnoordin. 1989. Bacteriological tests from the constructed wetland of the Big 5 Tunnel, Idaho Springs, Colorado, p. 550–557. *In* D. A. Hammer (ed.), *Constructed Wetlands for Wastewater Treatment*. Lewis Publishers, Chelsea, Mich.
Cloud, P. 1983. The biosphere. *Sci. Am.* **249(3)**:176–189.
Fogel, S., R. L. Lancione, and A. E. Sewall. 1982. Enhanced biodegradation of methoxychlor in soil under sequential environmental conditions. *Appl. Environ. Microbiol.* **44**:113–120.
Jannasch, H. W., and C. D. Taylor. 1984. Deep-sea microbiology. *Annu. Rev. Microbiol.* **38**:487–514.

Karl, E. M., C. O. Wirsen, and H. W. Jannasch. 1980. Deep sea primary production at the Galapagos hydrothermal vents. *Science* **207**:1345–1347.
Krieg, N. R., and J. G. Holt (ed.). 1984. *Bergey's Manual of Systematic Bacteriology*, 9th ed., vol. I. The Williams and Wilkins Co., Baltimore.
Langworthy, T. A. 1978. Microbial life in extreme pH values. *In* D. J. Kushner (ed.), *Microbial Life in Extreme Environments*. Academic Press, Inc., New York.
Lovley, D. R., and E. J. P. Phillips. 1986. Organic matter mineralization with reduction of ferric iron in anaerobic sediments. *Appl. Environ. Microbiol.* **51**:683–689.
Lundgren, D. G., and M. Silver. 1980. Ore leaching by bacteria. *Annu. Rev. Microbiol.* **34**:263–283.
Margulis, L. 1981. *Symbiosis in Cell Evolution: Life and Its Environment on the Early Earth*. W. H. Freeman, New York.
Pereira, W. E., C. E. Rostad, T. J. Leiker, D. M. Updegraff, and J. L. Bennett. 1988. Microbial hydroxylation of quinoline in contaminated groundwater: evidence for incorporation of the oxygen atom of water. *Appl. Environ. Microbiol.* **54**:827–829.
Power, U., and J. K. Collins. 1990. Tissue distribution of a coliphage and *Escherichia coli* in mussels after contamination and depuration. *Appl. Environ. Microbiol.* **56**:803–807.
Rice, D. D., and G. E. Claypool. 1980. Generation, accumulation, and resource potential of biogenic gas. *Bull. Am. Assoc. Petroleum Geologists* **65**:5–25.
Rice, R. R. 1988. The Safe Drinking Water Act amendments of 1986 and their impacts on the use of ozone for drinking water treatment in the United States. Proceedings of the International Ozone Association Conference, Monroe, Mich., 27–28 April 1988. International Ozone Association, Monroe, Mich.
Schneider, S. H. 1989. The changing climate. *Sci. Am.* **261**:70–79.
Schopf, J. W., and B. M. Packer. 1987. Early Archean (3.3-billion to 3.5-billion-year-old) microfossils from Warrawoona group, Australia. *Science* **237**:70–73.
Singer, M. E., and W. R. Finnerty. 1984. Microbial metabolism of straight chain and branched alkanes, p. 1–59. *In* R. M. Atlas (ed.), *Petroleum Microbiology*. Macmillan, New York.
Tomei, F. A., J. S. Maki, and R. Mitchell. 1985. Interactions in syntrophic associations of endospore-forming butyrate-degrading bacteria and H_2-consuming bacteria. *Appl. Environ. Microbiol.* **50**:1244–1250.
Trudinger, P. A. 1976. Microbiological processes in relation to ore genesis, p. 135–176. *In* K. H. Wolf (ed.), *Handbook of Strata-Bound and Uniform Ore Deposits*, vol. 2: *Geochemical Studies*. Elsevier, New York.
Wachtershauser, G. 1988. Before enzymes and templates: theory of surface metabolism. *Microbiol. Rev.* **52**:452–484.
Whittaker, R. H. 1969. New concepts in kingdoms of organisms. *Science* **163**:150–160.
Yarbrough, H. F., and V. F. Coty. 1982. Microbial enhanced oil recovery from the Upper Cretaceous Nacatoch formation, Union County, Arkansas, p. 149–153. *In* E. C. Donaldson and J. B. Clark (ed.), *Proceedings of 1982 Conference on Microbial Enhanced Oil Recovery, Afton, Okla*. U.S. Dept of Energy, Bartlesville, Okla.
Zhang, X., and J. Wiegel. 1990. Sequential anaerobic degradation of 2,4-dichlorophenol in freshwater sediments. *Appl. Environ. Microbiol.* **56**:1119–1127.

II. FATE IN THE SUBSURFACE WORLD

Chapter 2

Problems with Using Existing Transport Models To Describe Microbial Transport in Porous Media

Ruth A. Dickinson

Introduction 22
   Basics of Flow through Porous Media 22
   Fundamental Aquifer Hydraulic Parameters 23
   The Continuum Approach 24
Equations for Groundwater Flow 25
   Isotropy and Homogeneity 26
   Equation for Saturated, Confined Groundwater Flow 26
   Equation for Saturated, Unconfined Groundwater Flow 27
   Equation for Unsaturated Soil Moisture Flow 27
Equations for Contaminant Transport 28
   Hydrodynamic Dispersion 29
   Contaminant Decay Terms 30
   Colloidal Transport 30
Behavior of Microorganisms in Porous Media 31
   Microbial Kinetics 31
   Examples of Microbial Transport 32
   Microorganisms as Tracers of Contaminant Movement 34
Using an Existing Contaminant Transport Model 35
   An Unsaturated Zone Transport Model 36
   Results Using Contaminant Transport Model 37
Problems with Using Contaminant Transport Models 38
   Parameter Estimation and Variability 39
   Reporting Aquifer Hydraulic Parameters 40
   Colloidal Transport 43
   Transport Factors Not Included 43
Conclusions 44
   Possible Resolution of Problems 45
References 45

Ruth A. Dickinson • Camp Dresser & McKee Inc., Fort Myers, Florida 33919; formerly with Versar Inc., Springfield, Virginia 22151.

This chapter explores the concept of modeling microbial transport in porous media and presents the thesis that contaminant transport models can be adapted and used to model microbial transport. It addresses the feasibility of, and problems associated with, using groundwater flow and contaminant transport models to predict the movement of microorganisms (i.e., bacteria and viruses) in porous media, such as in soil and groundwater aquifers.

The approach presented in this chapter is based on the premise that the hydraulic gradient, and the resulting movement of water, is the principal driving force of the transport of microorganisms in porous media. In this approach, tropisms and mechanisms of microbial self-propulsion are assumed to be secondary in magnitude to movement driven by the hydraulic gradient in porous media. Thus, the modeling approach is that of microorganisms being transported with soil moisture in the unsaturated zone of soil and with groundwater flow in aquifers.

A number of computer models have been developed to describe flow and contaminant solute transport through porous media. Though microorganisms do not behave entirely like dissolved contaminants in a groundwater flow regime, the mathematical expressions used to describe their motion take on the same form as the equations used in the groundwater flow and contaminant (solute) transport models.

The problems associated with using existing contaminant transport models to describe microbial transport include the following: (i) important parameters are not directly measurable and/or are variable in the flow regime; (ii) aquifer hydraulic properties are not measured or reported for many microbial studies; (iii) microorganisms are not transported as contaminants in solution but as colloids; and (iv) important factors in the behavior of the microorganisms, such as local changes in porosity and permeability due to clogging by the microorganisms, are not included in the governing equations.

This chapter is composed of seven sections. The first section presents basic concepts of flow through porous media. The next two sections establish the terminology of flow and transport in porous media and a premise for the modeling approach. They are followed by a section that presents mathematical expressions for microbial behavior in porous media and a section on adapting an existing model to microbial transport. The sixth section discusses the problems associated with using contaminant transport models to describe microbial movement in porous media. The last section of this chapter recommends ways to avoid these problems or to minimize their impact.

BASICS OF FLOW THROUGH POROUS MEDIA

A porous medium is any solid that has interconnected pore spaces, or void spaces, within it. This general definition encompasses soil, fractured rock, membrane or fabric filters, trickling filters at wastewater treatment plants, activated carbon columns, and other such examples. The interconnected pore spaces allow fluids such as water or air to flow through the porous medium. The pore spaces may be connected by tortuous pathways, that is, indirect or circuitous routes through the porous medium (Dullien, 1979).

Water flows through the pore spaces, or voids, of the soil. The term porosity, represented by n, is defined as the ratio of volume of void spaces to the total volume (void spaces and soil) in a representative elemental volume. Generally the porosity of clayey soils exceeds that of sandy soils. However, the entire volume of the void spaces is not available for the flow of water or transport of particles or solutes. Therefore, another term, effective porosity, is introduced. Effective porosity, represented by n_e, is the ratio of the volume of void spaces that is available for flow to the total volume of a representative elemental volume. This value is always less than the porosity of the soil and, because it is highly dependent on the arrangement of the void spaces and pathway in the porous medium, is not directly measurable. The effective porosity of sandy soils exceeds that of clayey soils.

When the pore spaces are completely filled with fluid (in this case, water) the soil is referred to as a saturated porous medium; flow through it is known as saturated flow. This condition occurs in unconfined and confined aquifers. In most areas of the country, the uppermost aquifer below land surface is an unconfined aquifer and is known as the water table aquifer. A confined aquifer is one in which the water in the pore spaces is held under pressure, and the aquifer is confined (constrained) above and below by less permeable materials (aquitards). Between the water table aquifer and the ground surface the soil's pore spaces are not completely filled with water. This area is known as the unsaturated zone or the vadose zone. The pore spaces in the unsaturated zone are filled with water and air, and the movement of water is referred to as soil moisture flow.

Fundamental Aquifer Hydraulic Parameters

Three fundamental aquifer hydraulic parameters that are used to describe the flow of groundwater include aquifer hydraulic conductivity, represented by K; hydraulic gradient, represented by i; and the aquifer storage coefficient, called storativity (S) in a confined aquifer, and specific yield (S_y) in an unconfined aquifer. The hydraulic conductivity is a measure of the ease by which water can flow through the porous medium. It is a function of both the fluid (water) and the porous medium through which the fluid flows. The aquifer transmissivity, represented by T, is directly related to hydraulic conductivity. Aquifer transmissivity is the product of hydraulic conductivity and the thickness of the aquifer. The value for aquifer transmissivity is one of the parameters that may be calculated from aquifer pumping tests in confined aquifers, so it is often reported instead of hydraulic conductivity.

The hydraulic gradient is the change in hydraulic head over a linear distance. The hydraulic head at a point in an aquifer is a measure of the potentiometric energy at that point; it is the sum of the energy due to position (height above a reference point) and energy due to pressure. The hydraulic head can be measured as the height to which water will rise in an observation well that is screened in the aquifer of interest. In an unconfined aquifer, the energy due to pressure is zero, and the water in the observation well rises to the same height as in the aquifer. In a confined aquifer, the water in an observation well rises above the top of the aquifer because the water is under pressure. When the constraint of the overlying confining unit is removed, as in an observation well, the groundwater rises to a position that reflects the energy stored within the confined aquifer. Water in the unsaturated zone will not enter an observation well because water is held in the pore spaces by capillary forces. The pressure in the unsaturated zone is less than atmospheric and is referred to as matric pressure.

The potentiometric surface of an aquifer can be contoured or mapped to record the hydraulic conditions at a given time. In an aquifer, areas that exhibit equal hydraulic heads will lie on equipotential contours. The direction of groundwater flow is from areas of higher potentiometric head to lower potentiometric head along flowpaths perpendicular to the equipotential contours. The hydraulic gradient can be found as the difference between the hydraulic heads at two locations divided by the distance along a perpendicular flowpath between those locations.

The aquifer storage coefficient is defined as the volume of water that is released from an aquifer per unit head decline over a given unit area. It represents the amount of water that is released from storage when there is a decline in the potentiometric head caused by draining or by a pumping well. In a confined aquifer, the storage coefficient is called storativity. In an unconfined aquifer, the storage coefficient is called the specific yield. It represents the water released by dewatering the pore spaces, because water is not stored under pressure in the unconfined aquifer. The specific yield of an unconfined aquifer is equal to the effective porosity in that aquifer (Bear, 1979).

The Continuum Approach

To describe the flow of fluids in a porous medium, it is necessary to employ the concept of a continuum (Bear, 1979). Under a continuum approach, the microscopic flow pattern along tortuous pathways through individual pore spaces is neglected, and only the average flow through the medium is addressed. Either of two perspectives, the Eulerian approach or the Lagrangian approach, may be employed when developing hydrodynamic equations based on the conservation of mass. In the Eulerian approach, a control volume or control box concept is used and the hydrodynamics equations are written based on the mass volume (e.g., water) entering, leaving, or being stored in the control volume. In the Lagrangian approach, a fixed unit of mass is tracked and described mathematically as it moves through time and space.

The earliest developments in quantifying flow through porous media began with a laboratory experiment conducted by Henry Darcy in 1856 to evaluate flow of water through a sand filter. The approach used was an Eulerian approach by nature. That experiment considered a bulk mass of water flowing into and then out of a control volume (in this case, the sand filter) without consideration of the water movement within the pore spaces of the filter. Darcy observed that the water discharged through the filter, represented by Q, was proportional to the hydraulic gradient (difference in hydraulic heads over the length of the filter), represented by i; the cross-sectional area of the filter, represented by A; and a proportionality constant, represented by K. Darcy's Law is given by

$$Q = KiA \qquad (1)$$

where the proportionality constant has come to be known as hydraulic conductivity. The hydraulic conductivity is a function of the permeability of the porous medium and the fluid flowing through the medium. The hydraulic conductivity of porous media cannot be measured directly but can be determined in situ by a number of field tests and in laboratory column tests using samples. Darcy's Law is valid for laminar flow (nonturbulent flow) only. The simple Darcy's Law forms the basis for all groundwater flow derivations.

The discharge flow, Q, is sometimes expressed on a per unit area basis, Q/A. Then equation 1 becomes

$$q = Ki \qquad (2)$$

In equation 2, the term q is called specific discharge (preferred) or sometimes darcy velocity (Bear, 1979; Freeze and Cherry, 1979). Though specific discharge is described in units of length per time, it does not represent the velocity of the fluid; it is the discharge of the fluid per unit cross-sectional area of a porous medium. Use of this term is entirely consistent with the Eulerian point of view that is inherent in the concept of a continuum and is a basic part of the development of theories in groundwater flow.

It wasn't until over a century later that concern over the movement of contaminants in porous media gave rise to the need for a term to represent groundwater flow velocity. Tracking the movement of an individual parcel of water, or a particle such as a contaminant suspended in the water, necessitates a Lagrangian point of view (e.g., Bear, 1979). In a Lagrangian approach, a discrete mass of water or particle is followed as it moves along the tortuous pathways of the interconnected pore spaces in a porous medium. Individual particles will move at different rates through the porous medium depending upon the length of the tortuous path followed, the relationship of fluid flow to pore size and geometry, the turbulence within the pore spaces, and other factors present on a microscopic scale. The number of variables and the complex combinations and interrelationships, plus the problems with attempting to identify representative particles to extrapolate behavior of discrete particles to the whole population of particles, make it advisable to view transport phenomena using the continuum concept and an Eulerian approach.

The average rate of groundwater flow through a porous medium can be derived by borrowing a fundamental concept from hydrodynamics. The discharge rate of a fluid flowing in an open channel is the product of the velocity of the fluid and the cross-sectional area perpendicular to the flow direction and through which the fluid passes. In a uniform porous medium that has an effective porosity of n_e, the cross-sectional area of the pore spaces that are available for flow is $n_e A$. The average groundwater flow velocity, v_{gw}, can be derived as

$$v_{gw} = \frac{Q}{(n_e A)} = \frac{(Ki)}{n_e} = \frac{q}{n_e} \qquad (3)$$

This representation of the average groundwater flow velocity is used in determining the magnitude of advective and dispersive transport of a substance that is transported with groundwater in a porous medium.

EQUATIONS FOR GROUNDWATER FLOW

Governing equations for groundwater flow are generated by employing Darcy's Law and the equations of continuity for a representative volume of the aquifer. The concept of continuity is based on the conservation of mass. For applications of groundwater flow, continuity leads to the observation that the difference of water flowing into and out of a control (or representative) volume is equal to the time rate of change of water stored in the control volume. When the amounts of water entering and leaving the control volume are equal, there is no change in the volume of water in storage. This special case is known as steady-state flow.

The governing equation for a fluid in a porous medium is given by the following partial differential equation:

$$\text{div} \{\rho [K \text{ grad}(\phi)]\} = \partial (\rho n)/\partial t = \rho S_0 (\partial \phi / \partial t) \qquad (4)$$

where div and grad are mathematical operators that stand for divergence and grad, respectively; ρ is the density of the flowing fluid; K is the hydraulic conductivity of the medium; ϕ is the potentiometric head at a given location; n is the saturated porosity of the medium; S_0 is the specific storativity, equal to the storage coefficient divided by the aquifer thickness; and t is time.

Good explanations of the derivation of the governing equation can be found in chapter 2 of Remson et al. (1971) and in chapter 5 of Bear (1979). Equation 4 is appropriate for flow of a compressible or incompressible fluid in a nondeforming porous medium. It describes flow in a saturated and confined, a saturated and unconfined, or an unsaturated medium. However, other forms of the equation are derived to describe each of these special cases and are presented in sections that follow.

Isotropy and Homogeneity

Under particular flow conditions it is possible to simplify equation 4. The most common simplifying assumption for saturated flow is that the porous medium is isotropic and homogeneous. Isotropy is a condition under which the aquifer parameters are nondirectional; that is, they have the same value in the x, y, and z directions. For most geologic formations, the action of the geologic deposition causes anisotropic (varying with direction) conditions to be the norm, rather than isotropic conditions. However, when aquifer conditions allow, or when the paucity of aquifer-specific data dictates, the assumption of isotropy may be applied.

Homogeneity refers to the uniformity of the porous medium. If a representative volume is truly representative of the entire aquifer, then the entire flow medium will have the same values of porosity, hydraulic conductivity (it may be anisotropic), and storage coefficient, and the aquifer is homogeneous. The hydraulic parameters in nonhomogeneous, or heterogeneous, aquifers are spatially variable. Heterogeneity is more likely to be the norm in a geologic formation, but often the aquifer is assumed to be homogeneous to simplify calculations or because there are insufficient data to support a more detailed treatment. The concept of heterogeneity should not be confused with a system of multiple aquifers and confining beds which can be described mathematically, and modeled, as layered (or abutting) homogeneous, isotropic porous media.

Equation for Saturated, Confined Groundwater Flow

The governing equation of flow for an incompressible fluid in a nondeforming, isotropic, homogeneous, confined aquifer becomes

$$T\left(\frac{\partial^2\phi}{\partial x^2} + \frac{\partial^2\phi}{\partial y^2} + \frac{\partial^2\phi}{\partial z^2}\right) + N = S\left(\frac{\partial\phi}{\partial t}\right) \quad (5)$$

where T is the transmissivity of the aquifer; ϕ is the potentiometric head; N is a term for a sink or source; S is the aquifer storativity; and t is time.

The governing equation is a second-order partial differential equation. To solve this equation, it is necessary to define two boundary conditions and the initial conditions (the potentiometric head when time is zero). Under special conditions, such as a uniform flow field in one dimension, the governing equation can be simplified further, and in some instances an analytical, or directly solvable, solution may be developed.

Equation for Saturated, Unconfined Groundwater Flow

The governing equation of flow for an incompressible fluid in a nondeforming, isotropic, unconfined homogeneous aquifer becomes

$$K\left[\frac{\partial}{\partial x}\left(h\frac{\partial h}{\partial x}\right) + \frac{\partial}{\partial y}\left(h\frac{\partial h}{\partial y}\right) + \frac{\partial}{\partial z}\left(h\frac{\partial h}{\partial z}\right)\right] + N = S_y\left(\frac{\partial h}{\partial t}\right) \quad (6)$$

where K is the hydraulic conductivity of the aquifer; h is the potentiometric head; N is a term for a sink or source; S_y is the aquifer specific yield; and t is time.

The governing equation is a second-order partial differential equation that is nonlinear in h. To solve this equation, it is necessary to define two boundary conditions and the initial conditions (the potentiometric head when time is zero). Under special conditions, such as steady-state flow in a uniform one-directional flow field, the governing equation can be simplified further, and in some instances an analytical, or directly solvable, solution can be developed.

Equation for Unsaturated Soil Moisture Flow

The most common simplifying assumption employed to describe flow in the unsaturated zone is that flow is one-dimensional downward. This is a reasonable assumption, because the acceleration due to gravity, which is applied only in the z direction, is a predominant driving force. The one-dimensional, governing equation of flow for an incompressible fluid in a nondeforming, isotropic, homogeneous vadose zone becomes

$$\frac{\partial}{\partial z}[K(\theta)\partial\phi/\partial z] + N = \frac{(\partial\theta)}{(\partial t)} \quad (7)$$

where θ is the soil moisture content; $K(\theta)$ is the hydraulic conductivity of the vadose zone and is a function of the moisture content; ϕ is the potentiometric head, which is the sum of the matric and gravitational potentials; N is a term for a sink or source; and t is time.

This governing equation is not solvable in this form because it has two independent variables, ϕ and θ. To develop a solvable form, it is necessary to introduce a term called the coefficient of diffusivity, $D(\theta)$, which is a function of the moisture content. The coefficient of soil moisture diffusivity is defined as

$$D(\theta) = -K(\theta)(\partial\psi/\partial\theta) \qquad (8)$$

where ψ is the matric potential and is inherently negative and $(\partial\psi/\partial\theta)$ is the slope of the soil moisture characteristic curve for the soil.

With application of the chain rule, equation 7 becomes the well-known Darcy-Richards equation:

$$\frac{\partial}{\partial z}\left[D(\theta)\left(\frac{\partial\theta}{\partial z}\right) - K(\theta)\right] = \frac{\partial\theta}{\partial t} \qquad (9)$$

This governing equation is a second-order partial differential equation that is nonlinear in diffusivity and hydraulic conductivity. To solve this equation, it is necessary to define two boundary conditions and the initial conditions (the moisture content when time is zero). There are no analytical, or directly solvable, solutions for the governing equation in the vertical direction (some semianalytical solutions are available for hydraulic diffusion in the horizontal plane, but they are not helpful for the studies under question) (Hillel, 1980a, 1980b). The hydraulic conductivity and soil moisture diffusivity are functions of (vary with) moisture content and matric pressure, so the flow of soil moisture in the unsaturated zone is nonlinear. Iterative solution techniques, such as those employed by numerical methods (i.e., finite difference, finite element solutions), can be used to solve unsaturated flow in the unsaturated zone.

Only the term for moisture content in equation 9 is measurable in the field, and it will vary with depth. The soil moisture characteristic curve (matric potential and a function of moisture content) and the unsaturated hydraulic conductivity curve are usually developed under laboratory conditions. Several empirical and semiempirical equations have been developed to enable the calculation of the value for unsaturated hydraulic conductivity given the soil moisture or matric pressure (Rawls et al., 1982; Dickinson, 1985). Some of the equations include a term for the air entry pressure (or bubble pressure), which represents the matric pressure at which air will first enter a draining, saturated soil. Measured values of hydraulic parameters for common soils have been tabulated (e.g., Mualem, 1976), and statistical averages for some soil groups have been determined (Clapp and Hornberger, 1978; Rawls et al., 1982).

EQUATIONS FOR CONTAMINANT TRANSPORT

The hydrodynamic advection-dispersion equation for transport of a solute dissolved in an incompressible solvent (e.g., water) moving in a porous medium is a nonlinear, partial differential equation in the form of

$$(\theta)(\partial C/\partial t) = \text{div}(\theta D_h \cdot \text{grad}C) - (q)(\text{grad}C) - \theta\mu C \qquad (10)$$

where θ is the moisture content of the medium; C is the mass concentration of solute (e.g., a contaminant in solution with groundwater); D_h is the coefficient of hydrodynamic dispersion which includes dispersion and molecular diffusion; q is the specific discharge of groundwater; μ is a first-order decay constant; and t is time.

This equation is valid for both saturated and unsaturated groundwater flow. If the moisture content is constant, as it is for saturated flow, the transport equation can be expressed as

$$\partial C/\partial t = \text{div}(D_h \cdot \text{grad}C) - (v_{gw})(\text{grad}C) - \mu C \qquad (11)$$

In equation 11, the term for the groundwater flow velocity is the link between the governing equations for groundwater flow and contaminant transport. Because the equations are linked, or coupled, both the groundwater flow equation (either equation 5, 6, or 7) and the contaminant transport equation (either equation 10 or 11) must be solved to generate a solution to the contaminant transport equation.

As with equation 4, for groundwater flow, simplifying assumptions and conditions may be applied for special cases. Analytical solutions to equation 10 have been developed for common contaminant release scenarios for flow and transport in a uniform flow field where groundwater flow is steady state (e.g., Bear, 1979; Javandel et al., 1984). In a uniform flow field, the groundwater pore velocity is the same throughout, and under steady-state conditions the groundwater pore velocity is constant.

Numerical computer models of transport include separate solutions of each governing equation so that the groundwater flow, or velocity flow field, is solved for a given time throughout the entire flow domain and then the transport or advection-dispersion equation is solved for that same time period. Numerical methods, such as finite difference or finite element methods, can be used to solve the coupled, partial differential equations. The interested reader can find good summaries of contaminant transport models in Javandel et al. (1984) and in Van der Heijde et al. (1985).

Hydrodynamic Dispersion

Solutes in groundwater will spread into a porous medium by mechanical mixing and by diffusion. The mechanical mixing, or mechanical dispersion, is primarily due to the velocity variations as groundwater moves through tortuous pathways. Turbulent flow inside pore spaces can contribute to mechanical dispersion. On a macroscopic scale, the dispersive nature of a porous medium is represented as the product of the groundwater pore velocity and the coefficient of dispersivity. The coefficient of dispersivity is a function of the porous medium and is anisotropic. When reported, values are usually given for the longitudinal dispersivity which acts in the primary direction of groundwater flow. Sometimes values are also reported for the transverse dispersivity (in the horizontal plane but perpendicular to the primary flow direction) and the vertical dispersivity. The values of the dispersivity coefficients are not derived from other basic properties, are not directly measurable, and appear to be scale dependent (Gelhar et al., 1985).

Molecular diffusion is another mechanism by which a solute is spread in a porous medium. Molecular diffusion is driven by a concentration gradient within the flow regime. Flow occurs from areas of higher concentrations to those of lower concentrations according to Fick's Law of Diffusion (Bear, 1979). Molecular diffusion can occur when the groundwater pore velocity is zero and there is no fluid motion. The coefficient of molecular diffusivity is not derived from other basic parameters and is not directly measurable.

The term hydrodynamic dispersion encompasses both mechanical dispersion and molecular diffusion. However, the magnitude of molecular diffusion is usually much less than that of mechanical dispersion, and it can be neglected without affecting the transport calculations. Since neither the dispersivity nor the diffusivity is directly measurable, the values assigned for hydrodynamic dispersion in a porous medium are inherently uncertain. Fortunately, in most flow regimes the contribution due to dispersive flux of a contaminant is much less than that of advective flux, and in regional flow regimes dispersion can be neglected (Anderson, 1979).

Contaminant Decay Terms

The last term on the right-hand side of equations 10 and 11 includes the decay coefficient, which represents the production or decay of solute concentration within the porous medium. The decay coefficient is a rate constant that represents increasing concentration when it is negative in value and decreasing concentration when it is positive. The oft-cited parallel is the half-life decay rate of radioactive substances. For microorganisms the decay coefficients can represent bioactivity, including predation, die-off, and regrowth, which affects the number (population) present.

As depicted in equations 10 and 11, the decay coefficient is a first-order decay coefficient; the value of the quantity is dependent upon the concentration in solution. However, other decay or production coefficients could be introduced as sink or source terms. They could be classified as zero-order decay terms and are independent of the solute concentration in the porous medium.

Colloidal Transport

Many substances are transported through porous media as particles in suspension, i.e., colloids. The governing equation for colloidal transport is quite similar in form to those for mass (solute) transport (e.g., Nuttall, 1986). One potential difference in their treatments is the importance that variable velocity has in the transport of colloids. With a solute it is reasonable to assume that the concentration within the pore space is uniformly distributed, but that may not be true with colloids.

Colloidal transport models would be appropriate to model the movement of some microorganisms in the unsaturated and saturated zones because bacteria and viruses may attach to minute particles and are transported in porous media as colloids (McDowell-Boyer et al., 1985). Soil colloid particles have a very high surface-area-to-volume ratio that provides a large proportion of adsorption sites. Also, if the soil colloid particles are composed of clay, they will have a net negative charge which makes them more reactive (Hattori and Hattori, 1976). The presence of clay enhances the adsorption of viruses in soils.

BEHAVIOR OF MICROORGANISMS IN POROUS MEDIA

Mechanisms such as filtering, adsorption, and sedimentation influence the fate of microorganisms in porous media. Corapcioglu and Haridas (1984, 1985) identified key mechanisms that affect the fate of microorganisms in porous media and expressed them in mathematical terms that can be incorporated into physical transport models based on the advection-dispersion equation.

The modeling approach presented in this chapter assumes that microorganisms can be represented by the contaminant transport equations commonly used in the study of solute transport. There are precedents for this modeling approach in the literature (e.g., Matthess and Pekdeger, 1981; Haridas, 1983; Corapcioglu and Haridas, 1984, 1985). Matthess and Pekdeger (1981) established a quantitative description of the horizontal movement of microorganisms in a saturated groundwater flow system. In such a treatment, the hydrodynamic advection-dispersion equation is used to calculate the concentration of microorganisms as a function of location and time. In the approach of Matthess and Pekdeger, the adsorption of microorganisms is assumed to behave as a linear, equilibrium isotherm, so that its influence is taken into consideration by the inclusion of a retardation factor. These authors further identified first-order processes that reduce the microbial population due to physical, biological, and chemical conditions (e.g., temperature, predation, and pH, respectively) (Matthess and Pekdeger, 1981).

Microbial Kinetics

Kinetic and fate terms for microbial growth and death, aggregation and subsequent filtering in the porous media, adsorption, sloughing, and other kinetic mechanisms can be incorporated in the transport equation if appropriate mathematical expressions can be developed. Corapcioglu and Haridas (1984, 1985) have identified a number of kinetic mechanisms that occur during subsurface transport of microorganisms. They presented mathematical expressions for some kinetic mechanisms and provided reported values for some of the decay coefficients.

Corapcioglu and Haridas (1985) recognized that microorganisms are removed from the transporting medium (the liquid phase) by mechanisms of deposition including straining (or mechanical filtering), adsorption, and sedimentation. They identified an expression for the rate of deposition that is a function of (i) the rate at which microorganisms (e.g., aggregates) clog the pore spaces in the soil and (ii) the rate of declogging due to sloughing of microorganisms from clogged pore spaces. The latter reintroduces some microorganisms into the liquid (transporting) phase. The expression took on the form of

$$\text{Rate} = k_c(n - \sigma)C - k_y\rho\sigma^h \tag{12}$$

where k_c is the clogging rate; n is the porosity; σ is the volume of the deposited microorganisms per unit volume of the porous medium; C is the concentration in the liquid; k_y is the declogging rate; ρ is the density of the microorganism; and h is an experimentally determined constant that acts as a distribution coefficient.

In equation 12, the expression for a clogging factor was presented as a first-order function of concentration in the liquid (water) phase. The expression for a declogging factor was presented as a function of the adsorbed concentration, and the value of the exponent, h, was assumed equal to 1. Corapcioglu and Haridas (1985) provided some values for clogging and declogging rates for fecal coliforms and bacteriophages (viruses that prey upon enteric bacteria) that had been reported by other researchers.

The expression for microbial death or inactivation is a rate reaction that is a first-order function of concentration (Yates et al., 1987). The expressions developed by Corapcioglu and Haridas (1984, 1985) include terms for death of microorganisms in the free state (in pore water) and in the adsorbed state. The expressions for inactivation in the liquid phase and solid (adsorbed) phase take on the forms

$$\text{Rate}_{\text{liquid}} = -k_d \theta C \tag{13}$$

$$\text{Rate}_{\text{solid}} = -k_d \rho \sigma \tag{14}$$

where k_d is the decay, death, or inactivation coefficient. If the rate expressions are given a positive sign, they would represent an increase, or growth, in the population. The concentrations in both the free and adsorbed states were calculated in the Haridas (1983) saturated flow model.

The mathematical expressions developed by Corapcioglu and Haridas (1984, 1985) for deposition, death, and growth as first-order and zero-order functions of microbial concentrations in the liquid phase lend credence to the use of an existing transport model with the inclusion of decay terms and sink/source terms to account for microbial kinetics.

The rate of deposition is also dependent upon the changing effective porosity due to clogging. The Haridas (1983) model tracks the fraction of the source population of microorganisms clogged in the pore spaces and periodically calculates a new effective porosity. The effect of a decreasing porosity due to clogging may be a significant factor in the transport calculations, but its absence in models of unsaturated contaminant transport is prevalent.

Examples of Microbial Transport

This section summarizes only a selected few of the many research endeavors associated with microbial transport. It is meant to provide an understanding of the type of transport phenomena that can yield data and information to further model development and that may in turn be aided by establishment of accepted models for microbial transport. There are several areas of study where funding for research has historically been directed, including studies of releases from on-site disposal systems and from unit processes at wastewater treatment plants.

On-site disposal systems, which include septic tank systems for individual or groups of homes and businesses, have been a frequent topic of studies of microbial transport. Funding for such studies arises from concerns regarding the capacity of the unsaturated soil system to remove microorganisms by filtration or adsorption. Unit processes that involve flow through porous media at wastewater treatment plants include trickling filters and infiltration basins. The current research in transport involves infiltration basins or percolation ponds that dispose of treated effluent.

Field studies

A study of on-site disposal systems in Virginia generated results confirming decreases in fecal and total coliform concentrations up to 28 m from septic drainfields sited in marginally suited soils (Reneau and Pettry, 1975). The soils were considered marginally suitable because of high seasonal water table levels, the presence of restricting soil layers, or both. The researchers collected samples from nests of piezometers located at various distances (0.15 to 6 m, 13.5 m, or 28 m) away from existing drainfield sources at the three study sites. Samples were collected and analyzed over a 2-year period, but they were collected infrequently, usually with a 2- to 5-month interval between sampling events. The total release rate from the drainfield sources was estimated based on assumed daily disposal rates for the septic systems. The subsurface drainfields at each site were different, so the daily disposal rates per unit area were estimated for each study site. The researchers concluded that both total and fecal coliform bacteria from the drainfields were transported through groundwater, but that the concentrations of bacteria were significantly reduced during transport.

McCoy and Hagedorn (1979) studied microbial transport in a saturated flow system at a controlled site in Benton County, Oregon. The researchers introduced antibiotic-resistant strains of *Escherichia coli* into three soil horizons at a study site. The transport of bacteria was monitored by collecting samples from five nests of six piezometers each, installed from 2.5 to 20 m downgradient from the injection lines. Two of the three injection lines were situated in the unsaturated zone, but the third, installed at a depth of 60 cm below ground surface, may have been within the zone of saturation. The duration of the field test was 84 h, and samples were collected from the monitoring wells at systematic intervals which lengthened as time of the study increased. The researchers reported their results as the arithmetic mean of results obtained from the laboratory counts of the number of viable antibiotic-resistant bacteria present. Their results showed the reduction in relative concentration with increasing distance from the source and bore a striking similarity to the analytical solutions of the hydrodynamic advection-dispersion equation for contaminant transport.

The downgradient concentration results observed by McCoy and Hagedorn (1979) led to the calculation of an apparent horizontal groundwater flow velocity of 17 cm/min at the Oregon study site. Their calculation was based on observations of the time elapsed between release of bacteria from the source and their appearance at a downgradient sampling point at a known distance from the source. The actual velocity of bacteria moving through a porous medium differs from the apparent velocity primarily due to the tortuous path that the bacteria follow.

Schaub and Sorber (1977) performed a field study to evaluate the effectiveness of rapid infiltration beds in the removal of enteric bacteria and viruses. Their study used bacteriophage as a tracer of microbial movement in wastewater that was introduced through 0.324-ha (0.8-acre) infiltration beds at Ft. Devens, located northwest of Boston, Mass. The wastewater, with a known and constant concentration of f2 bacteriophage, was allowed to infiltrate for 7 days, followed by a 14-day dry period during which no water infiltrated into the study site. Groundwater elevations in monitoring wells at the study site were measured 11 times during the 22-day study. Groundwater samples were collected from downgradient monitoring wells and analyzed for the presence of f2 bacteriophage, indigenous fecal streptococci, and indigenous viruses. Concentrations of f2 bacteriophage observed in downgradient wells appeared to be very much like the theoretical solutions for the hydrodynamic advection-dispersion equation of contaminant transport.

Field studies performed by Gilbert et al. (1976a, 1976b) were among those conducted in conjunction with the Flushing Meadows Wastewater Renovation Project located west of Phoenix, Ariz. The Flushing Meadows project had six infiltration basins, each 6 by 210 m (about 20 by 690 ft) in size, that received effluent discharged following secondary treatment of sewage. The stated purpose of the Gilbert et al. (1976a, 1976b) study was to evaluate the effective removal of viruses and bacteria from secondary sewage effluent by land treatment. Gilbert et al. did detect the presence of bacteria and viruses in samples collected in four downgradient monitoring wells and calculated the percent reduction in concentrations (1976a, 1976b). They detected seasonal variations showing that the greatest concentrations of enteric viruses in monitoring wells occurred during the months of August and September and the lowest concentrations occurred in June and July (Gilbert et al., 1976b). The researchers concluded that human bacterial and viral pathogens are largely removed during transport through the soil and that observed concentrations decreased four orders of magnitude in some cases. The researchers recognized that microorganisms removed during transport may survive and continue to pose a potential health threat.

Microorganisms as Tracers of Contaminant Movement

McCoy and Hagedorn's (1979) observations of bacterial transport to identify groundwater flow rate (described above) led to a subsequent study comparing the use of antibiotic-resistant bacteria with that of fluorescein dye as a tracer of groundwater movement. Rahe et al. (1979) set up a controlled study at the Oregon study site in which antibiotic-resistant *E. coli* strains of known concentration were introduced into the subsurface through the horizontal injection lines. A constant infiltration rate of 1 cm/h at the ground surface was supplied by means of a sprinkler. Samples were collected from the downgradient piezometers at periodic intervals for 12 h after injection. Distinct measurements of *E. coli* concentrations over time at given downgradient distances at three soil depths provided pertinent data about groundwater flow and microbial transport below the site.

When Rahe et al. (1979) repeated the experiment using fluorescein dye, the dye solution at a known concentration was introduced through each injection line for three sampling runs. Samples were collected from the piezometers at intervals during a 12-h sampling period. No dye was visibly observed in the samples, but use of a spectrophotofluorometer revealed residual concentrations nearly six orders of magnitude smaller than the source strength, whereas the reduction in bacterial concentration was about five orders of magnitude. The researchers concluded that optical quenching of the dye and adsorption and scattering of fluorescence by clay and organic matter in the porous medium effectively diminished the dye within 2.5 m of its release. The researchers then questioned whether dyes, particularly fluorescein dye, should be used to evaluate whether a septic system is malfunctioning. Furthermore, they suggested that marked strains of bacteria could serve quite satisfactorily as tracers of groundwater movement and transport of pollutants.

Other researchers have investigated the application of microorganisms as tracers of groundwater movement. Wimpenny et al. (1972) had success using bacteriophage as groundwater tracers. Sinton (1980) performed a detailed experiment at the Burnham wastewater disposal site in New Zealand. In this study, *E. coli* and *Bacillus stearothermophilus* strains were injected into an injection well and downgradient concentrations were observed in samples collected over a 12-month period. Sinton injected suspensions of bacteria in one well and collected samples from four downgradient monitoring wells for 10 days after the injections. Samples were also collected from an upgradient well to take into account the background concentrations of naturally occurring strains. The author concluded that both microorganisms were able to travel considerable distances in an aquifer: up to 900 m (2,950 ft) was observed, with apparent mean velocities of 184 m/day (600 ft/day) for *B. stearothermophilus* and 199 m/day (650 ft/day) for *E. coli*. The study attempted to identify a correlation between precipitation and concentrations of bacteria in groundwater. Sinton (1980) suggested that wet weather could cause *B. stearothermophilus* to be washed from the soil and into the aquifer, thereby obscuring the data regarding transport of the injected organisms that were to act as tracers. Nevertheless, Sinton (1980) concluded that both species were successful as groundwater tracers.

USING AN EXISTING CONTAMINANT TRANSPORT MODEL

A number of existing transport models were reviewed to identify one acceptable to evaluate the potential for transport of genetically engineered microorganisms in hypothetical scenarios (Versar, 1986). It was found that most numerical transport and groundwater models are data-intensive and have not been validated for biological sources. However, a few computer models have been developed specifically for the purpose of modeling microbial transport. One was developed to describe the transport of enhanced oil recovery bacteria (Battelle, unpublished), and another was designed to model the fate of bacteria and viruses in saturated porous media (Haridas, 1983; Grondin, 1987; Grondin and Gerba, 1986). An unsaturated zone transport model, called SESOIL, has been validated for landspreading scenarios (Bonazountas et al., 1981), which are analogous to biotechnology scenarios that involve application of microorganisms to soil with mulching and subsequent cultivation. Several models for microbial transport in the saturated zone have been developed (e.g., Haridas, 1983; Corapcioglu and Haridas, 1984, 1985). These models were based upon the advection-dispersion equation for solute transport.

The implicit parallelism between microbial and solute transport assumes that populations of microorganisms in a porous medium can be described in terms of concentrations at a given time and location relative to the initial source strength (concentration) at which the microorganisms were introduced. Terms within the advection-dispersion equation could be modified to represent growth or death of microorganisms or removal of microorganisms from the transporting fluid (water) by straining, adsorption, predation, sedimentation, or filtering. Accepting the parallel to solute transport implies that the driving force for microbial movement is assumed to be water flowing into and throughout a porous medium. The movement of water, or soil moisture, in the unsaturated zone is caused by hydraulic gradients resulting from differences in matric pressure at adjacent locations in the soil and by the acceleration due to gravity. Because the vertical flow gradient is predominant in the vadose zone, it is appropriate to simplify the hydraulic equation to the one-dimensional, vertical flow version of the Darcy-Richards equation which is given above as equation 9.

Unsaturated-zone contaminant transport models developed by Van Genuchten (1978, 1986) were used in hypothetical scenarios to evaluate the potential transport of intentionally released genetically engineered microorganisms (Versar, 1986, 1987; Dickinson, 1988). Most of the scenarios involved land applications, so the use of an unsaturated flow and transport model was preferred. The objective was to determine whether the released microorganisms would be transported vertically downward and enter the groundwater table as viable organisms, or whether they would otherwise pose a concern due to potential transport of live or dormant organisms away from the intended area of application. Although Versar's work focused on recombinant (genetically engineered) microorganisms, the approach was pertinent to evaluations of naturally occurring microorganisms. Modeling efforts rely upon current knowledge regarding the existence and movement of naturally occurring microorganisms.

Because most of the hypothetical scenarios involved releases to the unsaturated zone, an initial goal was to identify an existing computer model for unsaturated moisture flow and microbial, or contaminant, transport.

Further criteria were applied to narrow the selection for the most appropriate existing computer model to the evaluation of hypothetical release scenarios (Versar, 1986, 1987). It was necessary to identify a model that would be applicable to many scenarios and that could accommodate a wide range of soil parameters, initial conditions, and boundary conditions. It was important that the model selected should have been tested and verified for solute contaminant transport and that it be fully documented. A finite element model for one-dimensional contaminant transport in the vadose zone, called Sumatra-1 (Van Genuchten, 1978), and a later revision called WORM (Van Genuchten, 1986), met those criteria and were used to describe possible releases and transport of microbial organisms for several hypothetical scenarios for land application of genetically engineered microorganisms (Versar, 1986, 1987).

An Unsaturated Zone Transport Model

In the Sumatra-1 model, the flux of contaminants moving vertically with water in an unsaturated porous medium is given by the hydrodynamic advection-dispersion equation of the form

$$\frac{\partial \theta C}{\partial t} = \frac{1}{R_d}\left[\frac{\partial}{\partial z}\left(\theta D_h \frac{\partial C}{\partial z} - qC\right) - (\alpha\theta + \beta\rho k_d)C + \gamma\theta\right] \quad (15)$$

where θ is moisture content; C is concentration of the contaminant; R_d is a retardation factor whose value is based on the soil bulk density, soil porosity, and the distribution coefficient; D_h is the coefficient of hydrodynamic dispersion; q is the specific discharge, or flux, of water; ρ is the soil bulk density; k_d is a laboratory-determined distribution coefficient which represents the contaminant's propensity to adsorb to soil; and α, β, and γ are coefficients of decay.

The decay coefficients can represent a decrease or increase (if the coefficients are negative) in the concentration of contaminants in the porous medium. Therefore, they may be used to describe functions that remove contaminants from transport, such as death of microorganisms or adsorption and filtering, and functions that increase contaminants in transport, such as growth of a microbial population. The decay coefficients α and β are first-order decay coefficients; that is, they are multiplied by the concentration in the liquid phase. Their contribution is a function of the amount (concentration) of the contaminant, or microorganisms, present in the liquid phase. The γ term is independent of the concentration and is called a zero-order decay coefficient. It has the effect of a sink or source term, that is, an introduction or removal of contaminants, or microorganisms, entirely independent of their amount (concentration).

In equation 15 the decay coefficients α and γ apply to concentration in the liquid (water) phase only. Thus, they are multiplied by the moisture content, represented by θ. The β decay coefficient applies to concentrations adsorbed to solid particles. One of the beneficial features of the Sumatra-1 model is this inclusion of decay functions for both the liquid and solid phases. If this model is used to calculate the transport of microorganisms, then fate (growth and death) parameters for microorganisms that have adsorbed to soil or been filtered out of the transporting fluid can be included.

The similarity between the mathematical expressions for the decay coefficients in equation 15 and the kinetics parameters presented by Corapcioglu and Haridas (1984, 1985) lends credence to the concept of applying contaminant transport models to describe the transport of microorganisms. Then if the fate parameters of microorganisms can be expressed as first-order and zero-order decay constants for the liquid and solid phases, models like Sumatra-1 may be directly applicable to describe problems of contaminant transport.

Improvements in the Sumatra-1 model have been made. The finite element solution technique employed by the code was improved to make the model more efficient. The revised numerical model was published under the name WORM (Van Genuchten, 1986). The revised code allowed for a hydrologic sink function in the Darcy-Richards equation representing uptake of soil moisture by plant roots. The WORM model handles extreme hydrologic boundary conditions more effectively than does Sumatra-1, and it tends to have greater numerical stability than the earlier version. The hydrodynamic dispersion-advection equation used by WORM is simpler than that coded in Sumatra-1. The governing transport equation in WORM is

$$\frac{\partial \theta C}{\partial t} = \frac{1}{R_d}\left[\frac{\partial}{\partial z}\left(\theta D_h \frac{\partial C}{\partial z} - qC\right) - \theta\mu C\right] \quad (16)$$

In this equation, all of the first-order decay coefficients are incorporated into one term represented by μ. The WORM model does not include a zero-order decay term and does not separate the solid and liquid phases of the fate parameters. These aspects make it less applicable to modeling transport of microorganisms than Sumatra-1. A critical comparison of the two models also helps to identify the strong features of each for consideration in creating new models for microbial transport.

Results Using Contaminant Transport Model

The use of numerical models like Sumatra-1 and WORM to describe the transport and fate of microorganisms in the vadose zone allows a detailed description of microbial movement in a vertical plane and incorporates both hydraulic and microbial (contaminant) terms. Both Sumatra-1 and WORM can handle layered soil profiles and diverse hydraulic properties. The user can establish any initial and boundary conditions appropriate for the hydraulic and concentration conditions of the site being studied.

The output from the Sumatra-1 and WORM models includes the soil moisture, matric pressure, and microbial concentration distributions through the soil profile at time periods requested by the user. The output is detailed because the solution technique generates values for each node, and values at all nodal points are printed. The Sumatra-1 and WORM models are one-dimensional, and their results illustrate how contaminant (microbial) concentration is distributed throughout depth in any given location. Not surprisingly, the results appear much like the movement of solutes in an unsaturated porous medium (Versar, 1986, 1987). The effect of the decay constants in reducing relative concentration by up to 5 orders of magnitude at 40 cm of depth after 60 days was evident in comparisons of computer runs with and without the first-order and zero-order decay coefficients. The magnitude of the reductions in relative concentrations when decay coefficients are introduced suggests the importance of acquiring laboratory and field data to assign appropriate values for the decay coefficients of microorganisms in differing soil environments.

PROBLEMS WITH USING CONTAMINANT TRANSPORT MODELS

In general, many of the studies associated with the movement of microorganisms in porous media have generated insufficient data to establish a predictive model. To calibrate a predictive model it is necessary to have enough observations to have two independent sets of data, one for the calibration and one for a verification. The final level in using models is the validation of that model as a predictive tool, which can only occur over a period of time during uses of the model and comparisons between the predicted and actual results. Discussion of the needs for general model calibration, verification, and validation can be found in several publications including McCuen (1985) and Versar (1988a). At present the microbial transport data available to attempt even a model calibration are lacking. A search for existing published data of microbial transport revealed that the most complete transport data sets to employ transport models were those studies in which the research was directed toward using microorganisms as tracers of contaminant movement (Versar, 1988b). This section will identify several more specific problems associated with adapting existing contaminant transport models for use in modeling microbial transport.

This section discusses the following four general areas of problems associated with adapting existing transport models to describe microbial transport. (i) Important parameters are not directly measurable and/or are variable (including the microbial source population) in the flow regime. (ii) Aquifer hydraulic properties are not measured or reported for many microbial studies. (iii) Microorganisms are not transported as contaminants in solution but as colloids. (iv) Finally, important factors in the behavior of the microorganisms are not included in the governing equations, such as local changes in porosity and permeability due to clogging by the microorganisms.

Parameter Estimation and Variability

A disadvantage of numerical models is that they are data-intensive, and the input data required for the Sumatra-1 and WORM models are highly site specific. Further work is necessary to determine how sensitive the calculation of microbial concentration is to the hydraulics of a flow system. Hydraulic information may be limited in actual field problems, and approximations or general values may have to be used. Some work has been done in identifying best parameters that can be assigned for spatially varying flow systems. Anderson (1979) reviewed the concept of lumped parameter models in which spatially averaged values are used. Techniques developed in geostatistics, such as kriging, can be used to estimate the values of parameters in a spatially varying flow regime (de Marsily, 1984).

Many of the terms present in the hydraulic flow and contaminant transport equations are not directly measurable. Techniques have been developed to determine the values of some aquifer parameters, such as aquifer transmissivity or hydraulic conductivity and storativity, but the parameters that must be known to solve the unsaturated flow equation or the contaminant transport equation are substantially greater in number and detail. For unsaturated flow, it is necessary to identify the soil moisture characteristic curve and the unsaturated hydraulic conductivity as a function of soil moisture. It may be necessary to calculate certain constants that relate the hydraulic conductivity to the soil moisture and, depending upon the internal model coding, the air entry (or bubble) pressure of the soil. The soil moisture characteristic curve and unsaturated hydraulic conductivity for some soils have been tabulated (e.g., Mualem, 1976). The extremes of the functions are not always reported, and the air entry pressure is not always known. It appears as if a numerical solution to the Darcy-Richards equation may be most sensitive to the value of the air entry pressure, which is the least well known of any of the unsaturated flow parameters (Dickinson, 1985).

The contaminant transport equation introduces other parameters that are not directly measurable, such as the groundwater pore velocity, dispersivity, diffusivity, and retardation coefficient; are measurable only in a laboratory, such as the contaminant distribution coefficient; or are highly variable in the flow regime, such as soil bulk density or effective porosity. The coefficient of dispersivity has been found to be scale dependent; that is, its value changes depending upon the size of the flow regime under study (e.g., Gelhar et al., 1985). An intensively monitored field study site at Columbus Air Force Base in northeastern Mississippi was used to identify the role of spatial variability of hydraulic conductivity in the hydrodynamic dispersion of solutes in a saturated groundwater flow regime (Gelhar, 1985). One objective of the study was to develop a practical method to estimate statistical parameters that reflect the spatial variability of hydraulic conductivity. The final phases of the study are to be completed in 1990.

Laboratory column studies are conducted using soil samples that are collected from a field study site and then transported to the laboratory for experimentation. Soil columns have the advantage of being more readily controlled than field studies, but they are not as representative of field conditions as in situ experiments. For example, physical size limitations of the columns, often less than 10 in. (ca. 25.4 cm) in diameter and 2 ft (ca. 61 cm) high, require that samples used in these experiments represent a relatively small volume of the study site. Even in cases where an undisturbed soil sample is collected with a thin-walled core sampler (such as a Shelby tube), transported to a laboratory, and transferred directly into a column for study, the sample as removed from its flow environment may yield results different from an in situ study. These limitations on sampling, and the fact that some (disturbed) samples may be repacked in the columns, mean that the samples may be more homogeneous than the field site. The process of collecting a soil sample and repacking it into a laboratory column can substantially reduce that sample's similarity to field conditions. While these factors may limit the ability to extrapolate experimental results to field conditions, they do provide an opportunity to generate the types of data needed for modeling.

Reporting Aquifer Hydraulic Parameters

The field studies in which microorganisms were used as tracers of groundwater movement were those which came the closest in generating sufficient data to meet the input data requirements of models (i.e., Rahe et al., 1979; Sinton, 1980). Experiments that are designed to test the effectiveness of microorganisms as tracers of groundwater movement may generate sufficient data to support a modeling effort because (i) the experiments are controlled, and (ii) usually a substantial amount of groundwater data is also collected in order to demonstrate the proposed tracer's effectiveness. In application, one of the benefits of an effective tracer is the ability to identify groundwater flow rates and directions without an exhaustive and detailed groundwater hydraulic evaluation. It is ironic, therefore, that the most appropriate studies for modeling microbial movement in groundwater are those which seek only to demonstrate the effectiveness of microorganisms as tracers.

The observation of incomplete hydrogeological data sets in most field studies of microbial fate in the subsurface points out the need to expand current studies to include collection of soil and/or aquifer hydraulic data so that more complete analysis of transport is possible. Studies of microbial fate in the unsaturated zone will present an even greater challenge than saturated flow studies in this regard because of the complexities of hydraulic movement under various moisture conditions, compounded by the heterogeneity of soil, pH, and nutrients that would affect the survival (including predation, die-off, and regrowth) and adsorption/desorption patterns of the microorganisms.

Selected research studies of microbial transport in laboratory columns or field tests were reviewed to evaluate whether the data generated for microbial transport could be used to develop a mathematical model of transport using the equations presented above (Versar, 1988b). For each of these studies, it was concluded that data were inadequate for a modeling approach.

Field studies

A field study performed by Reneau and Pettry (1975) provided some data essential to modeling efforts. They observed the movement of fecal and total coliform bacteria through soils at three sites on the coastal plain of Virginia. In their report they listed information taken from a soil survey, including a description of the soil physiography and classification of drainage. The researchers provided the particle size distribution, percent organic matter, and cation exchange capacity of samples from each soil horizon. Although such data reveal useful aspects about the nature of the soil, none of the parameters measured can be used directly in a modeling effort. Far more pertinent are the values of horizontal saturated hydraulic conductivity which the authors provided.

The paucity of observed concentration data and the uncertainty involved in the estimate of the release from the sources make the data of Reneau and Pettry unsatisfactory for modeling efforts. The researchers did not control the sources during the study, so it is impossible to know the amounts released, the period of the releases (intermittent or continuous), or the concentrations of the sources. This is important because the source data could vary from day to day because the on-site disposal systems were being used. Even if the daily release data could be pinpointed, samples from the piezometers would be difficult since piezometric readings were taken every several months, but releases occurred every day. It would not be possible to identify which releases caused corresponding observed concentrations in the groundwater.

Studies by McCoy and Hagedorn (1979) and Rahe et al. (1979) generated a nearly complete data set regarding the evaluation of concentrations as a function of space and time. Rahe et al. (1979) also included pertinent soil information, i.e., the soil profile description and the soil's particle size distribution, pH, percent organic matter, and cation exchange capacity. In an attempt to develop a mathematical relationship between observed concentrations and distance from the source, the authors did not use solutions to the transport equation but rather tried to establish existence of a linear relationship between mean maximum concentration and distance in the log-log space. The authors reported good correlation values, from 0.88 to 0.97, but that was true only in the log-log space (Rahe et al., 1979). Log transformations tend to inappropriately minimize unexplained error in the real space, making many relationships appear linear when, in fact, they may not be (McCuen, 1985).

Data collected by Schaub and Sorber (1977) related to saturated flow only and included some parameters that would be needed for modeling, such as the soil void ratio and one laboratory-measured value for vertical, saturated hydraulic conductivity (no values of horizontal hydraulic conductivity were provided). The value provided by the void ratio is not the same as that for soil porosity, but it could provide useful information to aid in estimating porosity. Also, the hydraulic gradient in the aquifer could be calculated from the water table contours provided by the researchers. It appears that the amount and quality of data provided by Schaub and Sorber (1977) would permit modeling, but the data are suitable only for generalized treatments.

Field studies performed by Gilbert et al. (1976a, 1976b) were among those conducted in conjunction with the Flushing Meadows Wastewater Renovation Project in Arizona. The researchers monitored releases from six infiltration basins which were subjected to intermittent wastewater flooding and drying periods of 14 days each. They estimated the annual infiltration of sewage effluent as 100 m/year per unit area, but it was also possible to establish infiltration rates for each event by considering the difference between inflow and outflow rates measured at flumes. The maximum depth of wastewater in the infiltration beds was controlled by overflow structures. It would be possible, then, to establish critical hydraulic data related to the source, and the concentrations of organisms at the source were measured.

However, the limited downgradient concentration data available from the study do not allow the site to be modeled. Samples were collected from only four wells every 2 months during week 2 of each flood period. The concentration data were reported as averaged values or as a range. Though this data manipulation makes the data manageable, it does not lend itself well to modeling efforts.

Laboratory column studies

Many of the laboratory column studies focus upon the role that soils perform in removing bacteria and viruses from sewage effluent during infiltration. Such studies report percentages of microorganisms retained or eluted and usually include information about the effluent used and its release rate or volume introduced to the column. Some of the studies also report pertinent soil characteristics such as soil pH, grain size, and soil bulk density.

Bitton et al. (1976) also reported the soil's water-holding capacity, which is related to soil porosity, and Smith et al. (1985) reported the initial water content in the soil columns at the start of their experiments.

Sobsey et al. (1980) measured the cation exchange capacity and organic matter content of soils in their study. These values are influential in adsorption and desorption phenomena, but the parameters are not directly usable in modeling unless expressed in terms of a distribution coefficient or retardation coefficient. In addition to soil organic carbon content data, Gerba and Lance (1978) developed coefficients for adsorption from batch experiments using soil from their study site west of Phoenix, Ariz. (the Flushing Meadows Wastewater Renovation Project), and primary-treated and secondary-treated sewage water appropriate to their laboratory column study. They found that the adsorption of polioviruses in the sewage water at that site fit a Freundlich isotherm relationship. This type of information is directly applicable to modeling approaches that incorporate retardation of contaminants or microorganisms.

Wang et al. (1981) provided data on the effects that flow rate has on the retention of poliovirus and echovirus in four different sandy or sandy-loamy soils. The study also evaluated the removal rate, or removal effectiveness, of the upper 17 cm of the soil column as compared to the lower 70 cm of the column. Wang et al. (1981) used saturated soil columns under different constant head boundary conditions. The data from this study and related studies from the Flushing Meadows Wastewater Restoration Project appear to be extensive and reliable. Such data may be appropriate for an evaluation of the applicability of models. However, Wang et al. (1981) did not report the soil hydraulic properties, such as hydraulic conductivity, so modeling would be limited to the given infiltration flow rates caused by changing the constant hydraulic heads at the top of the soil columns.

Microbial tracer studies

Sinton (1980) reported some data regarding aquifer properties and groundwater flow for the tracer study at the Burnham wastewater disposal site in New Zealand. The data available from both research efforts may provide enough information to perform a modeling study of transport of microorganisms in a saturated groundwater flow system. However, the Sinton study suffered from a limited number of downgradient observations. Samples from wells located closest to the injection well were collected seven times on the first day after injection and two to three times during the next 3 days, followed by once a day for the remaining 6 days of the sampling period. Samples were collected in the more distant wells only once or twice per day. All of these data are usable, but Sinton estimated travel times between the injection well and each of the monitoring wells as being halfway between the time of first observation and the previous sampling event because it was not possible to take more frequent samples. Sinton observed that this may have introduced considerable error in estimating travel times to the more distant monitoring wells. In fact, because of the infrequent sampling schedule, only three observations of *E. coli* in the most distant well, at 900 m (or 2,950 ft), and one observation of *B. stearothermophilus* on day 6 appear to be usable for that distance.

Colloidal Transport

The hydrodynamic advection-dispersion equation is the same form of a governing equation as was used in a model of colloidal transport in a porous medium (Nuttall, 1986). Since the basis for the transport is similar, an accurate representation of colloidal transport may be accomplished by addressing those aspects that make it unique. If one of the principal differences between solute and colloidal transport is the variability of the flow field, or velocity vector, it may be reasonable to use stochastic modeling approaches that allow introduction of parameter ranges. Bhattacharya et al. (1976) have added a statistical basis to the governing equations of flow through unsaturated porous media to address the impact of field variability of parameters. In a statistical treatment, flow and transport parameters are expressed as distributions that are assigned a given range, mean, and distribution type. The model calculations are made using random selection of parameters, often using a Monte Carlo technique (de Marsily, 1984). The output of such a model is also a range, with values falling within a prescribed confidence interval. Anderson et al. (1987) have used these concepts in applications for microbial transport in a saturated flow system.

Benefits of stochastic modeling are that it can address the effects of local variability in the velocity flow field and also address the problem of spatial variability of flow and transport parameters due to aquifer heterogeneity.

Transport Factors Not Included

Arguments in favor of modeling the transport of microorganisms using the advection-dispersion equation for contaminant transport are presented above. Researchers involved in studying the fate, transport, and survivability of microorganisms in porous media are evaluating a number of control factors that may or may not fit into the variables in the mathematical equations presented above. There is ongoing research to evaluate the significance of species-specific factors, such as hydrophobicity, flagellar movement, production of exopolysaccharides (capsular material), and surface electrical charge on the fate and transport of selected microorganisms in porous media. Such factors, if found to have a significant role in transport, may be incorporated into some of the sorption and first- or second-order decay (or growth) variables in the mathematical governing equations. There are other factors that would affect the structure of the governing equations, such as those described below.

Time-dependent porosity

The rate of bacterial deposition or straining is dependent upon the changes in the sizes of the pore spaces as clogging progresses. This could be tracked by allowing the effective porosity of the porous medium to be a variable rather than a constant value. The Haridas (1983) model tracks the amount of microorganisms clogged in the pore spaces and periodically calculates a new effective porosity. This feature is missing from the Sumatra-1 model (Van Genuchten, 1978) but is a modification that could be made to improve the model for use in microbial transport.

Geochemistry

There are other possible factors, such as soil and water chemistry, that may necessitate expanding the number of governing equations to include chemical balance. It has been suggested that the presence of clay particles in a porous medium may maintain the pH at suitable levels for microbial life (Hattori and Hattori, 1976).

An example of the potential importance of the influent water chemistry on the fate of sorbed microorganisms was given by research conducted by Landry et al. (1979). The researchers studied the adsorption of polioviruses, coxsackievirus, and echoviruses from sewage effluent as it percolated through columns of highly permeable sandy soil. The soil columns used were undisturbed samples collected from an operating recharge basin of a sewage treatment plant in Medford, N.Y., on Long Island. Sewage effluent containing known concentrations of viruses was allowed to infiltrate through each column. The filtrates collected after the infiltration of sewage were assayed for the presence of viruses, and the differences in the initial and final concentrations were assumed to be due to adsorption of viruses by the columns.

The columns were paired by infiltration rates. Tests for the elution of adsorbed viruses were conducted by subjecting each pair of columns to a subsequent infiltration of sewage effluent containing viruses and also subjecting the second column of the pair to an equal volume of simulated rainwater having the same anion and cation content as precipitation in the northeastern United States. Final elution was determined by destroying the soil core's structure and mixing the soil with an extract in a rotary shaker. Landry et al. (1979) found that 82 to 94% of poliovirus type 1 was adsorbed to the soil as the sewage water infiltrated through the columns, but 1 to 14% of the adsorbed viruses were eluted with subsequent applications of sewage water and 3 to 33% were eluted with applications of artificial rainwater. For other strains of viruses, 97 to 99% of poliovirus type 3 was adsorbed, but less than 1% was eluted with subsequent applications of sewage water, and 24 to 67% was eluted with applications of artificial rainwater. The researchers attributed the increased elution rates with artificial rainwater to its lower electrical conductivity (40 μmho/cm compared to 606 μmho/cm for sewage effluent) and, to a lesser extent, its lower pH value (4.4 for artificial rainwater compared to 7.5 for sewage effluent). In such cases, where the chemistry of the influent has a profound effect on the elution of adsorbed microorganisms, it is important to include chemistry-related factors in the governing equations. This is an area that warrants further research.

CONCLUSIONS

Observations of microbial transport and studies of survivability in porous media should be accompanied by measurement of the hydraulic and transport parameters needed in the governing equations. The hydrogeological data most needed to make reliable comparisons of subsurface field studies and existing models are hydraulic conductivity, hydraulic gradient, and aquifer porosity. Studies of microbial fate in the unsaturated zone also necessitate information on the soil moisture characteristic curve, that is, the value of hydraulic conductivity as a function of moisture content.

An order-of-magnitude comparison should be prepared to identify the critical driving forces in the transport of microorganisms. Sensitivity analyses should be performed to identify the principal parameters in the governing equations.

Possible Resolution of Problems

Further studies could employ statistical distributions of hydraulic and transport parameters and of the microbial behavioral characteristics to improve the universality of the solutions to the governing equations. Attention should be given to generating the appropriate statistical distribution functions.

The ideal model for microbial transport would include terms for the concentration of the microorganisms in the liquid state (in suspension) and in the solid state (adsorbed to the soil particles) and would allow for changes in porosity as the microorganisms clog and declog in the porous medium. The ideal model would also take into consideration the potential ranges of the values for all hydraulic, transport, and fate parameters. It may be a statistically based model that selects spatially and temporally distributed input data from probability distributions for each parameter and generates a probability distribution for the output solution. The ideal data set for a microbial transport study would include multiple measurements of all of the hydraulic, transport, and fate parameters and the hydraulic and concentration boundary conditions, as possible, as well as an estimate of the probable extremes of the parameters and boundary conditions for the flow regime.

REFERENCES

Anderson, D. L., J. M. Rice, M. L. Voorhees, R. A. Kirkner, and K. M. Sherman. 1987. Ground water modeling with uncertainty analysis to assess the contamination potential from onsite sewage disposal systems (OSDS) in Florida, p. 264–273. *In On-Site Wastewater Treatment*, vol. 5. Proceedings of the Fifth National Symposium on Individual and Small Community Sewage Systems. Publication 10-87. American Society of Agricultural Engineers, Chicago, Ill.

Anderson, M. P. 1979. Using models to simulate the movement of contaminants through groundwater flow systems. *Crit. Rev. Environ. Control* **9**:97–156.

Battelle. Evaluation of potential biotechnology models. Draft final report. *Prepared for* Contract no. 68-01-6721, U.S. Environmental Protection Agency, Washington, D.C.

Bear, J. 1979. *Hydraulics of Groundwater*. McGraw-Hill International Book Company, New York.

Bhattacharya, R. N., V. K. Gupta, and G. Sposito. 1976. On the stochastic foundations of the theory of water flow through unsaturated soil. *Water Resour. Res.* **12**:503–512.

Bitton, G., N. Masterson, and G. E. Gifford. 1976. Effect of a secondary treated effluent on the movement for viruses through a cypress dome soil. *J. Environ. Qual.* **5**:370–375.

Bonazountas, M., J. Wagner, and B. Goodwin. 1981. Evaluation of seasonal soil/groundwater pollutant pathways. *Prepared for* Contract no. 68-01-5949, U.S. Environmental Protection Agency, Washington, D.C.

Clapp, R. B., and G. M. Hornberger. 1978. Empirical equation for some soil hydraulic properties. *Water Resour. Res.* **14**:601–604.

Corapcioglu, M. Y., and A. Haridas. 1984. Transport and fate of microorganisms in porous media: a theoretical investigation. *J. Hydrol.* **72**:149–169.

Corapcioglu, M. Y., and A. Haridas. 1985. Microbial transport in soils and ground water: a numerical model. *Adv. Water Res.* **8**:188–200.

de Marsily, G. 1984. Spatial variability of properties in porous media: a stochastic approach, p. 719–770. *In* J. Bear and M. Y. Corapcioglu (ed.), *Fundamentals of Transport Phenomena in Porous Media*. Martinus Nijhoff Publishers, Boston.

Dickinson, R. A. 1985. Soil moisture flow in the unsaturated zone under hydraulic and thermal gradients. Thesis. University of Maryland, College Park.

Dickinson, R. A. 1988. Applications of an existing one-dimensional model to movement of microorganisms in the vadose zone. *In* Proceedings of a workshop: The Integration of Research and Predictive Model Development in Biotechnology Risk Assessment. U.S. Environmental Protection Agency, Office for Research and Development, Environmental Research Laboratory, Gulf Breeze, Fla.

Dullien, F. A. L. 1979. *Porous Media, Fluid Transport and Pore Structure*. Academic Press, Inc., New York.

Freeze, R. A., and J. A. Cherry. 1979. *Groundwater*. Prentice-Hall Inc., Englewood Cliffs, N.J.

Gelhar, L. W. 1985. Macrodispersion Experiment (MADE): design of a field experiment to investigate transport processes in a saturated groundwater zone. Report no. EA-4082. Research project 2485-5. *Prepared for* Electric Power Research Institute, Palo Alto, Calif.

Gelhar, L. W., A. Mantoglou, C. Welty, and K. R. Rehfeldt. 1985. A review of field-scale physical solute transport processes in saturated and unsaturated porous media. Report no. EA-4190. Research project 2485-5. *Prepared for* Electric Power Research Institute, Palo Alto, Calif.

Gerba, C. P., and J. C. Lance. 1978. Poliovirus removal from primary and secondary sewage effluent by soil filtration. *Appl. Environ. Microbiol.* **36**:247–251.

Gilbert, R. G., C. P. Gerba, R. C. Rice, H. Bouwer, C. Wallis, and J. L. Melnick. 1976a. Virus and bacteria removal from wastewater by land treatment. *Appl. Environ. Microbiol.* **32**:333–338.

Gilbert, R. G., C. P. Gerba, R. C. Rice, H. Bouwer, C. Wallis, and J. L. Melnick. 1976b. Virus removal by soil filtration. *Science* **192**:1004–1005.

Grondin, G. H. 1987. Transport of MS-2 and f2 bacteriophage through saturated Tanque Verde Wash soil. Masters thesis. University of Arizona, Tucson.

Grondin, G. H., and C. P. Gerba. 1986. Virus dispersion in a coarse porous medium. Meeting of the Arizona Section, American Water Resources Association, Glendale, Ariz., 1986.

Haridas, A. 1983. A mathematical model of microbial transport in porous media. Thesis. University of Delaware, Newark.

Hattori, T., and R. Hattori. 1976. The physical environment in soil microbiology: an attempt to extend principles of microbiology to soil microorganisms. *Crit. Rev. Microbiol.* **4**:423–461.

Hillel, D. 1980a. *Fundamentals of Soil Physics*. Academic Press, Inc., New York.

Hillel, D. 1980b. *Application of Soil Physics*. Academic Press, Inc., New York.

Javandel, I., D. Doughty, and C. F. Tsang. 1984. *Groundwater Transport: Handbook of Mathematical Models*. Water Resources Monograph Series no. 10. American Geophysical Union, Washington, D.C.

Landry, E. F., J. M. Vaughn, M. Z. Thomas, and C. A. Beckwith. 1979. Adsorption of enteroviruses to soil cores and their subsequent elution by artifical rainwater. *Appl. Environ. Microbiol.* **38**:680–687.

Matthess, G., and A. Pekdeger. 1981. Concepts of a survival and transport model of pathogenic bacteria and viruses in groundwater. *In* W. Van Duijvenbodden, P. Glasbergen, and H. Van Lelyveld (ed.), *Quality of Groundwater*. Proceedings of an International Symposium, Noordwijkerhout, The Netherlands. Studies in Environmental Science, vol. 17. Elsevier Scientific Publishing Company, The Netherlands.

McCoy, E. L., and C. Hagedorn. 1979. Quantitatively tracing bacterial transport in saturated soil systems. *Water Air Soil Pollut.* **11**:467–479.

McCuen, R. H. 1985. *Statistical Methods for Engineers*. Prentice-Hall, Inc., Englewood Cliffs, N.J.

McDowell-Boyer, L. M., J. R. Hunt, and N. Sitar. 1985. Particle transport through porous media. UCB-SEEHRL Report no. 85-12. Sanitary Engineering and Environmental Health Research Laboratory, University of California, Berkeley.

Mualem, Y. 1976. A catalogue of the hydraulic properties of unsaturated soils: development of methods, tools and solutions for unsaturated flow with application to watershed hydrology and other fields. Technion, Israel Institute of Technology, Haifa, Israel.

Nuttall, H. E. 1986. Population balance model for colloid transport. NNWSI Report no.

LA-OR 86-1914. Milestone R318. *Prepared for* Los Alamos National Laboratory, Los Alamos, N. Mex.

Rahe, T. M., C. Hagedorn, and E. L. McCoy. 1979. A comparison of fluorescein dye and antibiotic-resistant *Escherichia coli* as indicators of pollution in ground water. *Water Air Soil Pollut.* **11**:93–103.

Rawls, W. J., D. L. Brakensiek, and K. E. Saxton. 1982. Estimation of soil water properties. *Trans. ASAE Am. Soc. Agric. Eng.* **25**:1316–1320, 1328.

Remson, I., G. M. Hornberger, and F. J. Molz. 1971. *Numerical Methods in Subsurface Hydrology*. Wiley Interscience, New York.

Reneau, R. B., Jr., and D. E. Pettry. 1975. Movement of coliform bacteria from septic tank effluent through selected coastal plain soils of Virginia. *J. Environ. Qual.* **4**:41–44.

Schaub, S. A., and C. A. Sorber. 1977. Virus and bacteria removal from wastewater by rapid infiltration through soil. *Appl. Environ. Microbiol.* **33**:609–619.

Sinton, L. W. 1980. *Investigations into the Use of the Bacterial Species Bacillus stearothermophilus and Escherichia coli (H_2S Positive) as Tracers of Groundwater Movement.* Water and Soil Technical Publication no. 17. National Water and Soil Conservation Organization, Water and Soil Division, Ministry of Works and Development, Wellington, New Zealand.

Smith, M. S., G. W. Thomas, R. E. White, and D. Ritonga. 1985. Transport of *E. coli* through intact and disturbed soil columns. *J. Environ. Qual.* **14**:87–91.

Sobsey, M. D., C. H. Dean, M. E. Knuckles, and R. A. Wagner. 1980. Interactions and survival of enteric viruses in soil materials. *Appl. Environ. Microbiol.* **40**:92–101.

Van der Heijde, P., Y. Bachmat, J. Bredehoeft, B. Andrews, D. Holtz, and S. Sebastian. 1985. *Groundwater Management: the Use of Numerical Models*. Water Resources Monograph no. 5. American Geophysical Union, Washington, D.C.

Van Genuchten, M. T. 1978. Mass transport in saturated-unsaturated media: one-dimensional solutions. Water Resources Program, Department of Civil Engineering, Princeton University, Princeton, N.J.

Van Genuchten, M. T. 1986. A numerical model for water and solute movement in and below the root zone; model description and user manual. Draft report. U.S. Salinity Laboratory, U.S. Department of Agriculture, Riverside, Calif.

Versar Inc. 1986. Ambient exposure assessment methods for intentional releases of recombinant microorganisms. *Prepared for* Contract no. 68-02-3968, Task 161. Office of Toxic Substances, U.S. Environmental Protection Agency, Washington, D.C.

Versar Inc. 1987. Ambient exposures to recombinant microorganisms intentionally released to municipal and pulp and paper industry wastewaters. Draft final report. *Prepared for* Contract no. 68-02-4254, Task 45, Office of Toxic Substances, U.S. Environmental Protection Agency, Washington, D.C.

Versar Inc. 1988a. Guidance in assessing the validation of exposure models. Draft report. *Prepared for* Contract no. 68-02-4254, Task 84, Office of Health and Environmental Assessment, U.S. Environmental Protection Agency, Washington, D.C.

Versar Inc. 1988b. Evaluation of methods of assessing exposure to genetically engineered microorganisms—comparison of modeling and experimental data. Final report. *Prepared for* Contract no. 68-02-4254, Task 109, Office of Toxic Substances, U.S. Environmental Protection Agency, Washington, D.C.

Wang, D. S., C. P. Gerba, and J. C. Lance. 1981. Effect of soil permeability on virus removal through soil columns. *Appl. Environ. Microbiol.* **42**:83–88.

Wimpenny, J. W. J., N. Cotton, and M. Statham. 1972. Microbes as tracers of water movement. *Water Res.* **6**:731–739.

Yates, M. V., S. R. Yates, J. Wagner, and C. P. Gerba. 1987. Modeling virus survival and transport in the subsurface. *J. Contam. Hydrol.* **1**:329–345.

Chapter 3

Modeling Microbial Transport in the Subsurface: A Mathematical Discussion

Marylynn V. Yates and S. R. Yates

Introduction	48
Factors Controlling Microbial Fate in the Subsurface	50
Components of a Contaminant Transport Model	50
Survival	52
Transport	52
General Considerations	56
Application of Contaminant Transport Models to Microorganisms	56
Decay Rate	56
Advection and Dispersion	57
Adsorption	58
Examples of Virus Transport Models	58
Advection-Dispersion Model	58
Potential Flow Model	61
Discussion	66
Mathematical Discussion of the Advection-Dispersion Equation	66
Examples	69
Difficulties in Using the Advection-Dispersion Equation	70
Advection-Dispersion with Distance-Dependent Dispersivity	70
Conclusions	74
Glossary of Modeling Terms	74
References	75

The increasing number of reports of groundwater contamination by microorganisms has prompted federal, state, and local government agencies to propose and establish regulations to try to protect groundwater from microbial contamination. Examples of the types of regulations being proposed include the establishment of wellhead protection zones, drinking water standards for viruses and parasites, and

---

Marylynn V. Yates • Department of Soil and Environmental Sciences, University of California, Riverside, California 92521. *S. R. Yates* • U.S. Salinity Laboratory, U.S. Department of Agriculture/Agricultural Research Service, Riverside, California 92501.

septic tank placement ordinances. In many cases, the use of contaminant transport models is advocated to determine where sources of potential contamination should be placed in order to minimize the likelihood of microbial contamination of the underlying groundwater.

In November 1985, the U.S. Environmental Protection Agency proposed a Maximum Contaminant Level Goal for viruses in drinking water, setting a level of zero viruses (U.S. Environmental Protection Agency, 1985). For various technical and economic reasons, monitoring the water for the presence of viruses most likely will not be required; instead, various treatment techniques to eliminate or reduce virus contamination of drinking water were discussed. At that time, mandatory filtration and disinfection of surface water and disinfection of groundwater were discussed as possible treatment requirements. Since then, treatment techniques for surface water have been proposed, and in June 1989, the final rule requiring surface water sources to be filtered and disinfected was published. In April 1990, draft requirements for mandatory disinfection of groundwater sources of drinking water were published. Included in that rule were variance criteria for avoiding disinfection, one of which is to conduct a vulnerability assessment or sanitary survey of the source water. A microbial transport model similar to those described herein could be an integral part of such an assessment.

One result of these regulations is that microbiologists are being called upon more and more to work with the engineers, hydrologists, geologists, and soil scientists who are developing predictive models of contaminant transport. However, because most microbiologists have little, if any, training in modeling and the modelers have little or no knowledge of microbiology, communication problems can develop. Modelers require information about the factors that affect microbial transport, but many times that information is not available in a directly usable form. An understanding of the processes that are being modeled and the terminology used by the modelers will enable microbiologists to help in the development of models that are as accurate predictions of microbial transport as possible given the available data. It is intended that this discussion will provide a framework within which communications between microbiologists and modelers will be improved.

In most cases, the person using a contaminant transport model is trying to answer the question, "If we apply some quantity of microorganisms to the soil at point A, how many will make it to point B?" Point B may be a source of drinking water such as a well, where the presence of microorganisms in sufficient number would pose a potential threat to the health of the persons consuming the water. (The significance of that number in terms of whether it poses a health hazard must be answered by using some type of risk assessment.) Another application of microbial transport models is in the area of bioremediation. Some bioremediation processes involve the introduction of microorganisms to the soil surface and rely upon microbial movement to the contaminated site. Models can provide useful information about the extent of movement as well as the time required for the microorganisms to arrive at the desired location. While the microorganisms in these cases are not the contaminant, the same processes influence their survival and transport, so the concepts of transport modeling discussed here are applicable.

There are several pieces of information which are required before using a model to answer these questions. Generally, the more accurate the input data, the more accurate the prediction that can be made using a model.

FACTORS CONTROLLING MICROBIAL FATE IN THE SUBSURFACE

The fact that viruses remain infective long enough, and can travel far enough, in the subsurface to contaminate drinking water and cause waterborne disease outbreaks has led to attempts to develop predictive models of virus fate in the subsurface. To model the survival and transport of viruses in the subsurface, it is necessary to determine the factors which influence these processes. Over the past several years, a great deal of research has been done to determine the factors that influence how long viruses can survive in the environment as well as how far they can be transported in soils and groundwater. One of the provisions of the Safe Drinking Water Act states that "the administrator shall carry out a study of virus contamination of drinking water sources and means of control of such contamination" (section 1442). Research on virus contamination of drinking water has been ongoing in this country for over 20 years. A large amount of data has been collected by various researchers funded by local, state, and federal agencies, including the U.S. Environmental Protection Agency. The results of this research will be summarized here.

In the subsurface, two major factors control virus fate: survival and movement. Both factors must be considered when determining whether there is a hazard to human health associated with the contamination of groundwater by viruses. If a virus can survive for a long time in the subsurface, but cannot move through the soil very easily, it is not likely that it will pose a large threat to the groundwater. Similarly, if the virus is easily transported through the soil, but it does not survive for a very long period of time, it is probably not of much concern. However, if the virus can survive in an infective form long enough to be transported through the soil and into the groundwater, this may be cause for concern if the water is used for potable purposes.

In general, both survival and movement are controlled by the specific type of virus, the physical and chemical properties of the soil, and the climate of the environment. The susceptibility of viruses to different environmental factors varies considerably among different species as well as strains. The size and chemical composition of different viruses influence the extent to which they can travel in the subsurface. The soil properties play a major role in the survival and migration of bacteria and viruses. The texture of the soil and its pH, organic matter content, and moisture content all influence how long viruses can survive and how far they can travel in the subsurface. Temperature and rainfall are particularly important in determining microbial fate. Viruses can survive for extended periods of time at low temperatures; rainfall is important in that it can mobilize adsorbed viruses and promote their migration to the groundwater. A list of the factors important in controlling virus survival and movement in the subsurface is presented in Table 1.

COMPONENTS OF A CONTAMINANT TRANSPORT MODEL

Contaminant transport is commonly modeled using the advection-dispersion equation. There are four primary processes involved in contaminant transport which must be characterized quantitatively and accounted for by a model: decay, advection, dispersion, and adsorption. A mathematical description of the advection-dispersion model is presented in the next section. This discussion will be limited to a more general, less technical description of modeling.

Table 1. Factors influencing virus fate in the subsurface (Yates and Yates, 1988)

Factor	Influence on: Survival	Influence on: Movement
Temperature	Viruses survive longer at lower temperatures.	Temperature can influence water movement, which may affect virus movement.
Microbial activity	Some viruses are inactivated more readily in the presence of certain microorganisms; however, adsorption to the surface of bacteria can be protective.	Unknown.
Moisture content	Some viruses persist longer in moist soils than dry soils.	Generally, virus migration increases under saturated flow conditions.
pH	Most enteric viruses are stable over a pH range of 3 to 9; survival may be prolonged at near-neutral pH values.	Generally, low pH favors virus adsorption and high pH results in virus desorption from soil particles.
Salt species and concentration	Some viruses are protected from inactivation by certain cations; the opposite has also been observed.	Generally, increasing the concentration of ionic salts and increasing cation valences enhances virus adsorption.
Virus association with the soil	In many cases, survival is prolonged by adsorption to soil; the opposite has also been observed.	Virus movement through the soil is slowed or prevented by association with the soil.
Virus aggregation	Enhances survival.	Retards movement.
Soil properties	Effects on survival are probably related to the degree of virus adsorption.	Greater virus migration in coarse-textured soils; there is a high degree of retention by the clay fraction of soil.
Virus type	Different virus types vary in their susceptibility to inactivation by physical, chemical, and biological factors.	Virus adsorption to soils is probably related to physicochemical differences in virus capsid surfaces.
Organic matter	The presence of organic matter may protect viruses from inactivation; others have found that it may reversibly retard virus infectivity.	Soluble organic matter competes with viruses for adsorption sites on soil particles.
Hydraulic conditions	Unknown.	Generally, virus migration increases with increasing hydraulic loads and flow rates.

Survival

Information on the survival characteristics of the microorganism is usually included in a model in the form of a net decay rate. Decay, or inactivation, is the irreversible destruction of the contaminant by chemical, physical, or biological processes. Growth of microorganisms in effect offsets a portion of the decay and can be modeled by using the net decay rate (i.e., net decay = decay rate − rate of growth). In many models, the net decay rate is considered to be a constant; thus no allowance is made for the different decay rates of the microorganisms under changing environmental conditions.

Determination of the decay rate is a relatively simple process. The contaminant of interest is put into an environment in the laboratory which resembles field conditions as closely as possible. The concentration of the contaminant is measured at the beginning of the experiment and at several defined intervals for an appropriate time period. The concentration of the contaminant (or some transformation of the concentration, such as the logarithm) is then plotted as a function of time; the decay rate is the slope of this line.

Transport

The transport process can be described in terms of the three remaining factors, advection, dispersion, and adsorption. Each of these factors will be described separately.

Advection

Advection is the process by which the contaminant is transported with the bulk flow of the groundwater. For a simple model, the advection enters the model as the average velocity of the groundwater. Determination of velocity can be made in several ways. One common method is to conduct a field or laboratory tracer test in which a soluble, nonreactive substance such as chloride, bromide, or fluorescein (a colored dye) is added to the water. In the laboratory, water is applied to the top of a soil column, and samples of the outflow are collected. The samples are analyzed for the tracer, and the concentration of the tracer traveling through the column is determined as a function of time. Assuming that the tracer mimics the behavior of the water, the average velocity of the water can be calculated.

Field tracer tests are much more difficult to perform. Although an average value for the velocity can be obtained over a large area in the field, in many instances there may be a large variation in the velocity from location to location even when the distance separating two sites is just a few feet. There may also be changes in the measured velocity over time due to changes in the porous medium (e.g., shrinking and swelling materials), boundary conditions (e.g., application of water), and environmental setting (e.g., temperature changes or a lowering of the hydraulic head and therefore gradients).

Dispersion

Another process which is very important in describing transport is dispersion. As the water moves through the soil, it generally does not follow a straight path (unless it is flowing through a fracture). Two distinct processes are operating in dispersion: diffusion and mechanical mixing. Diffusion is the spreading out of a solute due to concentration gradients. This process is a relatively unimportant

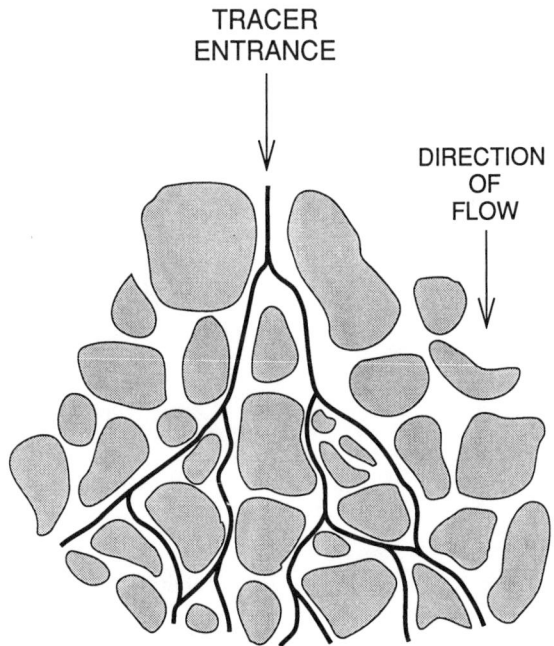

Figure 1. Mechanical dispersion of water and contaminant as water flows through the porous medium.

component of dispersion except at very low groundwater velocities. Under most circumstances, dispersion caused by the motion of the water (i.e., mechanical mixing) is the main process of interest. Mechanical mixing occurs when the water moving through the soil pores diverges around the soil particles. Sometimes the two streams of water will converge on the other side of the particle, and other times the two streams will impinge on another particle and separate again. After traveling around a number of particles, the initially concentrated flow pattern has spread out (Fig. 1). As a result, the water front, and thus the contaminant, spreads out as it moves through the soil. The amount of spreading depends upon the type of soil involved as well as on the velocity of the groundwater.

Dispersion can also occur in nonhomogeneous (nonuniform) porous media when a contaminant flows faster in one area than another. For example, given a homogeneous porous medium with a single fracture, contaminated water will flow faster in the fracture than through the porous medium. This causes the contaminant to move farther in the fracture in a given amount of time as compared with the contaminant in the porous medium. After the contaminant in the fracture has moved beyond the contaminated zone in the porous medium, it will begin to move into the porous medium in response to diffusion gradients as well as bulk flow. This faster-moving water causes a spreading of the contaminant far beyond what would occur if only the porous medium were considered. This process is especially important in field soils since heterogeneities (nonuniformities) such as old root

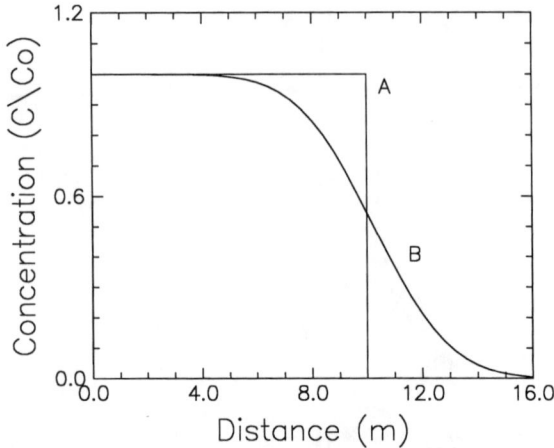

Figure 2. Breakthrough curve of a contaminant. A, Advection only; B, advection and dispersion.

channels or worm and ant holes can cause dispersion in a manner analogous to a fracture.

Like advection, dispersion can be measured using a tracer test as described above. Rather than calculating the average velocity of the water or contaminant as was done in the previous case, one is interested in the arrival history of tracer at the column outflow (i.e., the shape of the concentration-versus-time record). Dispersion causes the arrival of a fraction of the tracer at the column outflow to occur before the arrival of that fraction of the tracer which is traveling at the average velocity. If no dispersion is occurring, the breakthrough curve of the contaminant would look like curve A in Fig. 2; i.e., all of the contaminant travels through the column at the same rate. Dispersion causes the front to spread out, so that some of the contaminant is traveling faster than average and some is traveling slower. The breakthrough curve of a contaminant that has undergone dispersion is shown as curve B in Fig. 2.

There may be some problems encountered when determining dispersion values for use in models. It has been found that dispersion values calculated from soil column experiments generally are not applicable to field situations, where the scale of interest is much larger. Compounding this problem further, the dispersion of the contaminant changes as the water moves farther through the field. If the dispersion is determined after the water has traveled 1 m from the application site, one value is obtained. If dispersion is again determined 10 or 100 m from the point of application, generally the dispersion will be larger. This is because over the larger distances the tracer has experienced more and more heterogeneous features (i.e., worm holes, fractures, etc.) contained in the porous medium, and the scale of these features increases.

Most contaminant transport models assume that the dispersion is constant over the entire field of interest. The implications of this are shown in Fig. 3. If a constant value for dispersion is used, the model would calculate a distribution curve for the contaminant as shown by A. If, however, the dispersion is allowed to

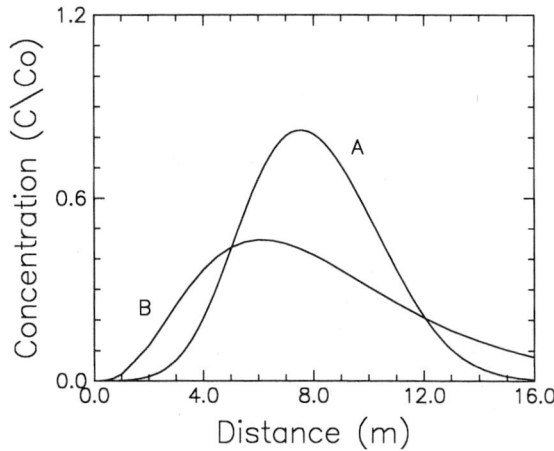

Figure 3. Effects of distance-dependent dispersion on contaminant transport. A, Constant dispersion; B, distance-dependent dispersion.

vary (and linearly increase with the average travel distance), the model would predict a much different distribution (curve B). It can be seen that, when using scale-dependent dispersion (curve B), the contaminant has been predicted to move much farther than when a constant value for dispersion is used. Thus, using a constant value would underestimate the time of arrival of the contaminant at a specified location. This underestimation could prove to be serious if the concentration of the contaminant in that leading edge is high enough to endanger health.

Adsorption

The third factor needed to describe the transport process is adsorption. Adsorption is the chemical binding of a contaminant to the surface of the solid medium (e.g., soil). The process of a chemical binding to a soil surface may be reversible or irreversible and equilibrium or diffusion controlled, depending upon the properties of the contaminant and the subsurface medium. Because adsorption removes contaminant from the fluid phase, even if only temporarily, it acts to slow the movement of the contaminant. The term "retardation" is commonly used to describe the effects of contaminant adsorption.

Adsorption is often measured in the laboratory by calculating a batch adsorption isotherm. The contaminant is added to a flask containing a mixture of soil and water which is generally agitated during the course of the experiment. After a sufficient period of time has passed to allow the system to come to equilibrium, the distribution of the contaminant in the adsorbed and free state is determined. This process is repeated for several contaminant concentrations (or several soil amounts) to determine the adsorption characteristics. The results of these experiments are usually described using either a Freundlich or Langmuir isotherm.

Many times, however, the adsorption information obtained in laboratory batch experiments is not particularly useful in flowing systems. In flowing systems, the adsorption process may be diffusion controlled; that is, the diffusion of the chemical from the flowing liquid to the liquid adjacent to the adsorption sites governs the adsorption process. For these conditions, it is desirable to use laboratory column or field tracer experiments rather than static batch experiments to determine the adsorption properties of the porous medium.

General Considerations

In general terms, characterization of the subsurface medium is the most important consideration for obtaining an accurate picture of the transport in porous media. Heterogeneities (nonuniformities) such as cracks, fractures, worm holes, and other large openings or soils with markedly different physical properties will have a profound effect on the flow of water, and hence the transport of any contaminant. Characterizing these nonuniformities in sufficient detail for modeling purposes is both time-consuming and very expensive, so other methods for modeling transport in heterogeneous porous media are currently being investigated.

Caution should be exercised before using a model like the one described above to predict the transport behavior in a field because it usually too simple a model to describe the complex nature of field sites. Such a model can be useful, however, for describing contaminant transport behavior in situations where the variation in physical properties is small or for situations where comparisons are to be made between different contaminants for hypothetical situations (e.g., screening models). The simple model described here cannot be used to predict transport in nonuniform subsurface media, although the concepts described are generally applicable to any model.

Although this discussion has been restricted to situations in which the chemical and hydrological parameters are constants, more comprehensive numerical models can be written so that these parameters can vary over space and time. While numerical models allow very complicated systems to be analyzed, they do suffer from many shortcomings. For example, numerical errors may result from the numerical approximations (i.e., discretization) used. In addition, they can be very time-consuming and expensive to use and may require extensive training in mathematics and computer techniques. Obtaining the required data is also expensive.

APPLICATION OF CONTAMINANT TRANSPORT MODELS TO MICROORGANISMS

Decay Rate

Of all of the information which must be input into the model, more is known about microbial decay rates than the other factors. There are many factors which affect the survival of microorganisms in the subsurface; these have been the subject of several recent reviews, including those by Sobsey (1983), Vaughn and Landry (1983), and Yates and Yates (1988). Some of the more important factors affecting survival include the soil texture (i.e., the amounts of sand, silt, and clay), the amount of organic matter in the soil, the amount of moisture in the soil, the pH of the soil, the temperature, and the chemical composition of the soil-water.

The particular microorganism of interest has a profound impact on the decay rate that should be used in the model. Microbial decay rates have been found to vary by several orders of magnitude. If the behavior of microorganisms in wastewater is of interest, several genera of bacteria, viruses, and parasites are involved. In such a situation, it may not be appropriate to use a decay rate for *Escherichia coli* alone. The decay rate of *E. coli* may not be representative of the decay rates of all microorganisms present. When a mixed population of microorganisms (i.e., bacteria, viruses, and parasites) is involved, a decision must be made about which decay rate should be chosen as representative of the group. It might seem most appropriate to choose the smallest decay rate, so that if an error is to be made, it would be on the side of safety. However, this could also result in predictions being made that would be so restrictive that they would be impractical to enforce in a real situation. If several different microorganisms are involved, it would also be possible to model the behavior of each microorganism separately and then combine the results to determine the overall concentrations at a specified time and location of interest.

Another problem that arises when deciding what decay rate to use is simply finding a value that is applicable to the particular situation of interest. Many of the decay rates in the literature have been obtained in the laboratory using a liquid medium at one or two different temperatures. Studies have shown that there are differences in microbial decay rates in liquid as compared with liquid-soil mixtures (Sobsey, 1983). Adjustments may have to be made so that the value is relevant to an environmental setting. In some instances, it may be necessary to conduct inactivation studies using the microorganisms of concern under conditions that resemble the environment of interest to obtain relevant decay rates.

Advection and Dispersion

Values for the velocity and dispersion of a contaminant are generally obtained from tracer tests. However, the use of a soluble chemical tracer may not give appropriate values for microorganisms. The failure of the solute transport equation to predict the extent of microbial movement has been observed in several laboratory and field studies (Krone et al., 1958; Wilson et al., 1984; Grondin, 1987; Champ and Schroeter, 1988; Matthess et al., 1988; Vaughn, 1988). In all of these studies, the average velocity of the microorganisms was greater than that of a conservative tracer such as chloride.

One possible explanation for these observations is a pore-size exclusion phenomenon. The advection-dispersion equation assumes that the contaminant is in solution and therefore (neglecting adsorption) has the same average velocity as the groundwater. If, however, the contaminant is a particle, it may not be able to move through all of the pores with the water. If the particle is large enough (such as a microorganism), it may be excluded from the smaller pores where, on the average, water travels more slowly, and thus be forced to travel through the larger pores where their average velocity would be greater than that of the medium taken as a whole. Thus, the microorganism would travel more rapidly through the medium than would a solute in the same porous medium. The situation is analogous to gel chromatography, where the chemicals are separated by size, with the largest molecules appearing at the end of the gel first. Under these circumstances, the microorganism would have moved more rapidly and to a greater extent than would have been predicted using a model whose coefficients have been determined using soluble chemicals. Therefore, tracer tests to determine the velocity and dispersion should be conducted using microorganisms or other particles with characteristics similar to those of the microorganisms of interest. If the microorganisms are large in comparison to the pores in the soil matrix (e.g., bacteria or parasites), filtration becomes an important removal mechanism.

Adsorption

There have been many studies of the adsorption of microorganisms to soil materials (Burge and Enkiri, 1978; Goyal and Gerba, 1979; Moore et al., 1981). The adsorption varies greatly depending upon the microorganism being studied and the soil used in the study. Attempts to predict the degree of adsorption on the basis of either soil properties (such as pH, organic matter content, and percent clay) or characteristics of the microorganism (such as its isoelectric point) have not been entirely successful. Therefore, it may be necessary to conduct experiments to determine the adsorption characteristics of the microorganism of interest using the soil at the site of interest.

The degree of adsorption of a particular microorganism may also affect the observed decay rate. The results of some studies have suggested that adsorption of some microorganisms to a solid surface may protect them from some types of inactivation or may slow the inactivation process in some way. Thus, there are some microorganisms that have two decay rates: one for the suspended state and another for the adsorbed state. As stated previously, the decay rate is required to be a constant in many contaminant transport models, so this factor would cause some error to be introduced into any predictions.

EXAMPLES OF VIRUS TRANSPORT MODELS

There have been a few models developed to describe virus transport with the goal of calculating safe distances between contaminant sources and drinking water wells. Two very different models will be described to illustrate the types of modeling approaches that can be used. The data requirements and limitations of each of the models will also be discussed.

Advection-Dispersion Model

The influence of advection, dispersion, adsorption, and decay on the transport of microorganisms (or any contaminant) is illustrated in Fig. 4. Each curve depicts the concentration distribution of the microorganisms in the liquid phase with respect to position of the microorganisms in the soil as affected by the various processes. For illustration purposes, assume that the microorganisms were added to the water for 12 h and that the water moves at a rate of 10 m day^{-1}. In curve A, the only process affecting transport is advection. In this case, after 1 day the first microorganisms have traveled 10 m, while those added at the end of 12 h have only traveled 5 m. The concentration of microorganisms in this region (i.e., 5 to 10 m from the source) is constant and is equal to the applied concentration ($C/C_0 = 1$). The concentration at 10 m will remain constant for 12 h (the time required for the microorganisms added last to travel 10 m). After 12 h, the concentration of

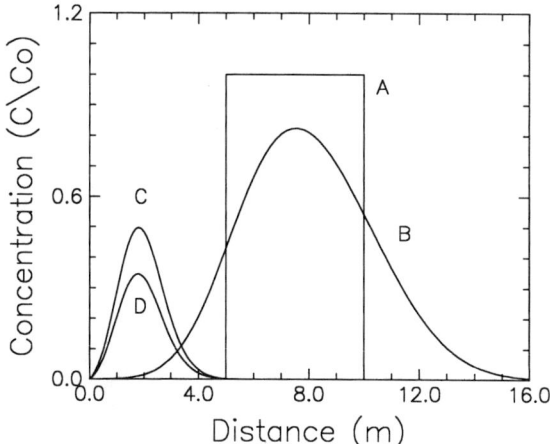

Figure 4. Effects of various processes on contaminant transport. A, Advection; B, advection and dispersion; C, advection, dispersion, and adsorption; D, advection, dispersion, adsorption, and decay.

microorganisms at 10 m will again drop to zero. This is because all of the microorganisms are traveling through the soil at the same velocity.

Curve B shows the distribution of microorganisms when both advection and dispersion are allowed to influence the transport process. In this case, the total area under the curve is the same as that in curve A, because all of the microorganisms are still present in the water. However, the microorganisms are not all traveling at the same rate through the soil. Some will travel farther than 10 m in 24 h (those traveling at a velocity greater than the average water velocity), while some will take longer than 24 h to move the same distance. This causes a spreading out of the distribution curve. The peak of the curve occurs at the halfway point of the advection curve; in other words, the average velocity of the microorganisms is the same as in the previous case, but some are traveling faster and others slower than that average. The height of the curve will change depending upon the amount of dispersion occurring in the particular system of interest, but the general shape of the curve and position of the peak will remain constant.

The influence of adsorption is added to the advection and dispersion processes for curve C. The area under this curve is reduced as compared to the previous ones because some of the microorganisms have been removed from the water phase and are adsorbed onto the solid medium. As stated previously, only the microorganisms in the liquid phase are shown in Fig. 4. If the total concentration of microorganisms in the system (both solid and liquid phases) were shown, the area under curve C would be the same as curve B, but the position of the curve would be closer to the source. Not only does adsorption act to decrease the total number of microorganisms in the water, it also slows the movement of the contaminant relative to what it would be if no adsorption occurred. The more strongly the microorganism is adsorbed, the longer will be the lag before the first microorganism reaches the 10-m point.

Table 2. Variables for which data are required for a transport model

Soil hydraulic properties	Hydraulic conductivity
	Moisture content
	Porosity
Transport characteristics	Diffusion^a
	Dispersion^a
	Adsorption capacity^a
	Mobile-immobile phase transport
Aquifer configuration parameters	Storage coefficient of the aquifer
	Radius of the well
	Location of top of aquifer
	Location of base of aquifer
	Location of water level
	Locations of observation wells
Initial and boundary conditions	Amount of recharge to aquifer
	Pumping rate of well
	Time of starting and cessation of pumping
	Initial virus concentration^a
	Location of contaminant^a
Virus properties	Virus inactivation (decay) rate^a
	Virus adsorption coefficient^a

^{a} Variables required for input in the advection-dispersion model used to generate Fig. 4.

In curve D, the effect of decay of the microorganisms during the transport process is added to the other factors. Because decay permanently removes microorganisms from the system, the area under this curve is smaller than that of curve C. Decay has no effect on the rate of movement of the microorganisms, so the peak of the curve is in the same place as the peak for the previous curve; i.e., the average velocity of the microorganisms is the same as when no decay is occurring. This curve is the most realistic in terms of what is actually occurring in a real-world situation, although dispersion has been considered to be a constant, which is probably not the case.

Data Input Requirements

The input parameters for a transport model based on the advection-dispersion equation vary widely from model to model. Models that give more accurate predictions will normally require more data than models that give more general results. It is desirable to know what model is going to be used prior to collecting data so that the proper measurements are made. Some of the important variables for which values may be required are given in Table 2.

Limitations

There are many limitations encountered in using an advection-dispersion model such as the one illustrated in Fig. 4 for predicting the behavior of viruses under "real-world" conditions, as follows. (i) The number of required input parameters is very high. (ii) Input values for many of the parameters are unknown, and therefore estimates must be made. For example, data on the virus inactivation rate and adsorption coefficient usually must be taken from the literature and not from experiments conducted under the site conditions used in the model. Obtaining actual values for many of the input parameters would be very costly. (iii) Only transport in the saturated zone is considered, so any removal of viruses in the unsaturated zone is not taken into account. (iv) Finally, the model illustrated in Fig. 4 is a site-specific application of the advection-dispersion equation which would not be applicable for regional screening purposes in its current form.

Potential Flow Model

Another type of transport model is a potential flow model. In this case, rather than considering water flow and transport of the microorganisms separately, the microorganisms are assumed to flow with the water. Only two of the four processes previously described are considered in this model, i.e., advection and decay. This type of model is useful for regional planning purposes in that areas in a community with relatively higher vulnerability to groundwater contamination can be distinguished from areas where contamination is less likely to be a problem.

A model of this type has been developed and used to predict setback distances between sources of viruses (in this case, septic tanks) and drinking water wells for a 200-km^2 area in the city of Tucson, Ariz. (Yates and Yates, 1989). Although septic tank setback distances are used here for illustrative purposes, this model could be used for determining separation distances between any potential source of contamination and a drinking water well.

Septic tanks are the most frequently reported causes of contamination in groundwater disease outbreaks associated with the consumption of untreated groundwater in the United States. The placement of septic tanks is generally controlled by county-wide or statewide regulations, with little consideration given to the local hydrogeologic, climatic, and land use conditions. This model illustrates the effects of including local variation in subsurface conditions when using geostatistics in the calculation of septic tank setback distances in a part of the city of Tucson, Ariz.

Data Input Requirements

The data required to use a model of this type are considerably different from those required for the previously discussed model. To use this geostatistics-based model, a large number of data points as well as their location in space are needed. Generally, at least 50 data points are required to determine whether the data are spatially correlated. Once the data are found to be spatially correlated, the correlation structure must be determined and input into the geostatistical model. In the Tucson model, data were available at 71 locations, so all 71 points were used in the calculations. Many large municipalities that use groundwater wells have this type of information readily available. Data required include the following. (i) Virus inactivation rates in the groundwater at 71 locations in the city. Groundwater temperatures may be used to calculate the virus inactivation rate using a simple linear regression equation. A comparison of the results of the Tucson model using actual virus inactivation rates versus virus inactivation rates estimated using groundwater temperatures can be found in Yates and Yates (1987). Figure 5 shows the relative locations of the samples used. (ii) Hydraulic conductivity of the aquifer at those 71 locations (at least). (iii) Hydraulic gradients at those 71 locations (at least).

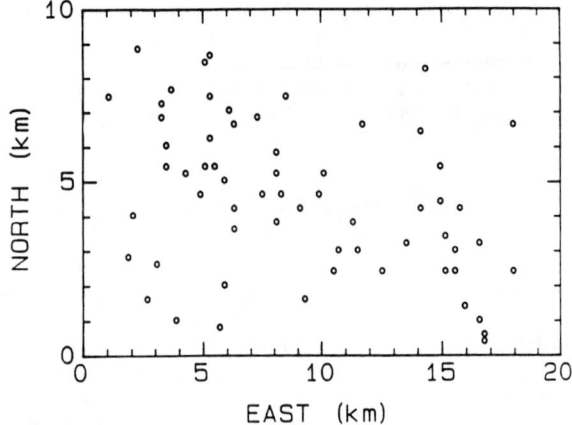

Figure 5. Sample collection sites.

Model Output

The output from the Tucson model is in the form of contour maps, which were generated using geostatistical programs. Figure 6 shows the distances between contamination sources (e.g., septic tanks) and drinking water wells that would be required to achieve a 7-order-of-magnitude reduction in virus number (i.e., the removal of 10×10^6 viruses) in the time necessary for the water to move that distance. To interpret this and following contour maps: if a septic tank is placed on a contour marked 30, this means that a well would have to be 30 m away for there to be a removal of 10×10^6 virus particles. (The model can be run for any amount of virus reduction desired.) A wide range of septic tank setback distances (from less than 15 m to greater than 75 m) was calculated for a part of the city of Tucson.

Using this model, one can also calculate the conditional probability that contamination will occur given a separation distance. In other words, the model can be used to answer the following questions: (i) given a setback distance (e.g., specified by regulation), what is the probability that this distance would be adequate to protect the groundwater from viral contamination at different locations

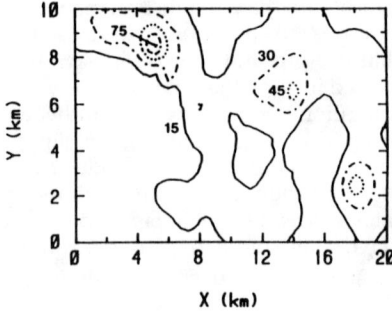

Figure 6. Septic tank setback distances (meters).

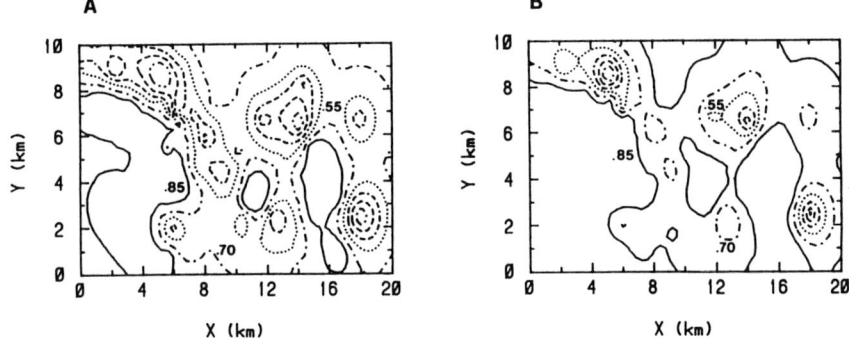

Figure 7. Contour diagrams for the conditional probabilities that the distances are greater than (A) 15 m or (B) 30 m.

in the city? and (ii) given a desired probability level, what setback distance would be necessary to be that confident that the groundwater would be protected from contamination by viruses?

Case (i): probabilities associated with specified setback distances. Probability maps were calculated for two setback distances for comparative purposes. Suppose that the local ordinance requires a minimum of 15 m separation distance between a septic tank and a drinking water well. Figure 7A shows the probability that there would be a 7-order-of-magnitude reduction in virus numbers in the time required for the water to travel 15 m. For the contour marked 0.85, there is an 85% probability that a 15-m separation distance will be adequate to meet the stated criterion of protection of the well water from virus contamination.

Figure 7B shows the probability contour map calculated using a 30-m separation distance between a septic tank and a well. Comparing this figure with Fig. 7A, it can be seen that the contour which had a 70% probability in Fig. 7A now has an 85% probability of meeting our criterion. This is because of the fact that 30 m has been set as the separation distance, which means that it will take longer for the viruses to travel to the well. The longer the travel time, the more inactivation of virus that will occur. Thus, it follows that the probability that a 30-m separation distance would be adequate is higher (85%) as compared with the probability estimated for a 15-m separation distance (70%).

Case (ii): setback distances associated with specified probabilities. In case (ii), rather than specifying a setback distance and calculating the associated probabilities, the desired probability level is specified and the associated setback distances are calculated. In the first example (Fig. 8A), a probability level of 0.9 was specified. In other words, what setback distance is necessary to be 90% certain that the actual setback distance required to achieve the criterion of 7-orders-of-magnitude reduction in virus number is less than or equal to that distance? In Fig. 8A, it can be seen that the required setback distances range from 20 to over 100 m. If one wanted to be 99% certain that the setback distances were adequate to prevent viral contamination, much larger separation distances are calculated (Fig. 8B). For example, in some areas a 100-m setback distance would be required rather than the 60 m calculated when a 90% probability of achieving the criterion was required.

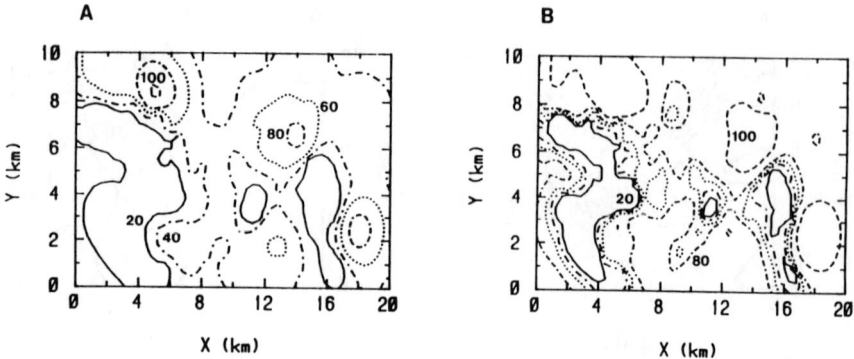

Figure 8. Contour diagrams for the setback distances (meters) given a conditional probability of (A) 90% or (B) 99%.

To demonstrate the effect of adding pumping wells to the regional groundwater flow in the model calculations, a simple one-well case may be used. The well chosen is pumped at a rate of 150 gal (ca. 568 liters) per min. In the former calculation, in which only regional groundwater flow is used in the setback distance calculation, this well is located on a 60-m contour (Fig. 6). When the 150-gal/min pumping rate is added to the travel time calculation, a setback distance of 156 m is required to achieve a 7-order-of-magnitude reduction in virus number (Fig. 9). If only 4 orders of magnitude of virus inactivation is required, the setback distance would be 93 m, which is still 1.5 times greater than that calculated without adding

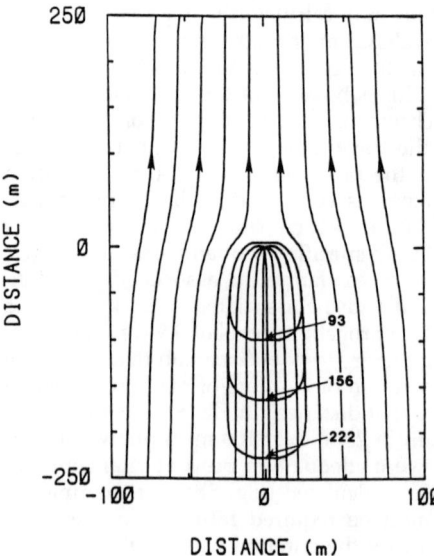

Figure 9. Setback distances (meters) calculated for a pumping well in a regional flow field.

the effects of pumping. The actual calculations would be more complicated than described here, as the interference effects from multiple well pumping would have to be included to get an accurate picture of the flow field in the Tucson Basin. This simple example does show, however, that pumping has a large impact on the travel time, and thus setback distance, calculations and must be considered if the method is to be used for municipal planning purposes.

With the appropriate modifications to model the specific situation of interest, the methods could be used for community planning purposes. The first case described, namely, calculating the conditional probabilities given a specified setback distance, would be useful in a situation where the minimum setback distance was specified by regulation. For example, a certain community has a regulation stating that 30 m is the minimum separation between a well and a septic tank. This model could be used to generate a conditional probability contour map. A decision to allow a septic tank to be placed in a certain location could then be based on the calculated probabilities. For example, it might be decided that if the probability was 75% or greater, a septic tank would be permitted on any lot, provided that soil percolation test requirements were met. If the probability was between 50 and 75%, soil percolation test requirements could be made more stringent or the minimum lot size could be increased for a septic tank permit to be issued. If the probability were less than 50%, it might be decided that septic tanks would not be allowed at all.

The approach described in the second case could also be used for community planning purposes, in that a desired probability level could be specified (e.g., in a regulation) and the setback distances necessary to achieve that level would be calculated. One advantage of using this method is that it would avoid the implicit assumption that the hydrogeologic characteristics of the area are constant. The regulations would only have to specify a probability level to be met in order to allow a septic tank permit.

Limitations

There are several limitations in the Tucson model in its current form which must be recognized when using it. These include the following.

(i) Only saturated zone transport is considered. There is no allowance for reduction in virus number as the water moves vertically through the unsaturated soil. This is a serious limitation because the greatest percentage of virus loss during subsurface transport is most likely to occur in the unsaturated zone. A model of unsaturated zone transport is currently being developed.

(ii) The influence of multiple pumping wells on the pattern and rate of groundwater flow has not been considered. Pumping wells act to increase the flow rate of water in some parts of the aquifer and slow it in others, and they may cause the direction of groundwater flow to be reversed in certain areas. This will have a profound effect on the calculated setback distances, as illustrated above. The effects of multiple pumping wells will change the flow field even more and must be considered in actual practice.

(iii) Inactivation of the viruses was the only removal mechanism included. From the previous discussion, it is obvious that adsorption to soil particles is an important removal mechanism, especially in the unsaturated zone.

(iv) This model used a bacterial virus, MS-2 coliphage, as a model for the behavior of human viruses. The viruses of concern may or may not behave in the same manner.

Discussion

Both of the models discussed above have several limitations, as noted. For example, in both cases a model virus was used to predict the behavior of all viruses of concern, and it has been well documented that there is a large degree of variation in the behavior of different viruses in the environment. Ideally, one would want to input data about how the particular virus(es) of concern survives and is transported in the particular soil and aquifer of interest under the environmental conditions present at the site. The more accurate the input data, the more accurate the predictions made by the model will be. In most cases, however, data are not available on the particular virus(es) of concern under the environmental conditions of interest. In such cases, there are several options, including using estimates from the literature, obtaining data using indicators, and performing experiments on the organism of interest.

Another point to consider when using virus transport models is that in many situations, our knowledge of the physical and hydraulic properties of the soil and groundwater systems used as input is as uncertain as that of the behavior of viruses. Thus, while there are differences in the length of time that various viruses remain infective and in the distances they can travel in soil, these differences are relatively small when considering the large variation in soil and aquifer properties that can occur in a relatively small area. In other words, the predictions made by a transport model will be affected much more by uncertainties in the soil and aquifer properties than by differences among viruses.

As stated previously, with the information that is available about the length of time viruses remain infective and the factors influencing the distances they can travel in the subsurface, model predictions can be made that would provide useful information in terms of protecting groundwater from contamination. Uncertainties in the soil and hydraulic properties have the potential to cause greater uncertainties in model predictions than does our knowledge of virus behavior in the subsurface.

If, however, the use of models to show that drinking water is likely to be free from viruses is advocated as a means whereby variances from a disinfection requirement are granted, there are several areas which need research to refine our capabilities to make accurate predictions of virus transport, as follows. (i) Virus behavior in the unsaturated zone should be experimentally determined. This is the portion of the subsurface where the greatest potential for removal exists. (ii) Methods need to be developed for the isolation and identification of Norwalk virus and other viruses which are important causes of waterborne diseases. Once we can work with these viruses in the laboratory, experiments need to be done to determine how they behave in the environment relative to the other viruses that have been studied and to any model virus. (iii) Risk assessment studies should be performed to determine the risks associated with the consumption of water containing low numbers of viruses. (iv) Predictive models of virus transport in unsaturated and saturated soils should be developed. Finally, (v) field studies are needed to tell us whether the predictive models that have been developed really tell us what happens in the environment.

MATHEMATICAL DISCUSSION OF THE ADVECTION-DISPERSION EQUATION

Contaminant transport can be modeled using the following form of the advection-dispersion equation:

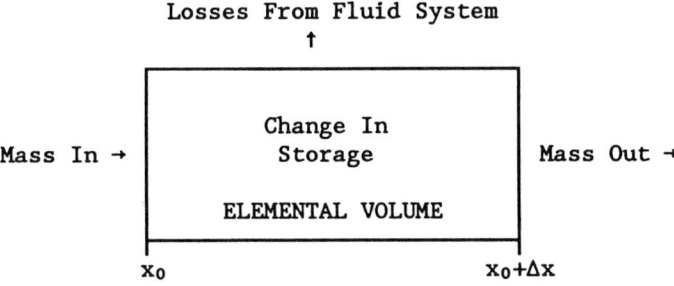

Figure 10. Diagram describing mass conservation for flow in a porous medium.

$$R \frac{\partial C}{\partial t} = \frac{\partial}{\partial x}\left[D \frac{\partial C}{\partial x}\right] - V \frac{\partial C}{\partial x} - \mu C - \Sigma Q_i \quad (1)$$

In this equation, R is the retardation coefficient describing the adsorption of the microorganism (note: although the word microorganism is used throughout, this equation can be used for any contaminant or solute) onto the solid particles; C is the concentration (or mass) of the microorganism at a place and time; x is the position; t is time; D is hydrodynamic dispersion, which describes mechanical mixing and diffusional processes; V is the average velocity of the water as it moves through the porous medium; μ is the decay rate of the microorganism; and Q_i is the source/sink term. In general, the coefficients R, D, V, and μ may depend on the position, the elapsed time, and the concentration as well as on a number of other equations that govern the transport of other variables, such as water and temperature. Under the most general conditions (i.e., when the coefficients depend on the concentration or when equation 1 depends on other transport equations), equation 1 is nonlinear and may be very difficult to solve.

Equation 1 is a statement of mass balance, where the term on the left-hand side describes the time rate of change in the concentration of the microorganism in an arbitrary elemental volume and is equal to the right-hand side, which characterizes movement of the microorganisms through the elemental volume, losses due to decay, and additions or losses due to inputs or outputs.

To derive an equation for the transport of microorganisms in porous media, a mathematical statement of mass conservation must be formulated. In other words, the amount of material leaving the system must be equal to the amount entering the system plus or minus any changes in storage in the system. Shown in Fig. 10 is a schematic of the movement through and changes in a quantity (mass) of contaminant in an elemental volume. It is assumed in this development that the porous medium is homogeneous, isotropic, incompressible, and fully saturated. This diagram (Fig. 10) states that the change in concentration in the elemental volume depends on the mass entering minus the mass leaving minus the quantity that changes state or that decays. Expressed as an equation, that is:

$$\text{mass in} - \text{mass out} - \text{mass lost} = \text{change in storage} \quad (2)$$

The next step in solving the problem is to express each term in the above equation in terms of appropriate fluxes. The flux is defined as the total mass of contaminant per unit cross-sectional area transported in the x direction per unit time:

$$F_x \Delta y \Delta z \bigg|_{x = x_0} - F_x \Delta y \Delta z \bigg|_{x = x_0 + \Delta x} - \left(\rho_b \frac{\partial S}{\partial t} + \eta \mu C \right) \Delta x \Delta y \Delta z$$

$$= \eta \frac{\partial C}{\partial t} \Delta x \Delta y \Delta z \quad (3)$$

where F_x is the flux in the x direction at the positions $x = x_0$ and $x = x_0 + \Delta x$; Δx, Δy, and Δz are the sizes of the elemental volume in the x, y, and z directions; ρ_b is the bulk density of the porous media; S is the concentration on the solid phase due to adsorption; η is the porosity of the porous media; and μ is the first-order decay coefficient. Before equation 3 can be written in a form that can be solved mathematically, some definition (i.e., an equation) must be found which describes the flux terms. A common equation for describing these fluxes uses Fick's Law and is

$$F_x = -\eta D \frac{\partial C}{\partial x} + V \eta C \quad (4)$$

where the first term on the right-hand side is Fick's Law and the second describes the mass flux due to advection. Additional equations for the y and z fluxes would be necessary if multidimensional transport was being considered. Equation 3 includes terms for describing the flow of the microorganism through the volume (terms 1 and 2 on the left-hand side of equation 3), for sorption of the microorganism on the soil matrix (i.e., $\rho_b \partial S/\partial t$), for first-order decay of the microorganism in the liquid phase (i.e., $\eta \mu C$), and for the net change in the concentration of the contaminant due to the above listed mechanism (the term on the right-hand side of equation 3).

The flux at $x = x_0 + \Delta x$ can be approximated by the flux at $x = x_0$ by using a Taylor series expansion, which is

$$F_x \bigg|_{x = x_0 + \Delta x} = F_x \bigg|_{x = x_0} + \frac{\partial F_x}{\partial x} \bigg|_{x = x_0} \Delta x + \frac{\partial^2 F_x}{\partial x^2} \bigg|_{x = x_0} \Delta^2 x + \ldots \quad (5)$$

Combining equations 3 and 5, cancelling terms, dividing through by $\Delta x \Delta y \Delta z$, and taking the limit as the elemental volume gets very, very small (mathematically, this means that Δx approaches 0 and provides a means for removing the derivatives of second or higher order from the equation) gives a transport equation written in terms of the flux:

$$\frac{\partial C}{\partial t} + \rho_b \frac{\partial S}{\partial t} = -\frac{\partial F_x}{\partial x} - \mu C \quad (6)$$

Using the equation for the solute flux given by equation 4 in equation 6; using the expression $S = k_d C$; expressing the retardation coefficient, R, as $R = 1 + \rho_b k_d / \eta$; and simplifying, gives equation 1.

Since equation 1 is relatively simple, many solutions can be obtained provided that the initial conditions (i.e., the initial distribution of the chemical in the porous medium at the starting time) and boundary conditions (i.e., the conditions that occur at a boundary such as the soil surface) are also relatively simple. Examples of the solutions that are commonly used are given below.

Examples

Consider the following situation: microorganisms will be applied to the soil at some site. The concentration of these particular microorganisms in the soil before application is zero. A known and constant concentration of microorganisms will be added to the soil for a known period of time at a particular location $x = 0$, known as the boundary. Mathematically, the initial and boundary conditions are as follows: the initial condition is $C(x, t = 0) = C_i$; the boundary conditions are $C(x = 0, t) = C_0$ and $C(x \to \infty, t)$ is finite. For this case the solution to equation 1, subject to the initial and boundary conditions, is

$$C(x, t) = C_i A(x, t) + C_0 B(x, t) \qquad (7)$$

where

$$A(x, t) = e^{-\mu t/R} \left\{ 1 - \tfrac{1}{2}\text{erfc}\left[\frac{Rx - Vt}{2\sqrt{DRt}}\right] - \tfrac{1}{2}e^{Vx/D}\text{erfc}\left[\frac{Rx + Vt}{2\sqrt{DRt}}\right] \right\} \qquad (8)$$

and

$$B(x, t) = \tfrac{1}{2}e^{(V-u)x/2D}\text{erfc}\left[\frac{Rx - ut}{2\sqrt{DRt}}\right] + \tfrac{1}{2}e^{(V+u)x/2D}\text{erfc}\left[\frac{Rx + ut}{2\sqrt{DRt}}\right] \qquad (9)$$

with

$$u = V\left[1 + \frac{4\mu D}{V^2}\right]^{1/2} \qquad (10)$$

where $\text{erfc}(x) = 1 - \text{erf}(x)$. It is also possible to obtain solutions for a constant and specified initial condition with a constant flux concentration boundary condition, that is, where the following flux is specified:

$$\left(-D\frac{\partial C}{\partial x} + V\right)\bigg|_{x=0} = VC_0 \qquad (11)$$

Solutions can be found for even more complicated initial and boundary conditions. Some of these can be found in van Genuchten and Alves (1982).

Difficulties in Using the Advection-Dispersion Equation

Although many of the models used to describe the transport of solutes in the subsurface are based on the advection-dispersion equation, models that assume that the coefficients are constant tend to be overly restrictive and often do not produce adequate comparisons with what is observed in the field. Recent field evidence suggests that the classical form of the advection-dispersion equation is inadequate for describing field-scale solute transport (Matheron and De Marsily, 1980; Sposito et al., 1986) due, in part, to an apparent increase in the dispersivity as a function of the travel distance. This has stimulated research to develop other methods for describing field transport behavior.

A variety of approaches are available for describing the transport of solutes in heterogeneous porous media. If it is possible to fully characterize the heterogeneous porous medium, then deterministic approaches may produce acceptable results. In a gross sense this may be accomplished if large-scale features such as layers, lenses, etc., can be located and the material and hydraulic properties can be determined. Major disadvantages in attempting to characterize the transport process in this manner include the extreme cost of sampling and data analysis as well as the difficulty in knowing the number and placement of samples and the level of sampling detail necessary to achieve a given accuracy in describing the porous medium. Such difficulties in describing field-scale solute transport have prompted investigators to use stochastic approaches (Gelhar et al., 1979; Jury, 1982; Dagan, 1984). Stochastic analyses have yielded important information concerning field-scale transport processes and, in particular, have provided a theoretical basis for the observation that field-scale dispersion often depends on the mean travel distance.

A possible alternative to stochastic approaches for modeling the field-scale transport process is to develop more accurate models for the field-scale dispersion process. Pickens and Grisak (1981) developed a finite element model which allows the dispersivity to depend on the mean travel time. They investigated several dispersivity functions: linear, parabolic, asymptomatic, and exponential. In each case, the dispersivity functions contained one or more constants which have to be determined for a given field location.

Advection-Dispersion with Distance-Dependent Dispersivity

Consider a one-dimensional physical system containing a solute that is moving with an average steady-state advective velocity, $\bar{v}$. It is assumed that any linear growth of the dispersion process is a direct consequence of the heterogeneous nature of the porous medium, that the observed growth is caused by fluctuations of the water velocity around an areal mean value, and that the growth of the dispersivity, $\alpha(x)$, as a function of distance can be adequately described using an equation of the form (see Pickens and Grisak, 1981)

$$\alpha(x) = ax \tag{12}$$

Since, to a first approximation, the hydrodynamic dispersion has been shown to be a function of the average flow velocity, $\bar{v}$, times the dispersivity plus a diffusion term, the hydrodynamic dispersion as a function of position can be written, using equation 12, as

$$D(x) = \alpha(x)\bar{v} + \text{diffusion} = (ax + Lb)\bar{v} \qquad (13)$$

where the diffusion is rewritten for convenience as $Lb\bar{v}$; L, b, and $\bar{v}$ are strictly constants; parameter a [l^0] is the slope of the dispersivity-distance relationship; parameter b [l^0] is a constant which characterizes the fluid diffusional processes; and L is a characteristic distance.

The transport of solutes in a one-dimensional system including linear equilibrium adsorption and first-order decay is

$$R\frac{\partial C}{\partial t} = \frac{\partial}{\partial x}\left[D(x)\frac{\partial C}{\partial x}\right] - \bar{v}\frac{\partial C}{\partial x} - \mu RC \qquad (14)$$

where C is the solute concentration as a function of space and time, R is the retardation coefficient, and μ is the first-order decay coefficient.

Although a numerical solution to equation 14 yields a complete solution to the distance-dependent problem, analytical solutions can be obtained for simplified initial and boundary conditions. One solution is for a boundary condition at $x = 0$ that has a constant concentration and the other has a constant flux condition. In both cases, as x gets large, the gradient $\partial C/\partial x$ is almost 0. The initial and boundary conditions can be stated mathematically as

$$C(x, 0) = 0; \quad C(0, t) = C_0; \quad \left.\frac{\partial C}{\partial x}\right|_{x \to \infty} = 0 \qquad (15)$$

for transport with a constant concentration boundary condition and

$$C(x, 0) = 0; \quad \left.-D(x)\frac{\partial C}{\partial x}\right|_{x=0} + \left.\bar{v}C\right|_{x=0} = \bar{v}C_0; \quad \left.\frac{\partial C}{\partial x}\right|_{x \to \infty} = 0 \qquad (16)$$

for a constant flux boundary condition. To obtain the analytical solutions, a number of steps are required, as given in detail by Yates (1990). Various transformations are used to obtain the following equation, which can be solved using a variety of methods, including Laplace transforms.

$$\xi^2 \frac{d^2\overline{C}}{d\xi^2} + (1 - 2/a)\xi\frac{d\overline{C}}{d\xi} - (2/a)^2(\beta + s)\xi^2\overline{C} = 0 \qquad (17)$$

Solution 1: Initial concentration zero and constant concentration boundary condition

If the initial and boundary conditions are

$$C(\xi_0, s) = C_U \text{ and } C(\xi, 0) = 0 \qquad (18)$$

then the solution is

$$C_c(\xi, \tau)/C_0 = \left[\frac{\xi}{\xi_0}\right]^\gamma \left[\frac{K_\gamma[2\sqrt{\beta}(\xi/a)]}{K_\gamma[2\sqrt{\beta}(\xi_0/a)]} - \frac{2}{\pi}I_c\right] \qquad (19)$$

where I_c is an integral for the constant concentration case on the interval $[\sqrt{\beta}, \infty]$ of Bessel functions of order γ and is

$$I_c = \int_{\sqrt{\beta}}^\infty \frac{\exp[-\chi^2\tau]}{\chi} \left[\frac{J_\gamma(\varepsilon)Y_\gamma(\varepsilon_0) - J_\gamma(\varepsilon_0)Y_\gamma(\varepsilon)}{J_\gamma(\varepsilon_0)^2 + Y_\gamma(\varepsilon_0)^2}\right] d\chi \qquad (20)$$

where

$$\varepsilon = 2\sqrt{\chi^2 - \beta}\,\xi/a \text{ and } \varepsilon_0 = 2\sqrt{\chi^2 - \beta}\,\xi_0/a \qquad (21)$$

and $J_\gamma(x)$ and $Y_\gamma(x)$ are Bessel functions of the first and second kind, respectively. As the first-order decay coefficient approaches zero, the ratio of modified Bessel functions in equation 19 will approach $(\xi_0/\xi)^\gamma$.

Solution 2: Initial concentration zero and a constant flux boundary condition

If the initial and boundary conditions in the Laplace domain are

$$\left.\frac{-a\xi}{2}\frac{\partial C}{\partial \xi}\right|_{\xi=\xi_0} + C\Big|_{\xi=\xi_0} = C_0 \text{ and } C(\xi, 0) = 0 \qquad (22)$$

then the solution is

$$C_f(\xi, \tau)/C_0 = \left[\frac{\xi}{\xi_0}\right]^\gamma \left[\frac{K_\gamma[2\sqrt{\beta}\,\xi/a]}{K_\gamma[2\sqrt{\beta}\,\xi_0/a] + \sqrt{\beta}\,\xi_0 K_{\gamma-1}[2\sqrt{\beta}\,\xi_0/a]} - \frac{2}{\pi}I_f\right] \qquad (23)$$

where I_f is an integral of Bessel functions of order γ on the interval $[\sqrt{\beta}, \infty]$ for the constant flux case and is

$$I_f = \int_{\sqrt{\beta}}^\infty \frac{\exp[-\chi^2\tau]}{\chi}$$

$$\cdot \left[\frac{J_\gamma(\varepsilon)[Y_\gamma(\varepsilon_0) - \phi(\chi)Y_{\gamma-1}(\varepsilon_0)] - Y_\gamma(\varepsilon)[J_\gamma(\varepsilon_0) - \phi(\chi)J_{\gamma-1}(\varepsilon_0)]}{[J_\gamma(\varepsilon_0) - \phi(\chi)J_{\gamma-1}(\varepsilon_0)]^2 + [Y_\gamma(\varepsilon_0) - \phi(\chi)Y_{\gamma-1}(\varepsilon_0)]^2}\right] d\chi \qquad (24)$$

where $\phi(\chi) = \sqrt{\chi^2 - \beta}\,\xi_0$ and ε and ε_0 are defined in equation 20. As the first-order decay coefficient approaches zero, the ratio of Bessel functions in equation 23 approaches $(\xi_0/\xi)^\gamma$.

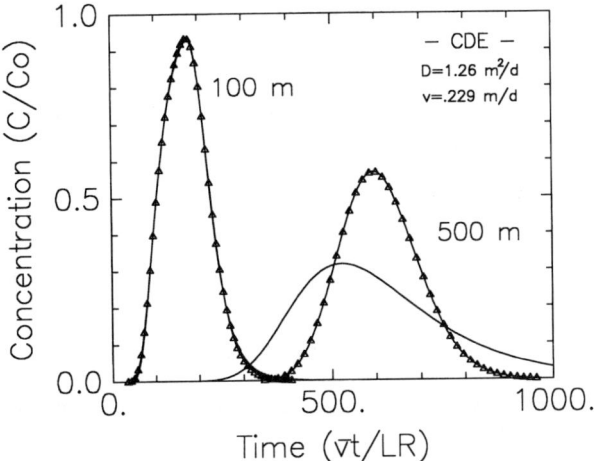

Figure 11. Concentration as a function of time for a pulse-type, constant concentration boundary condition at distances $x = 100$ m and 500 m. The solid line results from the distance-dependent solution, and the line with triangles is the advection-dispersion solution with constant coefficients that were obtained by fitting the model to the solid line at $x = 100$ m.

Example

To illustrate the distance-dependent solution, a hypothetical example was constructed. The values for R, b, v, L, and μ were 1.0, 0.01, 0.25 m/day, 1.0 m, and 0.0 day^{-1}, respectively.

Figure 11 shows a comparison between the distance-dependent solution described by equation 14 and the classical advection-dispersion solution (equation 1) with a constant dispersion coefficient (see van Genuchten and Alves, 1982) for a constant concentration boundary condition and a pulse input. The pulse boundary condition used in this example is

$$C(0, \tau) = 1 \quad \text{for } 0 \leq \tau \leq 125 \qquad (25)$$

$$C(0, \tau) = 0 \quad \text{for } \tau > 125$$

The concentration history is shown at two locations, $x = 100$ m and $x = 500$ m. This example shows the difference between predicted concentrations at the location $x = 500$ m using either the classical or the distance-dependent dispersion solution with model coefficients that produce essentially the same concentration curve at $x = 100$ cm. The values for the classical solute transport equation obtained from the inverse method of Parker and van Genuchten (1984) are $v = 0.229$ m/day and $D = 1.26$ m^2/day. For this example, the inverse method produces an estimated value for the average water velocity which is close to the aquifer value of 0.25 m/day. Comparing the results of the two solutions (i.e., see the curves marked "100 m" in

Fig. 11) demonstrates that the classical advection-dispersion equation can produce a curve with the same shape as a curve which results from a distance-dependent dispersion model. If the classical solution, with the parameter estimates obtained at $x = 100$ m, is used to predict the concentration-time curve at a later time, for instance at $x = 500$ m, a significant deviation from the distance-dependent solution results. The classical solute transport equation produces a higher peak concentration and less overall dispersion, but the mean values for the respective concentration distributions are approximately the same, due for the most part to the similar values for the average water velocity. Figure 11 demonstrates the need to adequately characterize the dispersion process whenever models are to be used to predict the spatial and temporal distribution of contaminants in soils.

CONCLUSIONS

The use of contaminant transport models to predict the movement of microorganisms for purposes of groundwater protection and bioremediation is likely to continue to increase in the future. The establishment of wellhead protection zones within which potential sources of contamination may not be placed is one area in which the use of such models will be necessary. Another function microbial transport models may serve is in granting variances from a mandatory groundwater disinfection requirement (to inactivate viruses and *Giardia* sp. in drinking water) under review by the U.S. Environmental Protection Agency. An understanding of some of the aspects of modeling will enable microbiologists to design experiments such that information critical to modeling efforts is obtained. It will also help to improve communications between microbiologists and contaminant transport modelers so that accurate models can be developed and model outputs can be interpreted correctly.

GLOSSARY OF MODELING TERMS

Advection: the process whereby contaminants are transported by the bulk motion of flowing groundwater

Boundary condition: the events at a boundary such as the soil surface where a source of contaminants resides

Breakthrough curve: the concentration of a contaminant plotted as a function of the time required for the contaminant to travel through a length of porous medium (e.g., a soil column)

Convection: sometimes used interchangeably with advection. Properly used, it refers to the transport of heat by flowing groundwater.

Decay: irreversible reduction in concentration of the contaminant due to chemical, physical, or biological processes

Dispersion (hydraulic dispersion, mechanical dispersion): the spreading out of the contaminant as it flows through the subsurface medium

Heterogeneous: describes a porous medium in which the hydraulic conductivity and other soil and hydraulic properties are dependent on the position within a geologic formation (i.e., the capacity to transmit water changes at different locations in the formation)

Homogeneous: describes a porous medium in which the hydraulic conductivity and other soil and hydraulic properties are independent of position (i.e., the capacity to transmit water is constant throughout the formation)

Hydraulic conductivity: a measure of the capacity of a porous medium to conduct water

Initial condition: the initial distribution of the contaminant in the porous medium

Retardation: the slowing of the movement of a contaminant front during flow due to adsorption onto the solid phase

Saturated medium: in which all of the pore spaces between solid particles are filled with water

Source, sink: additions or subtractions, respectively, of the contaminant in the system by means other than physical input or decay

Unsaturated medium: in which the pore spaces between solid particles are not completely filled with water; gas is contained within some of the pore spaces

REFERENCES

Burge, W. D., and N. K. Enkiri. 1978. Virus adsorption by five soils. *J. Environ. Qual.* **7**:73–76.

Champ, D. R., and J. Schroeter. 1988. Bacterial transport in fractured rock—a field scale tracer test at the Chalk River Nuclear Laboratories, p. 14:1–14:7. *In* B. H. Olson and D. Jenkins (ed.), Proceedings of the International Conference on Water and Wastewater Microbiology, Newport Beach, Calif., 8–11 February 1988.

Dagan, G. 1984. Solute transport in heterogeneous porous formations, groundwater. *J. Fluid Mech.* **145**:151–177.

Gelhar, L. W., A. L. Gutjahr, and R. L. Naff. 1979. Stochastic analysis of macrodispersion in a stratified aquifer. *Water Resour. Res.* **15**:1387–1397.

Goyal, S. M., and C. P. Gerba. 1979. Comparative adsorption of human enteroviruses, simian rotavirus, and selected bacteriophages to soils. *Appl. Environ. Microbiol.* **38**:241–247.

Grondin, G. H. 1987. Transport of MS-2 and f2 bacteriophage through saturated Tanque Verde Wash soil. Masters thesis. University of Arizona, Tucson.

Jury, W. A. 1982. Simulation of solute transport using a transfer function model. *Water Resour. Res.* **18**:363–368.

Krone, R. B., G. T. Orlob, and C. Hodgkinson. 1958. Movement of coliform bacteria through porous media. *Sewage Ind. Wastes* **30**:1–13.

Matheron, G., and G. De Marsily. 1980. Is transport in porous media always diffusive, a counter sample. *Water Resour. Res.* **16**:901–917.

Matthess, G., A. Pekdeger, and J. Schroeter. 1988. Persistence and transport of bacteria and viruses in groundwater—a conceptual evaluation. *J. Contam. Hydrol.* **2**:171–188.

Moore, R. S., D. H. Taylor, L. S. Sturman, M. M. Reddy, and G. W. Fuhs. 1981. Poliovirus adsorption by 34 minerals and soils. *Appl. Environ. Microbiol.* **42**:963–975.

Parker, J. C., and M. T. van Genuchten. 1984. Determining transport parameters from laboratory and field tracer experiments. Bulletin 84-3. Virginia Agricultural Experiment Station, Blacksburg.

Pickens, J. F., and G. E. Grisak. 1981. Modeling of scale-dependent dispersion in hydrogeologic systems. *Water Resour. Res.* **17**:1701–1711.

Sobsey, M. D. 1983. Transport and fate of viruses in soils, p. 175–197. *In* L. W. Canter, E. W. Akin, J. F. Kreissl, and J. F. McNabb (ed.), *Microbial Health Considerations of Soil Disposal of Domestic Wastewaters*. Publication no. 600/9-83-017. U.S. Environmental Protection Agency, Cincinnati, Ohio.

Sposito, G., W. A. Jury, and V. K. Gupta. 1986. Fundamental problems in the stochastic convection-dispersion model of solute transport in aquifers and field soils. *Water Resour. Res.* **22**:77–88.

U.S. Environmental Protection Agency. 1985. National primary drinking water regulations; synthetic organic chemicals, inorganic chemicals, and microorganisms. *Fed. Regist.* **50**: 46936–47022.

van Genuchten, M. T., and W. Alves. 1982. Analytical solutions of the one-dimensional convective-dispersive solute transport equation. Technical Bulletin no. 1661. U.S. Department of Agriculture, Washington, D.C.

Vaughn, J. 1988. Inactivation of human rotavirus by disinfectants. Final project report. *Prepared for* R. S. Kerr Environmental Research Laboratory, U.S. Environmental Protection Agency, Ada, Okla.

Vaughn, J. M., and E. F. Landry. 1983. Viruses in soils and groundwaters, p. 163–210. *In* G. Berg (ed.), *Viral Pollution of the Environment*. CRC Press, Boca Raton, Fla.

Wilson, L. G., C. P. Gerba, M. W. Bolton, and J. B. Rose. 1984. Subsurface transport of urban runoff pollutants, p. 158–160. *In* N. N. Durham and A. E. Redelfs (ed.), *Proceedings of the Second International Conference on Groundwater Quality Research*. University Printing Services, Oklahoma State University, Stillwater.

Yates, M. V., and S. R. Yates. 1987. A comparison of geostatistical methods for estimating virus inactivation rates in ground water. *Water Res.* **21:**1119–1125.

Yates, M. V., and S. R. Yates. 1988. Modeling microbial fate in the subsurface environment. *Crit. Rev. Environ. Control* **17:**307–344.

Yates, M. V., and S. R. Yates. 1989. Septic tank setback distances: a way to minimize virus contamination of drinking water. *Ground Water* **27:**202.

Yates, S. R. 1990. An analytical solution for one-dimensional transport in heterogeneous porous media. *Water Resour. Res.* **26:**2331–2338.

Chapter 4

Quantitation of Factors Controlling Viral and Bacterial Transport in the Subsurface

Charles P. Gerba, Marylynn V. Yates, and S. R. Yates

Introduction 77
Factors Influencing Microbial Transport 78
   Microbial Die-Off or Decay 78
   Filtration 80
   Adsorption 81
   Advection and Dispersion 84
Suggested Worst-Case and Default Values for Factors Describing
   Microbial Transport 86
References 87

The previous chapter discussed model equations to predict the survival and transport of enteric microorganisms in the subsurface. Such equations could be used to predict the concentration of microorganisms in the subsurface at any distance from the contaminant source. However, for a model to be useful, quantitative information on the various factors controlling transport is necessary. This information can only be obtained through laboratory and field experimentation. We have recently prepared several major reviews on this topic, and these were used as the primary source of information for the present chapter (Gerba and Bitton, 1984; Yates et al., 1987; Yates and Yates, 1988). This chapter reviews the current data base on factors which may be useful in models used to predict viral and bacterial transport in the subsurface.

In general, both the survival and migration of microorganisms are controlled by the specific microorganism type, the nature of the soil, and the climate of the environment. Specific factors affecting survival include temperature, organic matter, moisture content, pH, and the presence of other microorganisms. Migration is controlled by moisture content, pH, salt species and concentration, soil properties

---

Charles P. Gerba • Department of Microbiology and Immunology and Department of Soil and Water Science, University of Arizona, Tucson, Arizona 85721. *Marylynn V. Yates* • Department of Soil and Environmental Sciences, University of California, Riverside, California 92521. *S. R. Yates* • U.S. Salinity Laboratory, U.S. Department of Agriculture/Agricultural Research Service, Riverside, California 92521.

Table 1. Factors important in the transport of microorganisms through soil

Advection	The process by which contaminants are transported by the bulk motion of the flowing groundwater. Advection causes nonreactive contaminants to be carried at an average rate equal to the average velocity of the water as it passes through porous media.
Dispersion	The spreading out of the contaminant as it flows through the subsurface medium. Dispersion is the result of variations in the actual pore velocities as compared with the average velocity, and of the effect of Brownian motion on the microbial particles. Often dispersion is related to advection through $D = \alpha v$, where α is the dispersivity and v is the velocity.
Adsorption	The removal of microorganisms from groundwater flow through adhesion to soil particles.
Filtration	Removal by size exclusion.
Decay or die-off	The inactivation of microorganisms due to environmental stresses such as temperature or lack of nutrients.

(i.e., sand, silt, clay), organic matter, and hydraulic conditions. The importance of all of these factors has previously been reviewed in detail (Gerba and Bitton, 1984; Yates and Yates, 1988). While all of these factors have been shown experimentally to play a role in survival and transport of microorganisms in the subsurface, much of the existing data is described in a qualitative rather than in a quantitative sense. In some cases, too few data were generated to describe the results mathematically. In others, the results are so microorganism specific that they cannot easily be generalized to describe all situations.

FACTORS INFLUENCING MICROBIAL TRANSPORT

It is believed that there are five important factors which influence microbial transport in the subsurface. These are advection, dispersion, adsorption, filtration, and decay or die-off rate. Because of the small size of viruses, filtration is not believed to be important; however, it may be the most important factor in limiting the transport of bacteria. These factors are listed and defined in Table 1.

Microbial Die-Off or Decay

The soil is not the natural habit of enteric bacterial and viral pathogens, and eventually they will die or become inactivated. Under some situations enteric bacteria may demonstrate some regrowth, but eventually they will die out. The inactivation of microorganisms in water and soil has usually been described as a first-order reaction (Reddy et al., 1981; Crane and Moore, 1986; Yates and Yates, 1988). The inactivation rate is described by a first-order reaction rate expression:

$$\text{Inactivation rate} = \frac{\partial C}{\partial t} = K_i C \qquad (1)$$

where C is infective microorganism concentration at time t and K_i is the first-order inactivation constant (time^{-1}). Here K_i would be an expression of the sum total of all factors which influence microorganism survival. Measurement of virus inactivation has been conducted in a wide variety of surface waters, but such information on soils and groundwater has been limited until recently. As stated previously, many factors can be involved in controlling microbial survival in the subsurface. Of all of the factors controlling virus survival in the subsurface, temperature is probably the most important in the majority of subsurface environments. Yates et al. (1985) studied the effects of numerous physical-chemical and microbial properties of groundwater on virus survival. They found that groundwater temperature was the most important predictor of virus inactivation. This was found to be true for several enteroviruses and the coliphage MS-2.

The inactivation rate for coliphage MS-2 as a function of temperature was expressed by the following equation:

$$K_i = \text{inactivation rate } (\log_{10} \text{ day}^{-1})$$
$$= -0.181 + 0.0214 \times \text{temperature (°C)} \qquad (2)$$

Note that examination of this equation indicates that as groundwater temperatures approach 8°C, the rate of virus inactivation becomes negligible. Reddy et al. (1981) also attempted to model microbial survival in soil. They stated that the concentration of microorganisms in the soil at time t could be described using this first-order rate expression:

$$M_1 = M_0 \exp[(K_B - K_D)t] \qquad (3)$$

where M_1 is the microbial concentration at time t; M_0 is the initial microbial concentration after the waste application; K_B is the rate coefficient for the rate of division of the microorganism; and K_D is the rate coefficient for the die-off rate of the microorganism. This equation was simplified to

$$M_1 = M_0 \exp(-Kt) \qquad (4)$$

by defining $-K = K_B - K_D$.

Reddy et al. (1981) felt that temperature, pH, moisture, and the method of waste application were the most important factors controlling microbial inactivation. They reviewed the literature to find data which could be used to develop equations quantifying the influence of these factors on inactivation. Functional relationships were developed for each of the four factors, which were incorporated into the following expression, resulting in an effective die-off rate constant, K_2:

$$K_2 = K_1 \cdot F_T \cdot F_M \cdot F_{pH} \cdot F_{ma} \qquad t_{1/2} = 0.693/K_2 \qquad (5)$$

where the correction coefficients are F_T for temperature; F_M for soil moisture content; F_{pH} for soil pH; F_{ma} for method of application; and $t_{1/2}$ for the half-life for the survival of the organisms.

The usefulness of incorporating all of these factors was questioned by Crane and Moore (1986) and Yates and Yates (1988) because of the lack of literature data needed to estimate these coefficients. Examples given by Reddy et al. (1981) were based on a limited set of data, usually only involving one or a few microorganisms.

Table 2. Die-off rate constants for selected microorganisms in a soil-water-plant system[a]

Microorganism	Die-off rate constant (day^{-1})					
	Avg	Maximum	Minimum	± Standard deviation	% Coefficient of variation	No. of observations
Escherichia coli	0.92	6.39	0.15	0.64	179	26
Fecal coliforms	1.53	9.10	0.07	4.35	283	46
Fecal streptococci	0.37	3.87	0.05	0.69	188	34
Salmonella sp.	1.33	6.93	0.21	1.70	128	16
Shigella sp.	0.68	0.74	0.62	0.06	9	3

[a] Adapted from Reddy et al., 1981.

In addition, the data used to estimate the die-off rate constants and correction coefficients were not derived from studies on survival in soil or groundwater. Average die-off rate constants for selected enteric bacteria in soil groundwater environments and temperature correction coefficients are shown in Tables 2 and 3.

Filtration

Another factor which affects the transport of microorganisms through porous media is filtration. Filtration mechanisms include straining, sedimentation, inertial impingement, and diffusion. The straining mechanism occurs when the particle in suspension in the porous matrix cannot pass through a smaller pore opening or constriction (i.e., the wedge between two soil particles), and thus its transport is halted (Corapcioglu and Haridas, 1985). The relative magnitude of the effect of this process depends on many soil, water, and microbial factors. For small microbial particles (i.e., viruses) in coarse-grained material, filtration is probably negligible.

Table 3. Temperature correction coefficients for the survival of pathogens and indicator organisms in soil and water systems[a]

Type of organism	Temp range (°C)	Temp correction coefficient
Fecal coliforms	15–21	1.08
	10–20	1.05
Escherichia coli	5–10	1.09
	10–15	1.17
	15–20	1.15
	20–25	1.07
Salmonella typhimurium	10–20	1.05
Streptococcus faecalis	10–20	1.11
	4–10	1.02
	10–25	1.03
	25–37	1.06

[a] Adapted from Reddy et al., 1981.

For the large bacteria and virus aggregates, on the other hand, physical straining may be an important consideration. In general, under high flow velocities, the amount of bacteria filtered is less than for low flow velocities (Wollum and Cassel, 1978). This is probably due to the fact that a larger amount of the total flow quantity is derived from the larger pores, which will transmit a greater portion of the total number of bacteria present. The filter efficiency may also change with time: as the bacteria accumulate in the soil, they become part of the filter, thus increasing filter efficiency.

Sedimentation in the pores occurs where there is a density difference between the microorganism and water. If the microorganism is more dense than water and the flow properties are such that the tendency for gravitational settling is greater than the tendency to be resuspended into the flow stream, the bacteria may settle into quiescent parts of the porous matrix. Corapcioglu and Haridas (1985) analyzed the gravitational settling using Stokes' Law. A disadvantage of using this method is that Stokes' Law was derived for fluids at rest, which is generally not the case in groundwater systems. Also, Yao et al. (1971) noted that gravitational settlings play a significant role only in the capture of particles larger than 5 µm, which is larger than most enteric bacteria.

The filtration efficiency of a soil or aquifer can be simply defined as the removal to a certain flow length (Matthess et al., 1988):

$$C = C_0 \exp(-\lambda_f X) \tag{6}$$

where C_0 is the initial concentration of organisms; X is distance; C is the observed concentration of organisms; and λ_f is the filtration factor or coefficient.

Filter mechanisms depend on hydraulic conditions (flow velocity and flow direction). When these parameters change, the bacteria may be remobilized. The filter factor (λ_f) may also change with time; as the bacteria or other particles accumulate, a filter layer develops which further reduces the diameter of the pores available for microorganism movement (Krone et al. 1958). Also, very high microbial concentrations will induce flocculation and aggregation of microorganisms, which will further enhance clogging.

Matthess et al. (1988) reported filter coefficients of 10 to 44.6 m^{-1} for enteric bacteria in a sandy soil (<2% clay). Jang et al. (1983) reported filter coefficients of 40 to 93 m^{-1} for sandstone cores, depending on the type of bacteria.

In summary, the filter coefficient will depend upon the type of bacteria, water velocity, type of soil, number of bacteria, and water quality.

Adsorption

Adsorption is a factor in both bacterial and viral removal by soil. Numerous factors affect the adsorption of microorganisms in soil including nature of the soil, pH, organic matter, ionic strength, and the nature of the microorganism (i.e., isoelectric point and presence of slime layer) (Gerba and Bitton, 1984). For the purposes of developing a quantitative relationship for this removal process, both Langmuir and Freundlich isotherms have been used (Vilker and Burge, 1980; Vilker, 1981). The adsorption equilibrium of viruses and bacteria is established within 1 to 24 h (Gerba and Bitton, 1984; Matthess et al., 1988). Therefore, at flow velocities below 1 m/day adsorption equilibrium is probably attained (Matthess et al., 1988). At higher flow velocities the adsorption process may not be at equilibrium, which causes lower adsorption as compared with the theoretical equilibrium value. Microorganisms may also desorb from soil surfaces in response to changes in pH and lower salt concentrations in solution due to rainfall. Desorption rates are not well understood and need further research.

Generally, adsorption is quantified by the use of equilibrium adsorption isotherms which relate the concentration of the solute in the liquid to the concentration of the solute on the adsorbing particles under equilibrium conditions. Virus adsorption to soils can be characterized by a Freundlich isotherm:

$$A = K_A C^{1/n} \qquad (7)$$

where K_A and n are empirical constants; A is virus adsorbed to the solid phase; and C is virus concentration in solution.

For $n<1$, multilayer adsorption is the case; $n>1$ results in saturation adsorption; and $n = 1$ indicates single-layer adsorption, resulting in a linear isotherm. Multilayer adsorption occurs when more than one virus can adsorb to an adsorption site on the soil; single-layer adsorption indicates that only one virus can adsorb to an adsorption site on the soil. Saturation adsorption occurs when all the adsorption sites are occupied and reduced or no further adsorption can occur. Experimentally determined values of $1/n$ were summarized by Vilker and Burge (1980) and were found to range from 0.87 to 1.24.

Since reported values for $1/n$ are statistically close to 1, $1/n$ can often be replaced by 1 in the Freundlich equation, yielding $A = K_A C$, the linear isotherm. The value K_A, the linear adsorption factor, has been shown to be dependent on many factors. These factors include type of soil, type of virus, pH, chemical quality of the groundwater, and presence of soluble organic matter. A summary of K_A values which have been determined experimentally for poliovirus type 1 for various soils is shown in Table 4. As illustrated in this table, the K_A value will be greatly influenced by the solution in which the virus is suspended. Much greater adsorption of the virus occurs in distilled water than in the presence of sewage effluent. No information is available on which type of K_A value might be expected in leachates from sludge landfills. The K_A value for different types of viruses will vary greatly depending upon the type of virus. As shown in Table 5, some viruses exhibit little or no adsorption to some soils. Generally K_A values can be expected to be high in clay soil and low in sandy soils.

In modeling microbial transport, K_A can be used to determine the retardation factor (R). Retardation is the slowing of the movement of a contaminant relative to the bulk mass of the water due to absorption onto solids. It is related to the soil bulk density and the volumetric water content by

$$R = 1 + \frac{\rho_b}{\theta} \cdot K_A \qquad (8)$$

where ρ_b B is bulk density (grams per cubic centimeter); θ is the volumetric water content; and K_A is the linear adsorption coefficient.

For saturated soil water conditions, the volumetric water content is equal to the effective porosity. The relationship between retardation and volumetric water content and microbial retention has never been proven experimentally. Microbial

Table 4. Experimentally determined adsorption constants (K_A) for poliovirus type 1 adsorption to various soils[a]

Soil type	K_A (ml/g)	Water type	Remarks	Reference
Sand	5.8	Deep groundwater		Yeager and O'Brien, 1979
Sand	1.4	Shallow groundwater		Yeager and O'Brien, 1979
Sandy loam	499	Deep groundwater		Yeager and O'Brien, 1979
Sandy loam	66	Shallow groundwater		Yeager and O'Brien, 1979
Loamy sand	1,000	Distilled water	FM soil[b]	Goyal and Gerba, 1979
Loamy sand	142	Secondarily treated sewage effluent	FM soil	Goyal and Gerba, 1979
Sand	0.72	Deionized water	92% sand	Goyal and Gerba, 1979
Sandy loam	4.6	Deionized water	77% sand	Goyal and Gerba, 1979
Clay	1,000	Deionized water	54% clay	Goyal and Gerba, 1979
Sandy clay loam	99	Deionized water	28% clay	Goyal and Gerba, 1979
Sand and gravel	3	Tertiary treated sewage effluent		Landry et al., 1979
Sand	2.5	Phosphate-buffered saline	Ottawa sand	Vilker et al., 1983

[a] Modified from Grosser, 1985.
[b] FM soil, Soil from Flushing Meadows, Arizona.

movement under unsaturated flow conditions has only received limited study to date. Retardation factors determined experimentally in soil columns for MS-2 coliphage are shown in Table 6. In general this virus tended to move faster than the average groundwater flow. Bacteria and yeasts also have been observed to travel faster than conservative tracers such as bromine (Keswick et al., 1982; Bradford, 1987). There are several mechanisms which may explain this phenomenon. One is pore size exclusion, in which the microorganisms can only be transported through large pores, where the average pore water velocity is higher than the average for the entire porous medium. Another mechanism that could explain the apparent enhanced transport of microorganisms relative to the water is anion exclusion. In anion exclusion, the negatively charged microbial particles are pushed to the center of the pore where the velocity is, on the average, higher than that of the bulk medium. The effects of anion exclusion may result in an R value of less than 1.

Table 5. Adsorption (K_A) constants for various viruses to a loamy sand soil (Flushing Meadows)[a]

Virus type	K_A (ml/g)
f2	0
MS-2	0.20
Coxsackievirus type B4	0
Echovirus type 1	1.5
Poliovirus type 1	1,000
Rotavirus type SA-11	1.1

[a] Calculated from data of Goyal and Gerba, 1979.

Table 6. Hydrodynamic dispersion and retardation coefficients determined experimentally using MS-2 coliphage[a]

Soil name	Class	Soil content (%)				
		Gravel	Sand	Silt	Clay	Organic matter
Flushing Meadows	Loamy sand	0	89	8	3	0.9
Corolla	Fine sand	0	100	0	0	<0.1
Tanque Verde	Gravelly sand	57	43	0	0	<0.1

[a] Values from Grondin (1987) and Bradford (1987).

Many researchers have reported retardation factors of less than 1, suggesting that anion exclusion is occurring. For example, the retardation factor of native groundwater bacteria was observed to range from 0.8 to 1.0. Matthess et al. (1988) reported retardation values for *Escherichia coli* in laboratory and field experiments to range from 0.75 to 0.93. However, examination of reported values of K_A in Table 5 indicate that these data would result in R values less than or equal to 1 (see equation 8). This indicates that more research is needed to clearly define the mechanism(s) involved in the enhanced transport of microorganisms in the subsurface.

Advection and Dispersion

The transport of microorganisms by the flow of water is termed advection. Advection causes nonreactive (e.g., nonsorbing) substances to be transported through the subsurface at an average rate equal to the average velocity of the water. The spreading and dilution of a contaminated groundwater plume during its transport with the flowing groundwater are due to hydrodynamic dispersion, which occurs because of mechanical mixing during advection and because of Brownian diffusion. Mechanical dispersion is due to variation in pore water velocity and the tortuosity of the pore channels.

The larger the hydrodynamic dispersion coefficient the greater the distance over which the virus front will spread. The hydrodynamic dispersion coefficient D is generally dependent on groundwater flow velocity dispersivity of the soil and molecular diffusion (although molecular diffusion is generally much smaller than

Table 7. Worst-case and default values for parameters in microbial transport model

Parameter	Symbol	Worst-case value	Default value
Die-off or decay	μ	Virus = 0 $\log_{10}$/day; bacteria = 0.05 $\log_{10}$/day	Virus = 0.14 $\log_{10}$/day; bacteria = 0.37 $\log_{10}$/day
Retardation	R	0.5	0.75
Velocity	V	2.8×10^{-2} m/s	10^{-5} m/s
Hydrodynamic dispersion	D	0.01 and 10 cm^2/min	1 cm^2/min
Filtration (for bacteria only)	λ_f	0	20

Table 6—Continued

Bulk density (g/ml)	Porosity (%)	Dispersivity (cm)	Hydrodynamic dispersion coefficient (cm²/min)	Retardation factor
1.60	39.62	0.32438–0.92432	0.0552–0.1428	0.92–1.02
1.56	41.13	0.37656–1.2554	0.9522–3.2694	0.92–0.89
	25.89		0.8438–1.0417	0.537–0.610

Table 8. Explanation of worst-case values for parameters in transport model

Parameter	Symbol	Justification
Die-off or decay	μ	For viruses, worst-case conditions, i.e., the lowest decay rate, would occur at or below 8°C, where decay would not occur (Yates et al., 1985). In a review by Reddy et al. (1981) the lowest decay rate reported for an enteric bacteria in soil was 0.37 $\log_{10}$/day.
Retardation	R	Worst-case conditions assume that the microorganisms are moving faster than the average groundwater flow. The lowest retardation reported for a virus is 0.5 (Grondin and Gerba, 1986) and that for enteric bacteria is 0.75 (Matthess et al., 1988).
Velocity	V	The fastest movement of microorganisms reported is that for viruses in karst terrain at 2.8 $\times 10^{-2}$ m/s (Keswick et al., 1982). Faster velocities could be possible depending on gradients.
Hydrodynamic dispersion	D	A worst-case value is more difficult to suggest for this parameter. Dispersion of a high concentration of microorganisms over a large area could be considered worst case. However, little or no dispersion of a plume headed for a drinking water well could also be considered a worst-case situation. Thus it is suggested that a range of values be used such as 0.01 to 10 cm²/min.
Filtration	λ_f	Considered important only in the transport of bacteria. No filtration, such as might occur in karst terrain, would be considered a worst-case example.

Table 9. Explanation of default values for parameters in transport model

Parameter	Symbol	Justification
Die-off or decay	μ	Default value for virus represents that predicted from the equation of Yates et al. (1985) for 15°C. Groundwater temperatures approximate those of the mean annual surface air temperature, which would be about 15°C for middle latitudes in the United States. The 0.37 $\log_{10}$/day is the average value for fecal streptococci as determined by Reddy et al. (1981). Since fecal streptococci would be expected to be among the longer-surviving enteric bacteria in soil, an average value was chosen for these organisms to represent all enteric bacteria.
Retardation	R	Microorganisms move faster than or equal to the average groundwater flow. The value of 0.75 for *E. coli* as determined in laboratory and field studies by Matthews et al. (1988) was chosen. This falls midpoint between 0.5 and 1.0, which is the range of values reported in the literature (see Table 6 and Matthews et al., 1988).
Velocity	V	Velocity will be dependent not only on the nature of the soil but also on the gradient, which is site specific. A velocity of 10^{-5} m/s would be considered typical of silty sands.
Hydrodynamic dispersion	D	A value of 1 cm^2/min is typical of dispersion values obtained in field experiments for chemicals.
Filtration	λ_f	The median value reported by Matthews et al. (1988) for sandy soils was 20.

dispersion). Dispersivity is influenced by grain size, nonuniformity, grain form, pore size distribution, and structural features. Because the actual spread of a contaminant depends on the inhomogeneity of the substrata at various scales in addition to local variation, the dispersivity increases with the scale of the experiment. Dispersivity coefficients of porous media in laboratory experiments are on the order of 0.01 to 1 m, those in field experiments are on the order of 0.01 to 100 m, and those of fissured and karstic rocks in field experiments range from 10 to 1,000 m (Matthess and Pekdeger, 1981).

Information on the dispersion of microorganisms appears to be limited. Values for dispersivity and hydrodynamic dispersion obtained using coliphage MS-2 in sand and gravel soil columns are shown in Table 6.

SUGGESTED WORST-CASE AND DEFAULT VALUES FOR FACTORS DESCRIBING MICROBIAL TRANSPORT

Contaminant transport is generally modeled using a form of the advection-dispersion equation (see chapter 2, equation 1, in this volume). Suggested worst-case and default values for transport equations are given in Table 7. Explanations of how these values were chosen are given in Table 8 for worst-case values and in Table 9 for default values. Many of the values for the parameters in the transport equation are only available for a limited number of microorganisms and soil types. This is especially true for hydrodynamic dispersion and retardation. The effects of rainfall, pH, and dissolved solids could be better defined if values for desorption equilibrium were available. Clearly, more work is needed to better model microbial transport in the subsurface so that factors controlling transport can be quantified and compared with field observations.

REFERENCES

Bradford, A. W. 1987. Transport of MS-2 virus through saturated soil columns. M.S. thesis. University of Arizona, Tucson.

Corapcioglu, M. Y., and A. Haridas. 1985. Microbial transport in soils and groundwater: a numerical model. *Adv. Water Res.* 8:188–200.

Crane, S. R., and J. A. Moore. 1986. Modeling enteric bacterial die-off: a review. *Water Air Soil Pollut.* 27:411–439.

Gerba, C. P., and G. Bitton. 1984. Microbial pollutants, their survival and transport pattern to groundwater, p. 65–88. *In* G. Bitton and C. P. Gerba (ed.), *Groundwater Pollution Microbiology*. John Wiley & Sons, Inc., New York.

Goyal, S. M., and C. P. Gerba. 1979. Comparative adsorption of human enteroviruses, simian rotavirus, and selected bacteriophages to soils. *Appl. Environ. Microbiol.* 38:241–247.

Grondin, G. H. 1987. Transport of MS-2 and f2 bacteriophage through saturated Tanque Verde Wash soil. Masters thesis. University of Arizona, Tucson.

Grondin, J., and C. P. Gerba. 1986. Virus dispersion in a coarse porous medium. *Hydrol. Water Resour. Ariz. Southwest* 16:11–15.

Grosser, P. W. 1985. A one-dimensional mathematical model of virus transport, p. 105–107. *In* N. N. Durham and A. E. Redelfs (ed.), *Proceedings of the Second International Conference on Ground Water Quality Research*. University Center for Water Research, Oklahoma State Univ., Stillwater.

Jang, L. K., P. W. Chang, J. E. Findley, and T. F. Yen. 1983. Selection of bacteria with favorable transport properties through porous rock for the application of microbial-enhanced oil recovery. *Appl. Environ. Microbiol.* 46:1066–1072.

Keswick, B. H., D. Wang, and C. P. Gerba. 1982. The use of microorganisms as ground water tracers: a review. *Ground Water* 20:142–149.

Krone, R. B., G. T. Orlob, and C. Hodgkinson. 1958. Movement of coliform bacteria through porous media. *Sewage Ind. Wastes* 30:1–13.

Landry, E. F., J. M. Vaughn, M. Z. Thomas, and C. A. Beckwith. 1979. Adsorption of enterovirus to soil cores and their subsequent elution by artificial rainwater. *Appl. Environ. Microbiol.* 38:680–687.

Matthess, G., and A. Pekdeger. 1981. Concepts of the survival and transport model of pathogenic bacteria and viruses in groundwater. *Sci. Total Environ.* 21:149–159.

Matthess, G., A. Pekdeger, and J. Schroefer. 1988. Persistence and transport of bacteria and viruses in groundwater—a conceptual evaluation. *J. Contam. Hydrol.* 2:171–188.

Reddy, K. R., R. Khaleel, and M. R. Overcash. 1981. Behavior and transport of microbial pathogens and indicator organisms in soils treated with organic wastes. *J. Environ. Qual.* 10:255–266.

Vilker, V. L. 1981. Simulating virus movement in soils, p. 223–253. *In* I. K. Iskandar (ed.), *Modeling Wastewater Renovation*. John Wiley & Sons, Inc., New York.

Vilker, V. L., and W. D. Burge. 1980. Adsorption mass transfer model for virus transport in soils. *Water Res.* 14:783–790.

Vilker, V. L., J. C. Fong, and M. Seyyed-Hoseyni. 1983. Poliovirus adsorption to narrow particle size fractions of sand and montmorillinite clay. *J. Colloid Interface Sci.* 92:422–435.

Wollum, A. G., and D. K. Cassel. 1978. Transport of microorganisms in sand column. *Soil Sci. Soc. Am.* 42:72–76.

Yaeger, J. E., and R. T. O'Brien. 1979. Enterovirus inactivation in soil. *Appl. Environ. Microbiol.* **38**:694–701.

Yao, K. M., M. T. Habibian, and C. R. O'Melia. 1971. Water and wastewater filtration: concepts and applications. *Environ. Sci. Technol.* **5**:1105–1112.

Yates, M. V., C. P. Gerba, and L. M. Kelley. 1985. Virus persistence in groundwater. *Appl. Environ. Microbiol.* **49**:778–781.

Yates, M. V., and S. R. Yates. 1988. Modeling microbial fate in the subsurface environment. *Crit. Rev. Environ. Control* **17**:307–343.

Yates, M. V., S. R. Yates, J. Wagner, and C. P. Gerba. 1987. Modeling virus survival and transport in the subsurface. *J. Contam. Hydrol.* **1**:329–345.

Chapter 5

Parameters Involved in Modeling Movement of Bacteria in Groundwater

Ronald W. Harvey

Introduction	89
Approaches to Modeling	90
Modes of Macroscopic Migration	93
Continuous (Unretarded) Transport	94
Retarded (Intermittent) Transport	95
Tactic and Transverse Migration	95
Immobilization	96
Straining	97
Sorption	98
Growth	107
Survival	108
Groundwater Conditions	109
References	110

Continued widespread contamination of shallow drinking water aquifers by microbial pathogens and chemical wastes has resulted in an increased interest in the factors that control bacterial transport through groundwater. It has been hypothesized that degradation of highly mobile and persistent groundwater contaminants may be facilitated by cotransport of indigenous bacteria that have become acclimated to their degradation (Harvey and Barber, submitted). Recent advances in the application of molecular genetics have enhanced the feasibility of employing genetically engineered microorganisms for aquifer restoration. However, the success of many schemes that employ genetic technology in in situ treatment of organically contaminated aquifers involves, in part, the ability of the engineered organism to reach the contaminant-affected area in the aquifer.

The movement of nonindigenous bacteria through aquifers has long been a public health concern, since contamination of water supply wells by microbial pathogens has contributed significantly to the number of waterborne disease

---

Ronald W. Harvey • Water Resources Division, U.S. Geological Survey, Menlo Park, California 94025.

outbreaks (Keswick, 1984). A frequent problem relating to the domestic use of untreated groundwater involves transport through the aquifer of disease-causing bacteria from contamination sources upgradient, particularly from septic tanks, domestic waste lagoons, landfills, and on-land disposal facilities for domestic effluents. The reported ability of bacteria to move through limestone (Kingston, 1943) and fractured bedrock (Allen and Morrison, 1973) is not surprising. However, there are also a number of observations suggesting transport of pathogenic and enteric "indicator" bacteria over long distances (400 to 1,000 m, horizontally) through a variety of other aquifers, including pebbles, gravels, sand and gravel, stony silt loam, and even fine sand (Dappert, 1932; Merrell, 1967; Sinton, 1980; Anan'ev and Demin, 1971; Kudryavtseva, 1972; Martin and Noonon, 1977). Often, such observations are difficult to interpret because the source term(s) was ill-defined. Nevertheless, between 1971 and 1982, over 75% of the reported illnesses of known etiology relating to the consumption of inadequately treated or untreated groundwater involved pathogenic bacteria (Craun, 1985).

Unfortunately, the ability to model groundwater flow through the subsurface greatly exceeds our understanding of the factors which affect transport of indigenous and nonindigenous bacteria (Yates and Yates, 1988). Presently, predictive modeling of bacterial transport through contaminated groundwater is problematic and there is a scarcity of field data deriving from controlled, in situ experiments. It is clear that improvements in mathematical models that involve transport of bacteria in groundwater will require a better understanding of the various mechanisms and abiotic and biotic factors involved in bacterial migration through porous media. Transport of bacteria through contaminated groundwater is affected by a number of factors in addition to the geological and hydrological characteristics of the aquifer itself. These biotic and abiotic factors include bacterial growth, predation by eucaryotic microorganisms, parasitism by bacteriophages (bacteria-specific viruses) and predatory bacteria (*Bdellovibrio* sp.), lysis under unfavorable conditions, changes in bacterial size and propensity for attachment to solid surfaces in response to alterations in nutrient conditions, spore formation in the case of some gram-positive species, sorption and biological adhesion to solid surfaces, detachment from surfaces, and straining. This chapter discusses a number of these factors in the context of modeling.

APPROACHES TO MODELING

The modeling of bacterial transport is complicated by a number of factors that can substantively affect bacterial transport behavior. Several of these factors may be interrelated through other processes. For example, the chemical (including nutrient) conditions in the groundwater may influence several processes (e.g., survival, growth, attachment, and detachment) affecting abundance of bacteria traveling through the aquifer. Figure 1 depicts many of these factors and the manner in which they are interrelated. For the purposes of constructing a transport model, it is useful to consider the major processes that substantively change the abundance of unattached bacteria within a unit of aquifer sediment. These factors include interaction with solid surfaces (reversible and irreversible attachment and detachment), predation and parasitism, straining, transport (advective, diffusive, flagellar), and lysis (Fig. 2). A number of these factors (e.g., in situ predation and growth) remain poorly understood for groundwater environments, and how best

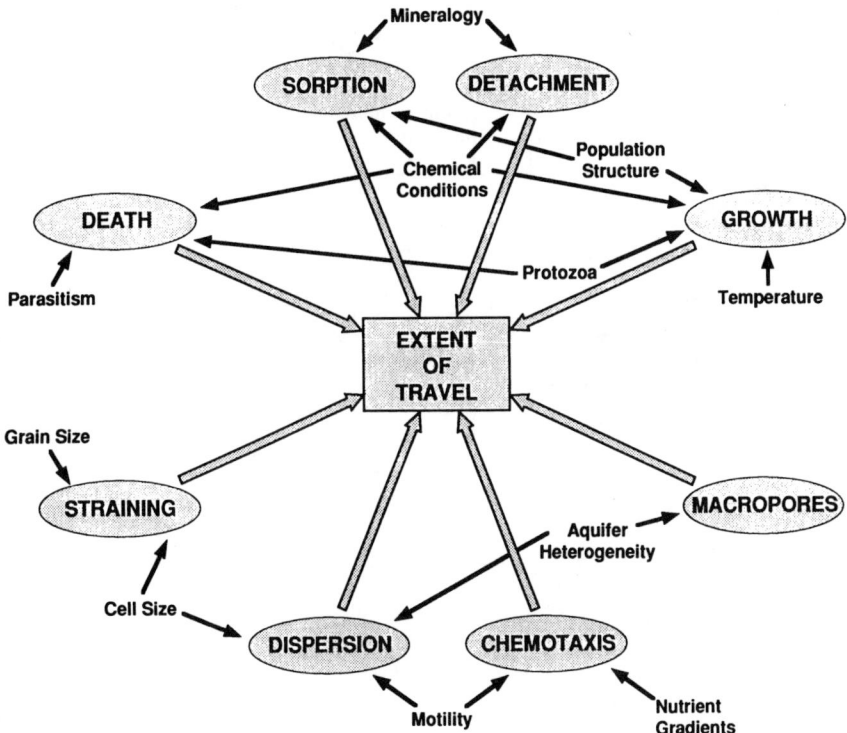

Figure 1. Schematic representation of the interrelationship of parameters and factors involved in the transport of bacteria through porous-medium aquifers.

to describe them mathematically is unclear. However, because of the complexity involved in integrating the various controlling factors into bacterial transport models, the modeling approach may be as important as the mathematical descriptions.

Models for describing subsurface transport of bacteria have employed both conceptual and data-based approaches. Recent theoretical models involving microbial transport through porous media (Corapcioglu and Haridas, 1984, 1985) have incorporated a number of governing processes depicted in Fig. 1 and 2. However, verification of existing theoretical models by field observation is problematic, owing to the complexity of the models and the number of parameters that need to be determined a priori. The predictive value of these models appears to be limited; in some cases, the degree of transport is significantly underpredicted, even when the model parameters are stressed to permissible limits (Germann et al., 1987). Nevertheless, comprehensive theoretical models serve to provide a conceptual framework for the development of more realistic models that are grounded with field observations. More accurate models will undoubtedly be developed as knowledge is gained about the parameters that govern bacterial transport through the aquifer.

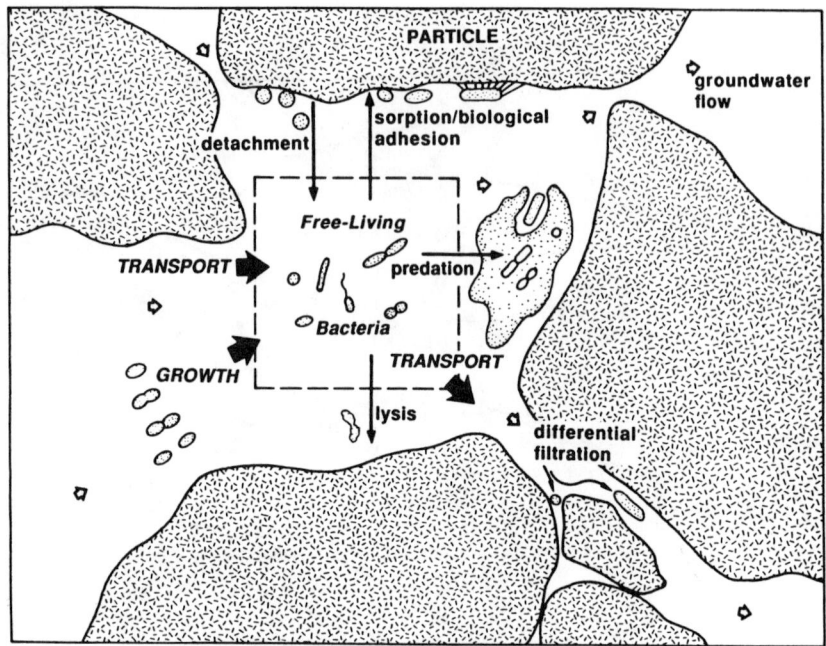

Figure 2. Schematic representation of factors affecting the mass balance of unattached bacteria in a unit of aquifer sediment.

As a result of both laboratory and field transport investigations, important advances have been made toward a better understanding of key processes governing microbial migration through aquifers. Columns packed with subsurface material allow experiments to be conducted with much greater control than is possible in field studies. Such experiments have allowed a clearer delineation of several factors governing microbial transport through the subsurface, including the effects of size-dependent exclusion from small porosity (Bales et al., 1989), survival (Bitton et al., 1979), sorption (Hendricks et al., 1979), and bacterial growth and taxis (Reynolds et al., 1989; Jenneman et al., 1985; Bosma et al., 1988). Column experiments have also proven useful in studying transport behavior of genetically engineered bacteria before they are released into the subsurface (Trevors et al., 1990). However, it has also been shown that transport of microbes through subsurface sediments that have been repacked into columns can differ from that observed in the field, even when flow velocity, porosity, and physicochemical conditions are similar (Harvey, 1988). Therefore, the overall mathematical descriptions of bacterial transport through aquifers may necessitate more information derived from field investigations, since repacking into columns can destroy the original secondary (preferred flowpath) pore structure (Smith et al., 1985).

There have been few reported attempts to model in situ observations of bacterial transport through a portion of an aquifer. One such model was used to simulate breakthrough curves for fluorochrome-labeled indigenous bacteria (whole population) that had been injected into sandy aquifer sediments 7 m upgradient from an evenly spaced network of multilevel observation wells (Harvey and Garabedian, in press). The instrumented site is part of the U.S. Geological Survey's groundwater contamination study area involving an unconfined glacial outwash aquifer near Otis Air Force Base on Cape Cod, Massachusetts. Construction of the hydrologic portion of the model was facilitated by the coinjection of a conservative, nonreactive tracer with the labeled bacteria. Employment of a solute transport model in the simulation of microbial transport through saturated sediment has also been employed in column experiments with nonsorbing viruses (Grondin and Gerba, 1986). The model described by Harvey and Garabedian (in press) includes terms for storage, advection, dispersion, and reversible and irreversible adsorption, and it handles physical heterogeneity within the bacterial population and the aquifer by superimposition of separate solutions. This deterministic approach was satisfactory over short distances (Harvey and Garabedian, in press) because the length of the layers having similar conductive properties (as delineated by hydraulic conductivity profiles) appears to be on the order of several meters (Hess, 1989). However, the modeling of transport over longer distances or thicknesses of the aquifer would necessitate a stochastic approach because it would be difficult to define aquifer structure deterministically at a larger scale. Both stochastic and deterministic approaches that attempt to account for physical heterogeneity are facilitated by information concerning the variability of hydraulic parameters within the aquifer.

Although the effect of aquifer pore structure upon microbial transport is beyond the scope of this chapter, it is clear that models describing bacterial transport in groundwater must take geohydrology into account. This is particularly important where preferred flow path (macropore) structure significantly influences the microbe's apparent downgradient mobility. In a number of groundwater tracer studies, microorganisms introduced into an aquifer appeared to travel faster than would be predicted from groundwater flow measurements, since their peak arrival at downgradient samplers occurred in advance of the conservative tracer. This phenomenon has been observed for yeasts (Wood and Ehrlich, 1978), for bacteria (Champ and Schroeter, 1988; Harvey et al., 1989), and, in sand-packed columns, for viruses (Grondin and Gerba, 1986). The apparent retardation of conservative tracer relative to bacteria in small-scale tracer tests was most significant in fractured rock (Champ and Schroeter, 1988) and least significant in sand and gravel (Harvey et al., 1989; Harvey and Garabedian, in press; Havemeister et al., 1985). Clearly, a key parameter governing transport of bacteria in groundwater involves the portion of the total porosity that is available for fluid flow, but unavailable for bacterial transport.

MODES OF MACROSCOPIC MIGRATION

Transport of bacteria through groundwater can involve several mechanisms. Three nondiffusive modes are continuous transport, discontinuous (intermittent) transport, and tactic migration (Fig. 3). All three modes of bacterial migration may be common to both indigenous and nonindigenous bacteria in groundwater. However, the contributions of each type are often poorly understood and depend upon the geohydrology, mineralogy, type of bacteria, and physical and chemical

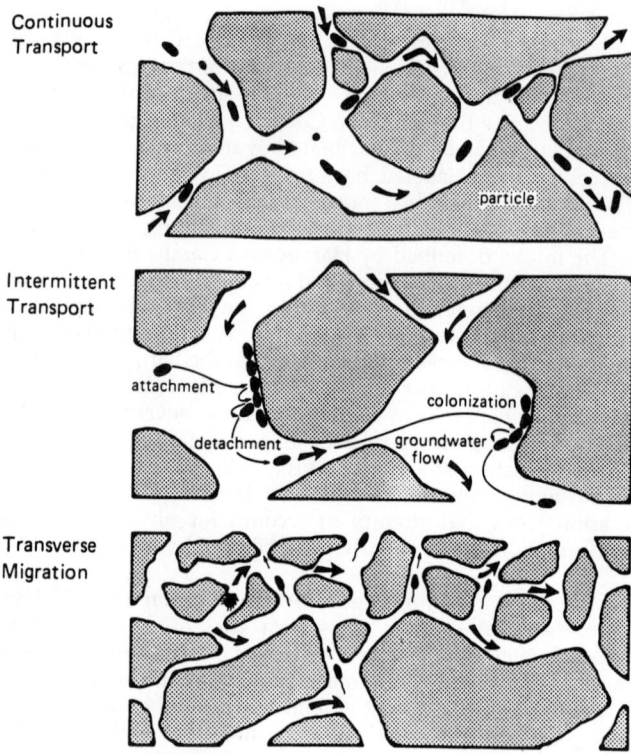

Figure 3. Schematic depiction of three modes of bacterial migration through aquifer sediments (from Harvey, 1989).

conditions of the aquifer. Facilitated migration of bacteria within a nonaqueous phase (e.g., micellular transport) is not considered here.

Continuous (Unretarded) Transport

"Continuous" is used here to refer to the unimpeded, advective movement of unattached bacteria through a section of the aquifer and implies an absence of retardation. Continuous transport is likely to be most significant within large fractures or macropores or in aquifers with high flow rates, elevated concentrations of readily degraded organic carbon, and low amounts of clay, fine sediments, and total dissolved solids. There are many conditions that affect the degree to which continuous transport contributes to the large-scale mobility of groundwater bacteria.

There is some evidence for continuous transport in several small-scale groundwater injection and recovery experiments, where breakthrough curves of bacteria at downgradient samplers were compared with those of conservative tracers (Champ and Schroeter, 1988; Harvey et al., 1989; Harvey and Garabedian, in press;

Havemeister et al., 1985). Because of the high degree of attenuation (immobilization of bacteria at solid surfaces), these experiments have been conducted over limited distances (2 to 30 m). The simultaneous arrival of the peak concentrations of conservative tracer and labeled bacteria at downgradient samplers in forced- and natural-gradient investigations performed at the Cape Cod groundwater study site (Harvey et al., 1989; Harvey and Garabedian, in press) suggests that many of the bacteria traveled through the sandy aquifer sediments in a relatively unimpeded manner. However, the appearance of some of the labeled bacteria after concentrations of the conservative tracer had declined to undetectable levels also suggested that at least some of the bacteria were subject to intermittent (retarded) transport, presumably as a result of interactions with solid surfaces. Furthermore, the concentration histories of the labeled bacteria at samplers downgradient in the latter experiments (Harvey and Garabedian, in press) were successfully modeled assuming both continuous and intermittent transport mechanisms were operative.

Retarded (Intermittent) Transport

Because of sorptive losses of unattached bacteria from moving groundwater to stationary surfaces, retarded transport would be expected to be more significant over longer distances, particularly in acidic or neutral-pH groundwater moving through relatively fine-grained sediments. Intermittent transport refers to the slow, discontinuous migration of bacteria through an aquifer. This can involve travel with groundwater flow, attachment to solid surfaces, and subsequent detachment and travel of at least a portion of the attaching population. However, adherent bacteria may become irreversibly adsorbed by an active process involving the use of highly surface-active, extracellular bridging polymers (exopolymer) (discussed under Sorption, below). Bacteria attached to solid surfaces with exopolymer may not be able to degrade their own exopolymer and are likely to be incapable of further transport. Therefore, detachment and subsequent transport of adherent groundwater bacteria likely involves weakly (reversibly) adsorbed cells or daughter cells in microcolonies that are associated with other surface-attached bacteria, but not directly with the surface.

The role of intermittent transport in the movement of bacteria through aquifers remains poorly understood. This type of transport can be substantially retarded due to long surface residence times. Retardation factors (ratio of groundwater flow velocity to bacterial migration velocity) as high as 10.0 have been reported for bacterial transport through sandy aquifers in Germany (Matthess et al., 1988). Although a much slower process, intermittent transport may be important both in the movement of subsurface bacteria over geological time and as a mechanism by which large areas of contaminated aquifers are eventually seeded with specific types of bacteria from the surface environment.

Tactic and Transverse Migration

Movement of motile groundwater bacteria in response to chemical gradients, herein referred to as "chemotactic migration," can involve movement via flagella toward or away from higher concentrations of a given nutrient or chemical. Many bacteria in aquifers appear to be motile (Harvey and Garabedian, in press) and presumably capable of directional locomotive response to certain chemical stimuli.

Chemotactic migration is proportional to the population density and to the concentration gradient of the attracting chemical. Growth of indigenous bacterial populations in uncontaminated groundwater environments is often limited by a lack of sufficient dissolved organic carbon that can be readily utilized as an energy source. This also appears to be the case for at least some contaminated aquifers such as the one at Cape Cod (Smith and Duff, 1988). The presence of spatial gradients in organic pollutants may be important to the net movement of bacteria through contaminant plumes, particularly in the vertical direction, where there is typically a low degree of dispersivity.

Bacterial movement in porous media due to chemotaxis can be substantially faster than that due to random thermal (Brownian) motion. Jenneman et al. (1985) demonstrated that the rate of bacterial penetration through Berea sandstone cores in the presence of a sucrose gradient was three- to eightfold higher for a motile bacterium, *Enterobacter aerogenes*, than for a nonmotile bacterium, *Klebsiella pneumoniae*. Observed rates of penetration by chemotactic bacteria in the sandstone core experiments ranged up to 0.11 m/day. This rate of movement suggests that motility may be an important mechanism in the movement of bacteria through contaminated, porous rock aquifers with low flow rates. A model which combines mathematical descriptions for chemotactic response with those for microbial growth by Monod kinetics was used by Bosma et al. (1988) to successfully simulate the degradation of 1,2-dichlorobenzene in soil columns inoculated with *Pseudomonas* sp. strain P52. The term used in their model to describe chemotactic movement of the pseudomonads is

$$\frac{\partial X}{\partial t} = -K_c \frac{\partial}{\partial z} \left[X \frac{\partial C}{\partial z} \right] \tag{1}$$

where K_c is the chemotactic coefficient, which has the units of centimeters squared per liter per microgram per hour; X is bacterial abundance; z is the longitudinal space coordinate; C is the concentration of the organic substrate; and t is time. Recently it was reported that penetration of motile strains of *Escherichia coli* through Ottawa sand-packed cores under anaerobic, nutrient-saturated conditions occurred four times faster than for nonmotile mutants (Reynolds et al., 1989). However, motile, nonchemotactic mutants penetrated the cores faster than did the chemotactic parental strain, suggesting that chemotactic response may not be required for enhanced transport of motile bacteria through unconsolidated porous media. In the absence of chemical gradients, taxis would be random and should also have the effect of increasing the apparent dispersion of the bacterial population being transported.

IMMOBILIZATION

In many situations involving transport of bacteria in groundwater, the extent of horizontal migration will be governed largely by the degree to which the bacteria become immobilized at solid surfaces. This is particularly true for indigenous groundwater bacteria that are well adapted to the low-nutrient conditions and, therefore, can survive for long periods of time in the near-absence of organic substrate. The two mechanisms responsible for the permanent immobilization of unattached bacteria from groundwater moving through porous media are straining and irreversible adsorption. Although the two mechanisms are often treated the same in microbial transport models, they can lead to quite different results. From a physical point of view, straining preferentially removes the larger bacteria, whereas the opposite is true for sorption. Also, the manner in which the capture efficiency of the sediment for bacteria, via straining and sorption, changes over time as the aquifer sediment "filter" develops can be substantially different. Therefore, it is desirable to treat straining and sorption separately in modeling transport of bacteria through porous media. This may require a priori knowledge of the grain and bacterial size distributions and the degree of physical heterogeneity of the aquifer.

Straining

Straining refers to the immobilization of suspended bacteria that occurs when they get caught in pore openings smaller than their limiting dimensions. It has been observed in porous media column experiments that strained particles can be dislodged and resuspended by flow reversal, but not by increases in flow velocity (Sakthivadivel, 1966). Straining is thought to be an important mechanism of immobilization in porous media when the average diameter of the free-living bacterial population is greater than 5% of that for the stationary sediment particles. This is based, in part, upon results of flow-through column experiments in which the effect of grain diameter upon removal of fine suspensions was assessed (Herzig et al., 1970). For ratios of media grain diameters to suspended particle diameters (d_m/d_p) of 20 and 50, it was observed that only 0.53% and 0.053%, respectively, of the bed volume would be filled by strained particles.

In highly heterogeneous media, such as aquifer sediments, criteria for straining that are based solely upon average grain and colloid diameters can be inaccurate, since sediments are typically characterized by a wide distribution of intergranular pore sizes. It has been suggested that for bacterial transport through sediments, the ratio of the bacterial diameter (d_p) to the critical pore size in sediments must be greater than 1.5 for straining to occur (Matthess and Pekdeger, 1985), i.e.:

$$\frac{d_p}{F_s d_K} \geq 1.5 \qquad (2)$$

where F_s is the empirical transit factor for suffusion and is related to the heterogeneity of the porous media, and d_K is the hydraulic equivalent diameter of pore canals. The latter parameter is equivalent to $0.455\ U^{1/6} e d_{17}$, where U is the uniformity coefficient (calculated as d_{60}/d_{10}) and e is the ratio of void to solid volume in the medium. The parameters d_{17}, d_{60}, and d_{10} are the grain diameters at which 17, 60, and 10%, respectively, of the particulate mass is of a smaller size. For sediments with uniform grain diameters, straining of most bacteria would not be predicted even for coarse silt (0.02- to 0.06-mm grain size), since the critical pore size may be estimated to be ~7 μm. However, it is estimated that over 10% of the pores in some heterogeneous sands are small enough to interfere with bacterial transport (Matthess and Pekdeger, 1985).

In field experiments, the contribution of straining in immobilization of bacteria traveling through heterogeneous, sandy aquifer sediments is often difficult to ascertain. In small-scale (2 to 7 m), forced- and natural-gradient tracer experiments (Harvey et al., 1989), involving various sizes of bacteria-sized microspheres moving through sandy aquifer sediments (0.59 mm, median grain size), straining did not appear to be a significant determinant of immobilization. The observed increase in the average diameter of the polydispersed suspension of microspheres during transport downgradient could be accurately simulated by sorptive-filtration theory (Harvey and Garabedian, in press) and was inconsistent with what would be expected with straining. In contrast, straining is likely to be an important cause of immobilization of bacteria traveling through silt (2 to 50 μm) and, in the absence of fractures, reasonably precludes bacterial transport through clayey layers in an aquifer. However, straining of bacteria traveling through heterogeneous, highly stratified sandy aquifers cannot be overlooked, particularly if longer travel distances (e.g., hundreds of meters) are considered. This is because bacteria may encounter a number of conductive layers, some of which can contain substantial amounts of silt. Straining may also be a factor in wastewater injection, since buildup of biofilm may restrict pore sizes in porous media adjacent to the well screen. A thorough discussion of straining and clogging is given by McDowell-Boyer et al. (1986).

Sorption

Mechanistic modeling of bacterial sorption that occurs during transport through an aquifer is complicated by the existence of different mechanisms for attachment, changes in bacterial surface properties and size in response to alterations in environmental conditions, changes in the nature of attachment due to biological processes, and the heterogeneous and complex nature of both bacterial and aquifer surfaces.

Surface properties

A primary determinant of the degree of bacterial attachment during transport through sandy or gravelly aquifers involves the characteristics of the surfaces involved. Several studies involving bacterial attachment to synthetic material have focused upon the surface characteristics of the abiotic solid. The two parameters of the solid phase which appear to be the more important controls of bacterial attachment are hydrophobicity (which involves the unfavorable reorientation of water molecules around nonpolar molecules) and surface charge. Fletcher and Loeb (1979) demonstrated that both hydrophobic and electrostatic interactions were involved in bacterial attachment. However, the degree of substratum hydrophobicity was the more important determinant of sorption onto hydrophobic surfaces, whereas the surface charge appeared to control sorption onto hydrophilic surfaces. Paul and Jeffrey (1985) suggested separate mechanisms governing attachment of *Vibrio proteolytica* onto hydrophilic and onto hydrophobic surfaces.

Many recent investigations involving the controls of bacterial sorption in aquatic environments have focused on the role of surface properties characterizing both microbe and abiotic solid. Bacterial cell hydrophobicity has been observed to affect bacterial attachment to a variety of materials, including glass (Kjelleberg and Hermansson, 1984), polystyrene (McEldowney and Fletcher, 1986; van Loosdrecht et al., 1987b; Rosenberg, 1981), mineral particles (quartz, albite, feldspar, and

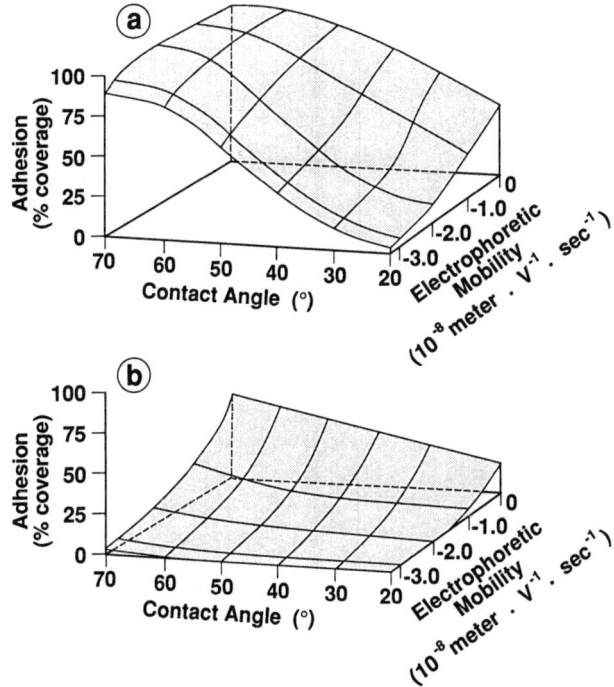

Figure 4. Relationship between bacterial adhesion to (a) sulfated polystyrene (hydrophobic) or (b) glass (hydrophilic) and bacterial surface characteristics as determined by contact angle measurement and electrophoretic mobility (reproduced with permission from van Loosdrecht et al., 1990b).

magnetite) (Stenstrom, 1989; Stenstrom and Kjelleberg, 1985), and whole sediments (van Loosdrecht et al., 1990b). Cell hydrophobicity may also have a role in determining the kinetics of initial bacterial attachment onto some surfaces. In one study involving sorption of 15 different isolates of *Pseudomonas aeruginosa* onto stainless steel, cell hydrophobicity correlated strongly with the sorption rate constant (Vanhaecke et al., 1990). A number of bacterial surface components are thought to contribute to cell hydrophobicity, including fimbriae, proteins A and M, prodigiosin (a pigment), core oligosaccharides, or outer membrane lipids. Bacterial surface hydrophobic components (hydrophobins) and the role of hydrophobic interactions in bacterial adhesion are reviewed in detail by Rosenberg and Kjelleberg (1986).

Several studies have demonstrated correlations between the degree of bacterial attachment to solid surfaces and the magnitude of charge on the bacterial surface (Feldner et al., 1983; van Loosdrecht et al., 1987b). However, only recently has the combined effect of bacterial cell hydrophobicity (which can be inferred from water drop contact angle measurement) and surface charge (which can be estimated from electrophoretic mobility data) upon the propensity of bacteria for attachment to both hydrophobic and hydrophilic surfaces been put into a proper perspective (van Loosdrecht et al., 1987a, 1987b; van Loosdrecht et al., 1990b). The relationship between the combined effects of a bacterium's electrophoretic mobility and hydrophobicity upon fractional adsorption is depicted in Fig. 4 for hydrophobic polystyrene (Fig. 4a) and hydrophilic glass (Fig. 4b). It is clear that the relative importance of hydrophobicity and surface charge with regard to bacterial sorption is different for hydrophobic and hydrophilic surfaces. However, maximal sorption onto both types of surfaces in the investigations of van Loosdrecht et al. (1990b) was evident for uncharged, highly hydrophobic cells.

There is a dearth of information relating bacterial and grain surface characteristics to bacterial sorption behavior in the aquifer. Most surfaces in the aquifer would be expected to be hydrophilic or only slightly hydrophobic. Scholl et al. (in press) reported that, in the absence of organic carbon, the surface charge of various minerals (quartz, muscovite, limestone, and iron hydroxide-coated quartz and muscovite) was the primary determinant of sorption for a slightly hydrophilic (contact angle measurement of 30°) bacterium (Lula-D) isolated from a sandy aquifer in Oklahoma. However, bacterial hydrophobicity may have a significant role in the initial attachment of bacteria moving through aquifer sediments, particularly where organic matter is present on mineral surfaces. Indeed, both hydrophobic attraction and electrostatic repulsion were necessary to explain the sorptive tendencies of a variety of bacteria in the presence of Rhine River sediment (0.05% organic carbon) (van Loosdrecht et al., 1990b).

There is growing evidence suggesting that extracellular appendages (fimbriae, flagella, and holdfasts) can facilitate sorption of bacteria, sometimes under conditions that would otherwise be energetically unfavorable. Adhesion of *Salmonella typhimurium* to a variety of mineral surfaces (albite, biotite, feldspar, magnetite, and quartz) was observed to be markedly enhanced by the presence of fimbriae (Stenstrom and Kjelleberg, 1985), and Rosenberg et al. (1982) demonstrated that thin fimbriae on *Acinetobacter calcoaceticus* enhanced its attachment to polystyrene. The role of fimbriae in the adhesion of bacteria to surfaces is thought to involve localized patches of positive charge and the apparent ability of the fimbriae to penetrate the electrostatic energy barrier between the bacterium and a like-charged surface (Isaacson, 1985). Fimbriae are found almost exclusively among the gram-negative bacteria and may be common features of many indigenous and "pollutant" bacteria found in contaminated aquifers. Flagella also appear to facilitate sorption of bacteria to at least some inorganic surfaces (Fletcher, 1979). Although the mechanism is not clear, flagella-induced motion (motility) appears to be insufficient to overcome the electrostatic barrier adjacent to a like-charged surface (Marshall et al., 1971), but flagella, like fimbriae, can apparently directly sorb to surfaces (Doetsch and Sjoblad, 1980). The relevance of morphologically distinct attachment structures called holdfasts to bacterial attachment in the groundwater environment is not known. However, a few species known to have holdfast-type structures are found in the subsurface. A review of the older (pre-1980) literature involving the role of microbial surface structures in attachment to surfaces is provided by Corpe (1980).

The role of extracellular polymers (exopolymer) in the initial attachment of bacteria is somewhat unclear. Allison and Sutherland (1987) reported that an exopolymer-producing bacterial isolate attached to glass to the same degree as a non-polymer-producing mutant. In contrast, exopolymer produced by a marine *Pseudomonas* sp. measurably affected its degree of adhesion to hydrophobic surfaces

(Wrangstadh et al., 1986). Undoubtedly, the role of exopolymer in the initial attachment of bacteria to solid surfaces depends upon the nature of the exopolymer, both surfaces, and the physicochemical conditions. However, it is clear that exopolymer has a key role in the time-dependent irreversible adhesion of bacteria to surfaces, which will be discussed in more detail in a subsequent section.

Models for bacterial attachment

An important control of bacterial immobilization occurring in porous aquifers involves the rate at which bacteria encounter stationary surfaces. For this reason, a number of transport models involving subsurface transport of microbes have incorporated filtration theory into descriptions of immobilization. The colloid filtration model of Yao et al. (1971), often employed to describe removal of colloidal particles during deep-bed filtration in water treatment applications, was used to account for immobilization in theoretical models describing bacterial transport through porous media (Corapcioglu and Haridas, 1984; Matthess and Pekdeger, 1981) and, more recently, in in situ bacteria transport experiments involving a contaminated, sandy aquifer (Harvey and Garabedian, in press). The colloid filtration model has the advantage of being relatively simple, accounting for the abiotic mechanisms by which bacteria contact stationary solid surfaces, and reasonably predicting the effect of cell size upon the rate of bacterial deposition. However, there is some uncertainty involving buoyant densities of bacteria in the aquifer, the effect of bacterial motility, and how to mathematically treat some of the physical differences between situations of groundwater flow through sandy aquifers and rapid, packed-bed filtration for which the model was developed. In addition, filtration theory does not account for differences in surface characteristics among bacteria. However, filtration theory appears to be useful in a multicomponent description of immobilization in transport models involving bacteria in sandy aquifers (Harvey and Garabedian, in press).

The filtration model considers three mechanisms by which suspended particles come into contact with stationary solid surfaces, i.e., Brownian motion (diffusion), sedimentation due to gravity, and physical interception resulting from the sizes of the stationary and suspended particles. An important parameter is the single collector efficiency, η, which is the rate at which suspended particles strike a single porous medium grain divided by the rate at which they approach the surface. The collector efficiency represents the physical factors determining microbial contact with grain surfaces and is the algebraic sum of collector efficiency caused by diffusion, interception, and settling (Fig. 5). If close-approach effects are neglected, collector efficiency may be estimated by the equation of Yao et al. (1971):

$$\eta = \eta_D + \eta_I + \eta_G = 0.9 \left[\frac{kT}{\mu d_p d v}\right]^{2/3} + 1.5\, (d_p/d)^2 + \frac{(\rho_p - \rho)g d_p^2}{18\, \mu v} \quad (3)$$

where η_D is the colloid-collector collision caused by Brownian motion; η_I is the colloid-collector collision caused by interception; η_G is the colloid-collector collision caused by settling; k is the Boltzman constant; T is the solute temperature; μ is the fluid viscosity; d_p is the bacterial diameter; ρ is the fluid density; ρ_p is the bacterial density (specific gravity of bacterial biomass); v is the fluid velocity; and g is the gravitational constant. An equation describing η that utilizes a similar trajectory

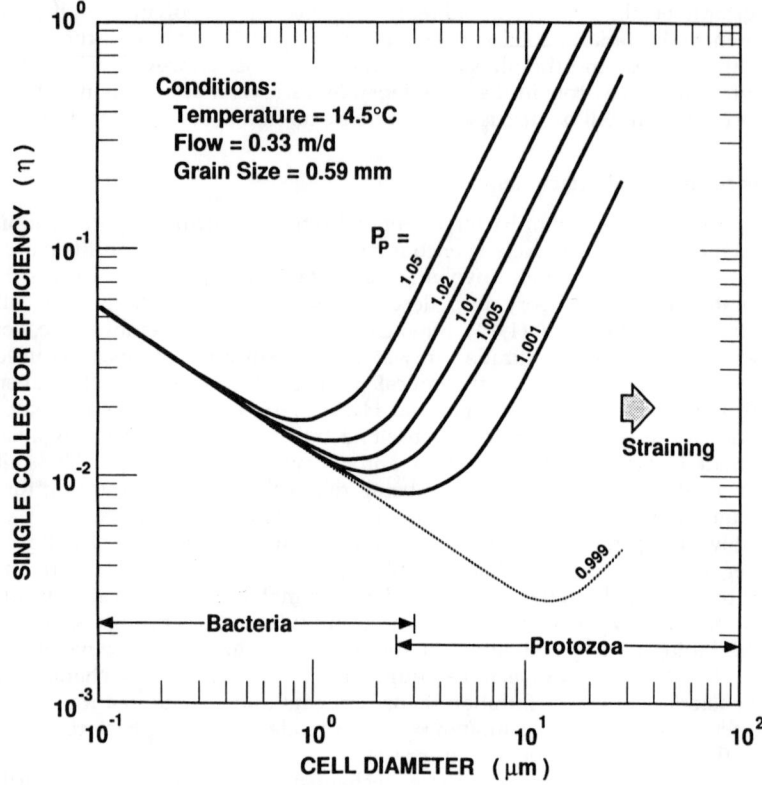

Figure 5. Theoretical estimates for collector efficiency as a function of cell diameter for microbes traveling through Cape Cod aquifer sediments. Different curves represent different cell buoyant densities (reproduced with permission from Harvey and Garabedian, in press).

approach, but includes hydrodynamic retardation and London-van der Waals interaction forces, is provided by Rajagopalan and Tien (1976).

For mixed bacterial populations, η must be determined separately for each size class. The effect of buoyant density and cell size upon collection efficiency for bacteria moving through sandy aquifer sediments is depicted in Fig. 5. As indicated by the minima in the η-versus-size function in Fig. 5, the optimal size for transport of bacteria with near-neutral buoyancy through the sandy sediments of the Cape Cod aquifer is likely 2 to 3 μm (diameter). In general, diffusion predominates for the 0.2- to 1-μm size classes of bacteria typically found in groundwater environments. The model predicts that smaller bacteria within this size range will be immobilized on solid surfaces at faster rates than will larger bacteria. This prediction is consistent with results of a recent groundwater injection and recovery experiment in which 0.2-, 0.7-, and 1.3-μm (diameter) microspheres were employed (Harvey et al., 1989).

Collector efficiency, η, is related to the flux of bacteria that become irreversibly sorbed in the porous media they are moving through by the so-called collision efficiency, α. α, which represents the physicochemical factors determining bacterial attachment to grain surfaces, is generally determined experimentally based upon observed immobilization rates of the suspended particles and theoretical values of η (Tobiason and O'Melia, 1988). For a pulse injection of bacteria and conservative tracer into a sandy or gravelly aquifer where the longitudinal dispersivity is small relative to the travel distance, α may be estimated by the following (Harvey and Garabedian, in press):

$$\alpha = \frac{-d \ln (RB)}{1.5(1 - \theta) \eta x_1} \tag{4}$$

where d is the median grain diameter; θ is the porosity; x_1 is the travel distance; and RB is the ratio of relative (injectate-normalized) masses of bacteria to conservative tracer appearing at the sampling point (Harvey et al., 1989). However, α is probably not constant and may vary spatially within the aquifer. This is because α depends, in part, on the aquifer characteristics, which may have substantial spatial variability. Although the value of α might be handled stochastically, the use of experimentally determined data obtained from small-scale in situ transport experiments would be helpful.

Models used to describe the initial attachment of bacteria to solid surfaces in aqueous systems have generally invoked either a surface free energy (Absolom et al., 1983; Busscher et al., 1984) or a colloid stability theory (Marshall et al., 1971; Rutter and Vincent, 1984; van Loosdrecht et al., 1989) approach. Although generally considered to be fundamentally different, there are important interrelationships between the two models (Pethica, 1980), and some recent progress has been made toward unifying the two approaches (van Loosdrecht et al., 1990b). However, the surface free energy approach, which involves "short-range" interactions (i.e., interfacial tensions), assumes that the bacteria are in direct contact with the solid surface, whereas the colloid stability theory is best used where there is a separation distance of 1 nm or greater between bacterium and surface (van Loosdrecht et al., 1989). The separation assumption is necessary because the colloid stability model is based upon so-called "long-range" (electrostatic and van der Waals) interactions. At shorter distances, the colloid stability model is complicated by the existence of other shorter-range interactions (e.g., steric effects, hydrogen bonding).

For direct bacterial contact on the solid surface, the extent of attachment is controlled by the surface properties of all three phases (liquid, solid, and bacteria). The surface thermodynamic approach employed by Absolom et al. (1983) involves a balance of surface tensions when a new phase boundary (bacterium-solid) is created because of attachment, i.e.:

$$\Delta F^{adh} = \gamma_{BS} - \gamma_{BL} - \gamma_{SL} \tag{5}$$

where ΔF^{adh} is the free energy of adhesion and γ_{BS}, γ_{BL}, and γ_{SL} are the interfacial tensions between bacterium and solid, bacterium and liquid, and solid and liquid, respectively. The difficulty with this approach involves obtaining accurate values for the various interfacial tensions. From theoretical considerations, bacterial

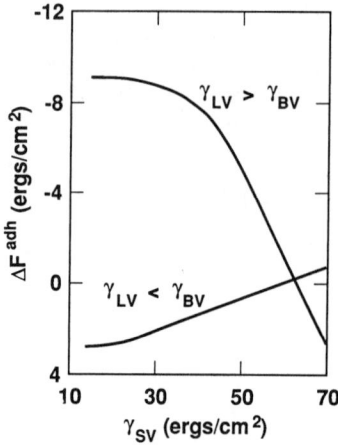

Figure 6. Free energy of adhesion calculated for *E. coli* 2617 as a function of substratum surface tension for liquid surface tension of 72.8 millinewton/m (upper curve) and 64.0 millinewton/m (reproduced with permission from Absolom et al., 1983).

adhesion increases with increasing surface tension of the solid (γ_{SV}) if the bacterium-vapor surface tension is greater than the liquid-vapor surface tension, i.e.: $\gamma_{LV} < \gamma_{BV}$. If, however, $\gamma_{LV} > \gamma_{BV}$, then bacterial adhesion would be expected to decrease with increasing surface tensions of the solid (higher-energy surfaces) (Absolom et al., 1983). The manner in which free energy of adhesion for *E. coli* varied as a function of the solid-vapor surface tension in the Absolom et al. study is depicted in Fig. 6. A potential problem in using this approach to predict the degree of bacterial sorption in a freshwater aquifer is that most bacteria attaching to a like charge surface would be held at a finite distance from the surface due to electrostatic repulsion.

The interactions between bacteria and particles have most often been explained in terms of long-range interactions by employing the theories of Derjaguin and Landau and Verwey and Overbeek, known collectively as the DLVO theory of colloid stability. In the absence of steric effects, the total energy of interaction as a function of separation distance between bacterium and a like-charged surface may be estimated as the algebraic sum of the repulsive (electrostatic) and attractive (van der Waals) forces. Although there are complications in applying the DLVO theory to sorption of bacteria (Pethica, 1980), it seems to explain the initial phase of bacterial adhesion in a number of studies (Marshall, 1980, 1985; van Loosdrecht et al., 1989, 1990a, 1990b).

An important parameter in the DLVO model for bacteria-surface interaction is the concentration of electrolyte. This is because the magnitude of electrostatic repulsion as a function of separation distance is controlled by the thickness of the diffuse double layer (electrical double layer) of counterions in solution, which in turn is a function of ionic strength (Shaw, 1976). The manner in which the total energy of interaction between bacteria and a negatively charged surface varies with distance at different concentrations of electrolytes is depicted in Fig. 7.

In highly saline aquifers, the electrical double layer is compressed to the point where the electrostatic energy barrier near the surface no longer exists and, therefore, the London-van der Waals attractive force dominates the overall bacteria-surface interaction energies at all separation distances (Fig. 7A). This may allow for

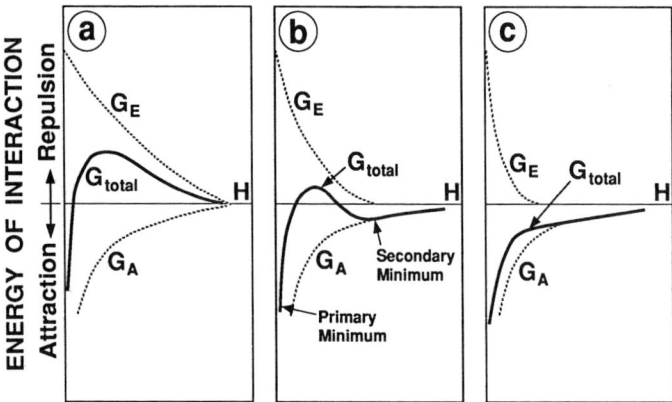

Figure 7. Schematic representation of Gibbs energy of interaction between a colloid-sized sphere and a like-charged surface as function of separation distance (H) for low (a), intermediate (b), and high (c) concentrations of electrolyte. G_{total}, Total energy of interaction; G_E, electrostatic interaction; G_A, van der Waals interaction. (Reproduced with permission from van Loosdrecht et al., 1990a.)

direct bacteria–surface contact. Since the interaction energy between a bacterium and a like-charged surface is at a minimum right at the surface, bacteria in direct contact with the surface are considered to be irreversibly adsorbed. In drinking water aquifers, the low ionic strength allows for a substantial electrical double layer (on the scale of nanometers in thickness), and electrostatic repulsion dominates such that an energy barrier separates approaching bacteria from the surface. At extremely low ionic strength (e.g., 10^{-4} M), there is a net repulsion that extends for a considerable distance from the surface (Fig. 7C). At intermediate ionic strengths (e.g., 10^{-2} M), the repulsive and attractive forces interact in a manner which allows for a secondary minimum of interaction energy at a finite distance (several nanometers) from the surface (Fig. 7B). Although there is a "primary" energy minimum right at the surface, approaching bacteria lack sufficient kinetic energy to penetrate the repulsive barrier. Bacteria which become trapped at the secondary minimum are reversibly sorbed and can be easily dislodged by their own motility or by changes in flow (Marshall, 1985).

In natural systems, application of the DLVO model can be complicated by presence of exopolymers. The presence of polymer on the solid surface as well as the bacteria can lead to additional interactions (steric and bridging) as well as alterations of van der Waals and electrostatic forces (van Loosdrecht et al., 1990b). Although entrapment in the secondary minimum, as predicted by the DLVO theory, is generally invoked to explain the ability of sorbing bacteria to exhibit a variety of movement at short distances from the surface (e.g., Marshall et al., 1971), there is at least some evidence to the contrary. Abbott et al. (1983) observed that about half of the *Streptococcus mutans* organisms sorbing onto clean and protein-coated glass oscillated a short distance from the surface, but apparently were not held in a secondary minimum. These authors suggested that the bacteria were sorbing to the surfaces by means of exopolymers, which can occur over a much longer distance than is possible with either van der Waals or electrostatic interactions (Rutter, 1980).

Reversibility of attachment

Irreversible sorption can be the most significant control of the extent of bacterial transport. This is particularly true for indigenous bacteria that are capable of surviving for long periods of time under typical groundwater conditions. Since reversible sorption is what controls retardation (slowing down of bacterial transport relative to groundwater flow), it is important to differentiate the two types of adsorption in transport models. However, such differentiation would be difficult in the absence of a priori information deriving from sorption experiments with core material and the bacteria of interest. Also, although bacteria that are reversibly sorbed can be removed by the shearing effects of a water jet or by sharp changes in chemical conditions (ionic strength, pH, or competing ligand concentrations), their "surface residence times" on aquifer solids may be quite long. This is because drastic changes in chemistry or in hydraulic conditions in the groundwater environment are rare. Therefore, it may be useful to incorporate a time frame into criteria used to differentiate reversibly sorbed from irreversibly sorbed bacteria in an aquifer.

Irreversible sorption or adhesion of bacteria onto solid surfaces has been defined as a time-dependent process in which bacteria are no longer subject to Brownian motion and cannot be recovered by washing (Marshall et al., 1971). Irreversible sorption requires some sort of bacterium-surface contact. Since bacteria in freshwater aquifers would be unable to contact most surfaces because of the formidable energy barrier created by the overlapping of bacterial and mineral electrical double layers, the initial adhesion would be largely reversible. Exopolymer is thought to be the primary mechanism responsible for permanent adhesion, and its role in the "cementing" of aquatic bacteria to solid surfaces is well documented (Marshall et al., 1971; Corpe, 1974). Although carbon limited, aquifer bacteria appear to channel a disproportionately large amount of their assimilated carbon into the production of extracellular and storage polymers. Adherent aquifer bacteria are often observed under electron microscopy to be anchored by large amounts of exopolymer (Ghiorse and Wilson, 1988).

Reversible sorption of bacteria to negatively charged surfaces involves the instantaneous attraction by so-called long-range forces near, but not directly at, the surface. Bacteria which are reversibly sorbed continue to exhibit Brownian motion and can be easily dislodged (Marshall, 1985). In general, the initial sorption of bacteria in a freshwater aquifer at low ionic strength is, by definition, reversible and can be an important determinant of bacterial transport. However, it is not always clear how best to represent reversible sorption in models. In the Cape Cod study, it was found that the breakthrough curves could be accurately simulated by using either of two approaches to adsorption (Harvey and Garabedian, in press). The first assumed that adsorption is instantaneous and linear, but that the propensity of the bacteria for solid surfaces differs among segments of the population. The second employed a kinetic approach that assumed that all the bacteria are equally capable of interacting with solid surfaces, but that the rates of adsorption and desorption are different. It is likely that sorption of a bacterial population in heterogeneous aquifer sediment involves both a kinetic limitation for adsorption and a difference among cells in their surface activity. It is clear that detailed studies are needed to more accurately describe sorption of bacteria onto aquifer material.

GROWTH

Growth of many pathogens and displaced bacteria in groundwater environments is likely to be negligible, since copiotrophic bacteria typically shift their metabolic capabilities from growth to starvation survival in response to low-nutrient conditions (Kurath and Morita, 1983). However, growth rates of indigenous bacteria moving through aquifers can be significant and must be accounted for in describing the population's movement downgradient (Harvey and George, 1987). Several models involving transport of bacteria or biodegradable contaminants have employed Monod kinetics to describe bacterial growth dynamics and utilization of organic carbon (Corapcioglu and Haridas, 1985, 1984; Molz et al., 1986). The Monod relationship is a hyperbolic function in which growth rate is related to the concentration of a limiting substrate and to two kinetic "constants," i.e., an asymptotic value defined as the maximum growth rate and the saturation constant. The latter is equivalent to the substrate concentration that will yield half the maximum growth rate. Monod kinetics were developed using experimental data derived from laboratory batch systems in which pure bacterial cultures and well-defined substrates were used. Although Monod kinetics have been successfully used to describe organic substrate utilization and bacterial growth in aquifer-derived laboratory microcosms (Godsy et al., 1990), their applicability to groundwater habitats has not been demonstrated.

The Monod relationship is used in the description of growth in subsurface transport models because it generally works well for bacterial populations having low saturation constants for organic substrates, as would be the case in most aquifers. However, a complication in this approach is that the kinetic "constants" in the Monod equation can be subject to change in response to changes in temperature, the nature of the substrate, and other factors (Gaudy and Gaudy, 1980). Therefore, caution should be used in the application of Monod growth kinetics to problems involving bacterial transport in contaminated groundwater, particularly where there are temporal and spatial changes in nutrient and physical conditions or in the bacterial populations themselves.

Growth of heterotrophic bacteria in the aquifer typically results from utilization of a variety of organic substrates. Therefore, in accounting for bacterial growth in transport models, it may be necessary to employ an aggregate measure of biodegradable organic material, such as chemical oxygen demand or dissolved organic carbon, instead of concentrations of an individual substrate. Rittmann et al. (1980) suggested this approach to account for changes in the nature of the growth-limiting carbon source as preferred substrates in a complex contaminant mixture are used up. However, the "aggregate approach" could benefit from the development of better methods for quantifying low levels of readily degraded dissolved organic carbon, since spatial changes in the nature of this substrate often occur in zones of contaminated groundwater (Harvey and Barber, submitted).

Before bacterial growth can be accurately accounted for in models involving bacterial transport in groundwater, more information on the growth characteristics of groundwater bacteria is needed. Accurate estimates of in situ growth rates for aquifer bacteria can be difficult to obtain owing to problems in obtaining uncontaminated samples, in the methods (frequency of dividing cells, tritiated thymidine uptake, closed-bottle incubations) themselves, and in using a number of inherent assumptions that may not be applicable to the aquifer (Harvey and George, 1987). Nevertheless, it appears that growth rates of bacteria in uncontaminated aquifer sediments are at least several orders of magnitude lower than those observed for surface sediments (Thorn and Ventullo, 1988). However, bacterial growth in oligotrophic aquifers can still affect the extent of bacterial transport over long periods of time, since groundwater flow can be quite slow (<1 m/day) and bacterial movement by intermittent transport can be even slower. Bacterial growth in organically contaminated groundwater can be quite rapid and approach that of at least some productive surface waters (Harvey and George, 1987). The contribution of growth to the overall down-gradient movement of indigenous bacteria in some contaminant plumes would appear to be significant (Harvey and Barber, submitted).

SURVIVAL

Since the initial attachment of bacteria sorbing to solid surfaces in freshwater systems is generally reversible (Marshall, 1985), the ultimate limitation of bacterial pathogen migration in groundwater may involve the duration of viability. For some bacterial pathogens, mortality in the groundwater environment can be quite significant. Survival of displaced bacteria in the aquifer can involve a number of factors, including the bacterial species involved; temperature; the presence of microbial competitors, predators, and parasites; groundwater chemistry; organic matter; and the degree of attachment. Generally, there is an initial period of relatively constant abundance. This is followed by a period in which temporal changes in abundance may be modeled by exponential decline. The initial period before exponential decline may vary from days in highly contaminated water to months under oligotrophic conditions (Matthess et al., 1988). For some displaced bacteria, die-off times in groundwater appear to be slow and a 1-order-of-magnitude reduction in abundance can involve several weeks or longer (Crane and Moore, 1984; Bitton et al., 1983). A review of the factors involved in survival of displaced bacterial populations in the subsurface is provided by Yates and Yates (1988).

Lysis of bacteria in the groundwater environment can result from unfavorable conditions, such as a lack of sufficient nutrients needed for cell maintenance or significant changes in salinity, or from parasites (e.g., bacteriophages or *Bdellovibrio* sp.). Bacteria that are indigenous to groundwater environments seem to be well adapted to low-nutrient conditions. Therefore, oligotrophic populations are likely to survive for long periods of time and even proliferate under conditions of severe nutrient limitation found in most contaminated groundwater. The extent of mortality due to parasitism in populations of indigenous groundwater bacteria is not known, but could be substantial in aquifers that are highly contaminated with organic material. The ability of genetically engineered bacteria to survive in the aquifer is also unclear and promises to be an important topic of future research.

Until recently, it was assumed that bactivorous protozoa were likely to be absent in most aquifers, since eucaryotic microorganisms were typically not detected in direct microscopic examinations of aquifer core material (Ghiorse and Balkwill, 1983; Wilson et al., 1983; Harvey et al., 1984). Also, bacterial abundances in groundwater appeared to be too low to support growth of a protozoan population. However, it now appears that this assumption is incorrect, at least for a number of aquifers. Sinclair and Ghiorse (1987) recently reported substantial numbers of protozoa in uncontaminated aquifer sediments that could be detected by using a most-probable-number technique, and Kinner et al. (1990), using both culturing and direct counting methods, have detected protozoa in contaminated aquifer sediments. Where present, protozoa are likely to constitute a significant sink for bacteria being transported in groundwater and, consequently, should be accounted for in transport models.

GROUNDWATER CONDITIONS

Groundwater conditions can have a significant effect upon the extent and reversibility of bacterial sorption onto solid surfaces in the aquifer. In general, bacterial attachment to stationary surfaces in an aquifer can be predicted to increase with increasing ionic strength. Increasing propensity for bacterial attachment with increasing ionic strength has been demonstrated for a variety of surfaces, e.g., muscovite (Scholl et al., in press), hydroxyapatite (Gordon and Millero, 1984), quartz sand (Sharma et al., 1985), and glass (Marshall et al., 1971). A decrease in ionic strength to the point where bacteria are repelled from the surfaces causes release of those cells which were reversibly sorbed (e.g., Marshall et al., 1971). Ionic composition is also important in bacterial attachment; calcium and magnesium ions appear to be particularly important and have been shown to be a requirement for irreversible sorption of a *Pseudomonas* sp. onto glass (Marshall et al., 1971).

The pH of the groundwater can have a significant effect upon bacterial attachment. In general, bacteria seem to sorb more readily to most surfaces under slightly acidic conditions, although exceptions to this rule have been reported. Scholl et al. (in press) reported greater sorption of a groundwater isolate onto quartz at pH 5 than at pH 7. In contrast, maximal adhesion of *P. aeruginosa* onto stainless steel occurred under near-neutral conditions. Abbott et al. (1983) reported that sorption of *S. mutans* onto clean and protein-coated glass surfaces was relatively insensitive to pH. Therefore, the role of pH upon bacterial sorption behavior may depend not only upon the chemistry of the groundwater, but the nature of the surfaces involved.

Dissolved organic matter can affect bacterial transport in groundwater by sorbing to abiotic surfaces. By sorbing to mineral surfaces in the aquifer, organics can substantively alter their sorptive characteristics. Sorbed proteins have generally been observed to have an inhibitory effect upon bacterial attachment (Pratt-Terpstra et al., 1987; Pringle and Fletcher, 1986). Decreased sorption of bacteria onto hydrophobic surfaces has been shown to occur with surfactants (Goldberg et al., 1990), and fulvic acid has been shown to decrease bacteriophage sorption onto soils (Bixby and O'Brien, 1979).

In addition to the effect upon bacteria growth rate, dissolved organic matter may also affect a bacterium's transport behavior in groundwater by causing alteration in cell size and hydrophobicity. Starvation is known to cause dwarfing in a number of bacteria, particularly in copiotrophic species (Kjelleberg, 1984; Humphrey et al., 1983). Because smaller bacteria exhibit greater Brownian motion, they would have a greater tendency to become immobilized in sandy aquifer sediments than do larger bacteria possessing similar surface characteristics (see Immobilization, above). Furthermore, it has been observed that starvation conditions can also cause an increase in cell hydrophobicity and, consequently, an increased propensity for attachment, at least in some species (Kjelleberg, 1984; Dawson et al., 1981). However, the relationship between starvation conditions and bacterial tendency for attachment is not entirely clear. There is at least one report suggesting that starvation conditions can lead to a decrease in hydrophobicity (Wrangstadh et al., 1986). In recent investigations involving organically contaminated aquifers, it has been observed that bacterial partitioning to the solid phase increases with increasing distance from the source of contamination (Godsy et al., submitted; Harvey and George, 1986). This suggests that unattached bacteria that are further away from their major source of organic substrate may have a greater tendency for attachment. However, it is clear that the role of organic matter in bacterial transport in groundwater needs further study.

REFERENCES

Abbott, A., P. R. Rutter, and R. C. W. Berkeley. 1983. The influence of ionic strength, pH and a protein layer on the interaction between *Streptococcus mutans* and glass surfaces. *J. Gen. Microbiol.* **129**:439–445.

Absolom, D. R., F. V. Lamberti, Z. Policova, W. Zingg, C. J. van Oss, and A. W. Neumann. 1983. Surface thermodynamics of bacterial adhesion. *Appl. Environ. Microbiol.* **46**:90–97.

Allen, M. J., and S. M. Morrison. 1973. Bacterial movement through fractured bedrock. *Ground Water* **11**:6–10.

Allison, D. G., and I. W. Sutherland. 1987. The role of exopolysaccharides in adhesion of freshwater bacteria. *J. Gen. Microbiol.* **133**:1319–1327.

Anan'ev, N. I., and N. D. Demin. 1971. On the spread of pollutants in subsurface waters. *Hyg. Sanit.* **36**:292.

Bales, R. C., C. P. Gerba, G. H. Grondin, and S. L. Jensen. 1989. Bacteriophage transport in sandy soil and fractured tuff. *Appl. Environ. Microbiol.* **55**:2061–2067.

Bitton, G., J. M. Davidson, and S. R. Farrah. 1979. On the value of soil columns for assessing the transport pattern of viruses through soils: a critical outlook. *Water Air Soil Pollut.* **12**:449–457.

Bitton, G., S. R. Farrah, R. H. Ruskin, J. Butner, and Y. J. Chou. 1983. Survival of pathogenic and indicator organisms in groundwater. *Ground Water* **213**:405–410.

Bixby, R. L., and D. J. O'Brien. 1979. Influence of fulvic acid on bacteriophage adsorption and complexation in soil. *Appl. Environ. Microbiol.* **38**:840–845.

Bosma, T. N. P., J. L. Schnoor, G. Schraa, and A. J. B. Zehnder. 1988. Simulation model for biotransformation of xenobiotics and chemotaxis in soil columns. *J. Contam. Hydrol.* **2**:225–236.

Busscher, H. J., A. H. Weerkamp, H. C. van der Mei, A. W. J. van Pelt, H. P. de Jong, and J. Arends. 1984. Measurement of the surface free energy of bacterial cell surfaces and its relevance for adhesion. *Appl. Environ. Microbiol.* **48**:980–983.

Champ, D. R., and J. Schroeter. 1988. Bacterial transport in fractured rock—a field scale tracer test at the Chalk River Nuclear Laboratories, p. 14:1–14:7. In Proceedings of the International Conference on Water and Wastewater Microbiology, Newport Beach, Calif., 8–11 February 1988.

Corapcioglu, M. Y., and A. Haridas. 1984. Transport and fate of microorganisms in porous media: a theoretical investigation. *J. Hydrol.* **72**:149–169.

Corapcioglu, M. Y., and A. Haridas. 1985. Microbial transport in soils and ground water: a numerical model. *Adv. Water Res.* **8**:188–200.

Corpe, W. A. 1974. Periphytic marine bacteria and the formation of microbial films on solid surfaces, p. 397–417. In R. R. Colwell and R. Y. Morita (ed.), *Effect of the Ocean Environment on Microbial Activities*. University Park Press, Baltimore.

Corpe, W. A. 1980. Microbial surface components involved in adsorption of microorganisms onto surfaces, p. 105–144. *In* G. Bitton and K. C. Marshall (ed.), *Adsorption of Microorganisms to Surfaces*. John Wiley & Sons, Inc., New York.

Crane, S. R., and J. A. Moore. 1984. Bacterial pollution of ground water: a review. *Water Air Soil Pollut.* **22:**67–83.

Craun, G. F. 1985. A summary of waterborne illness transmitted through contaminated groundwater. *J. Environ. Health* **48:**122–127.

Dappert, A. F. 1932. Tracing the travel and changes in composition of underground pollution. *Water Works Sewerage* **79:**265–274.

Dawson, M. P., B. Humphrey, and K. C. Marshall. 1981. Adhesion: a tactic in the survival strategy of a marine vibrio during starvation. *Curr. Microbiol.* **6:**195–198.

Doetsch, R. N., and R. D. Sjoblad. 1980. Flagellar structure and function in eubacteria. *Annu. Rev. Microbiol.* **34:**69–108.

Feldner, J., W. Bredt, and I. Kahane. 1983. Influence of cell shape and surface charge on attachment of *Mycoplasma pneumoniae* to glass surfaces. *J. Bacteriol.* **153:**1–5.

Fletcher, M. 1979. The attachment of bacteria to surfaces in aquatic environments, p. 87–108. *In* D. C. Ellwood, J. Melling, and P. R. Rutter (ed.), *Adhesion of Microorganisms to Surfaces*. Academic Press, Inc., New York.

Fletcher, M., and G. I. Loeb. 1979. Influence of substratum characteristics on the attachment of a marine pseudomonad to solid surfaces. *Appl. Environ. Microbiol.* **37:**67–72.

Gaudy, A. F., and E. T. Gaudy. 1980. *Microbiology for Environmental Scientists and Engineers*. McGraw-Hill Book Co., New York.

Germann, P. F., M. S. Smith, and G. W. Thomas. 1987. Kinematic wave approximation to the transport of Escherichia coli in the vadose zone. *Water Resour. Res.* **23:**1281–1287.

Ghiorse, W. C., and D. L. Balkwill. 1983. Enumeration and morphological characterization of bacteria indigenous to subsurface environments. *Dev. Ind. Microbiol.* **24:**213–224.

Ghiorse, W. C., and J. T. Wilson. 1988. Microbial ecology of the terrestrial subsurface. *Adv. Appl. Microbiol.* **33:**107–173.

Godsy, E. M., D. F. Goerlitz, and D. Grbic-Galic. Submitted for publication.

Godsy, E. M., L. M. Law, C. D. Fraley, and D. Grbic-Galic. 1990. Kinetics of phenolic compound degradation by aquifer derived methanogenic microcosms. Abstract, American Geophysical Union Winter Meeting, 3-7 December, 1989, San Francisco. *Eos* **71:**1319.

Goldberg, S., Y. Konis, and M. Rosenberg. 1990. Effect of cetylpyridinium chloride on microbial adhesion to hexadecane and polystyrene. *Appl. Environ. Microbiol.* **56:**1678–1682.

Gordon, A. S., and F. J. Millero. 1984. Electrolyte effects on attachment of an estuarine bacterium. *Appl. Environ. Microbiol.* **47:**495–499.

Grondin, G. H., and C. P. Gerba. 1986. Virus dispersion in a coarse porous medium. *Hydrol. Water Resour. Ariz. Southwest* **16:**11–15.

Harvey, R. W. 1988. Transport of bacteria in a contaminated aquifer, p. 183-188. *In* U.S. Geological Survey Water Resources Investigations Report no. 88-4220. U.S. Geological Survey, Reston, Va.

Harvey, R. W. 1989. Considerations for modeling transport of bacteria in contaminated aquifers, p. 75-82. *In* L. Abriola (ed.), *Groundwater Contamination*. Publication no. 185. IAHS Press, Wallingford, Oxfordshire, United Kingdom.

Harvey, R. W., and L. B. Barber. Submitted for publication.

Harvey, R. W., and S. P. Garabedian. Use of colloid filtration theory in modeling movement of bacteria through a contaminated sandy aquifer. *Environ. Sci. Technol.*, in press.

Harvey, R. W., and L. H. George. 1986. Bacterial distribution and transport in an organically-contaminated aquifer in Cape Cod, MA, p. B25–B26. *In* U.S. Geological Survey Open-File Report #86-481. U.S. Geological Survey, Reston, Va.

Harvey, R. W., and L. H. George. 1987. Growth determinations for unattached bacteria in a contaminated aquifer. *Appl. Environ. Microbiol.* **53:**2992–2996.

Harvey, R. W., L. H. George, R. L. Smith, and D. R. LeBlanc. 1989. Transport of microspheres and indigenous bacteria through a sandy aquifer: results of natural and forced-gradient tracer experiments. *Environ. Sci. Technol.* **23:**51–56.

Harvey, R. W., R. L. Smith, and L. H. George. 1984. Effect of organic contamination upon microbial distributions and heterotrophic uptake in a Cape Cod, Mass., aquifer. *Appl. Environ. Microbiol.* **48:**1197–1202.

Havemeister, G., R. Riemer, and J. Schroeter. 1985. Feldversuche zur Persistenz und zum Transportverhalten von Bakterien. *Versuchsfeld Segeberger Forst Umweltbundesamt Materialien* **2/85:**49–56.

Hendricks, D. W., F. J. Post, and D. R. Khairnar. 1979. Adsorption of bacteria on soils. *Water Air Soil Pollut.* **12:**219–232.

Herzig, J. P., D. M. Leclerc, and P. Le Goff. 1970. Flow of supensions through porous media—application to deep filtration. *Ind. Eng. Chem.* **62:**8–35.

Hess, K. M. 1989. Use of a borehole flowmeter to determine spatial heterogeneity of hydraulic conductivity and macrodispersion in a sand and gravel aquifer, Cape Cod, MA, p. 497-508. *In* F. J. Molz, J. G. Melville, and O. Guven (ed.), *Proceedings of the Conference on New Field Techniques for Quantifying the Physical and Chemical Properties of Heterogeneous Aquifers*, Dallas, Tex., 20–23 March 1989. National Water Well Association, Dublin, Ohio.

Humphrey, B., S. Kjelleberg, and K. C. Marshall. 1983. Responses of marine bacteria under starvation conditions at a solid-water interface. *Appl. Environ. Microbiol.* **45:**43–47.

Isaacson, R. E. 1985. Pilus adhesins, p. 307–336. *In* D. C. Savage and M. Fletcher (ed.), *Bacterial Adhesion*. Plenum Press, New York.

Jenneman, G. E., M. J. McInerney, and R. M. Knapp. 1985. Microbial penetration through nutrient-saturated Berea sandstone. *Appl. Environ. Microbiol.* **50:**383–391.

Keswick, B. H. 1984. Sources of groundwater pollution, p. 59-64. *In* G. Bitton and C. P. Gerba (ed.), *Groundwater Pollution Microbiology*. John Wiley & Sons, Inc., New York.

Kingston, S. P. 1943. Contamination of water supplies in limestone formation. *J. Am. Water Works Assoc.* **35:**1450–1456.

Kinner, N. E., A. L. Bunn, L. D. Meeker, and R. W. Harvey. 1990. Enumeration and variability in the distribution of protozoa in an organically-contaminated subsurface environment. Abstract, American Geophysical Union Winter Meeting, 3-7 December, San Francisco. *Eos* **71:**1319.

Kjelleberg, S. 1984. Effects of interfaces on survival mechanisms of copiotrophic bacteria in low-nutrient habitats, p. 151-159. *In* M. J. Klug and C. A. Reddy (ed.), *Current Perspectives in Microbial Ecology*. American Society for Micriobiology, Washington, D.C.

Kjelleberg, S., and M. Hermansson. 1984. Starvation-induced effects on bacterial surface characteristics. *Appl. Environ. Microbiol.* **48:**497–503.

Kudryavtseva, B. M. 1972. An experimental approach to the establishment of zones of hygienic protection of underground water sources on the basis of sanitary-bacteriological indices. *J. Hyg. Epidemiol. Microbiol. Immunol.* (Prague) **16:**503–511.

Kurath, G., and R. Y. Morita. 1983. Starvation-survival physiological studies of a marine *Pseudomonas* sp. *Appl. Environ. Microbiol.* **45:**1206–1211.

Marshall, K. C. 1980. Adsorption of microorganisms to soils and sediments, p. 317–329. *In* B. Bitton and K. C. Marshall (ed.), *Adsorption of Microorganisms to Surfaces*. John Wiley & Sons, Inc., New York.

Marshall, K. C. 1985. Mechanisms of bacterial adhesion at solid-water interfaces, p. 133–161. *In* D. C. Savage and M. Fletcher (ed.), *Bacterial Adhesion*. Plenum Press, New York.

Marshall, K. C., R. Stout, and R. Mitchell. 1971. Mechanism of the initial events in the sorption of marine bacteria to surfaces. *J. Gen. Microbiol.* **68:**337–348.

Martin, G. N., and M. J. Noonon. 1977. *Effects of Domestic Wastewater Disposal by Land Irrigation on Groundwater Quality of Central Canterbury Plains*. Water Soil Technology Publication no. 7. Water and Soil Division, Ministry of Work and Development, Wellington, New Zealand.

Matthess, G., and A. Pekdeger. 1981. Concepts of a survival and transport model of pathogenic bacteria and viruses in groundwater. *Sci. Total Environ.* **21:**149–159.

Matthess, G., and A. Pekdeger. 1985. Survival and transport of pathogenic bacteria and viruses in ground water, p. 472-482. *In* C. H. Ward, W. Giger, and P. McCarty (ed.), *Ground Water Quality*. John Wiley & Sons, Inc., New York.

Matthess, G., A. Pekdeger, and J. Schroeter. 1988. Persistence and transport of bacteria and viruses in groundwater—a conceptual evaluation. *J. Contam. Hydrol.* **2:**171–188.

McDowell-Boyer, L. M., J. R. Hunt, and N. Sitar. 1986. Particle transport through porous media. *Water Resour. Res.* **22**:1901–1921.
McEldowney, S., and M. Fletcher. 1986. Variability of the influence of physicochemical factors affecting bacterial adhesion to polystyrene substrata. *Appl. Environ. Microbiol.* **52**:460–465.
Merrell, J. C., Jr. 1967. *The Santee Recreation Project, Santee, Calif.* Water Pollution Research Series Publication no. WP-20-7. Federal Water Pollution Control Administration, Cincinnati, Ohio.
Molz, F. J., M. A. Widdowson, and L. D. Benefield. 1986. Simulations of microbial growth dynamics coupled to nutrient and oxygen transport in porous media. *Water Resour. Res.* **22**:1207–1216.
Paul, J. H., and W. H. Jeffrey. 1985. Evidence for separate adhesion mechanisms for hydrophilic and hydrophobic surfaces in *Vibrio proteolytica*. *Appl. Environ. Microbiol.* **50**:431–437.
Pethica, B. A. 1980. Microbial and cell adhesion, p. 19–46. *In* R. C. W. Berkeley, J. M. Lynch, J. Melling, P. R. Rutter, and B. Vincent (ed.), *Microbial Adhesion to Surfaces*. Horwood, Chichester, United Kingdom.
Pratt-Terpstra, I. H., A. H. Weerkamp, and H. J. Busscher. 1987. Adhesion of oral streptococci from a flowing suspension to uncoated and albumin-coated surfaces. *J. Gen. Microbiol.* **133**:3199–3206.
Pringle, J. H., and M. Fletcher. 1986. Influence of substratum hydration and adsorbed macromolecules on bacterial attachment to surfaces. *Appl. Environ. Microbiol.* **51**:1321–1325.
Rajagopalan, R., and C. Tien. 1976. Trajectory analysis of deep-bed filtration with the sphere-in-cell porous media model. *J. Am. Inst. Chem. Eng.* **22**:523–533.
Reynolds, P. J., P. Sharma, G. E. Jenneman, and J. J. McInerney. 1989. Mechanisms of microbial movement in subsurface materials. *Appl. Environ. Microbiol.* **55**:2280–2286.
Rittmann, B. E., P. L. McCarty, and P. V. Roberts. 1980. Trace organics biodegradation in aquifer recharge. *Ground Water* **18**:236–243.
Rosenberg, M. 1981. Bacterial adherence to polystyrene: a replica method of screening for bacterial hydrophobicity. *Appl. Environ. Microbiol.* **42**:375–377.
Rosenberg, M., E. A. Bayer, D. Delarea, and E. Rosenberg. 1982. Role of thin fimbriae in adherence and growth of *Acinetobacter calcoaceticus* RAG-1 on hexadecane. *Appl. Environ. Microbiol.* **44**:929–937.
Rosenberg, M., and S. Kjelleberg. 1986. Hydrophobic interactions: role in bacterial adhesion. *Adv. Microb. Ecol.* **9**:353–393.
Rutter, P. R. 1980. The physical chemistry of the adhesion of bacteria and other cells, p. 103–135. *In* A. S. G. Curtis and J. D. Pitt (ed.), *Cell Adhesion and Motility*. Cambridge University Press, Cambridge.
Rutter, P. R., and B. Vincent. 1984. Physicochemical interactions of the substratum, microorganisms, and the fluid phase, p. 21–38. *In* K. C. Marshall (ed.), *Microbial Adhesion and Aggregation*. Springer-Verlag, New York.
Sakthivadivel, R. 1966. Theory and mechanism of filtration of non-collidal fines through a porous medium. Report no. HEL 15-5. Hydraulic Engineering Laboratory, University of California, Berkeley.
Scholl, M. A., A. L. Mills, J. S. Herman, and G. M. Hornberger. The influence of mineralogy and solution chemistry on the attachment of bacteria to representative aquifer materials. *J. Contam. Hydrol.*, in press.
Sharma, M. M., Y. I. Chang, and T. F. Yen. 1985. Reversible and irreversible surface charge modifications of bacteria for facilitating transport through porous media. *Colloids Surf.* **16**:193–206.
Shaw, D. J. 1976. *Introduction to Colloid and Surface Chemistry*, 2nd ed. Butterworth Inc., Boston.
Sinclair, J. L., and W. C. Ghiorse. 1987. Distribution of protozoa in subsurface sediments of a pristine groundwater study site in Oklahoma. *Appl. Environ. Microbiol.* **53**:1157–1163.
Sinton, L. W. 1980. Investigations into the use of the bacterial species *Bacillus stearothermophilus* and *Escherichia coli* (H_2S positive) as tracers of ground water movement. Water and

Soil Technology Publication no. 17. Water Soil Division, Ministry of Works and Development, Wellington, New Zealand.
Smith, M. S., G. W. Thomas, R. E. White, and D. Ritonga. 1985. Transport of Escherichia coli through intact and disturbed soil columns. *J. Environ. Qual.* **14**:87–91.
Smith, R. L., and J. H. Duff. 1988. Denitrification in a sand and gravel aquifer. *Appl. Environ. Microbiol.* **54**:1071–1078.
Stenström, T. A. 1989. Bacterial hydrophobicity, an overall parameter for the measurement of adhesion potential to soil particles. *Appl. Environ. Microbiol.* **55**:142–147.
Stenström, T. A., and S. Kjelleberg. 1985. Fimbriae mediated nonspecific adhesion of *Salmonella typhimurium* to mineral particles. *Arch. Microbiol.* **143**:6–10.
Thorn, P. M., and R. M. Ventullo. 1988. Measurement of bacterial growth rates in subsurface sediments using the incorporation of tritiated thymidine into DNA. *Microb. Ecol.* **16**:3–16.
Tobiason, J. E., and C. R. O'Melia. 1988. Physicochemical aspects of particle removal in depth filtration. *J. Am. Water Works Assoc.* **80**:54–64.
Trevors, J. T., J. D. van Elsas, L. S. van Overbeek, and M.-E. Starodub. 1990. Transport of a genetically engineered *Pseudomonas fluorescens* strain through a soil microcosm. *Appl. Environ. Microbiol.* **56**:401–408.
Vanhaecke, E., J.-P. Remon, M. Moors, F. Raes, D. De Rudder, and A. Van Peteghem. 1990. Kinetics of *Pseudomonas aeruginosa* adhesion to 304 and 316-L stainless steel: role of cell surface hydrophobicity. *Appl. Environ. Microbiol.* **56**:788–795.
Van Loosdrecht, M. C. M., J. Lyklema, W. Norde, G. Schraa, and A. J. B. Zehnder. 1987a. The role of bacterial cell wall hydrophobicity in adhesion. *Appl. Environ. Microbiol.* **53**:1893–1897.
Van Loosdrecht, M. C. M., J. Lyklema, W. Norde, G. Schraa, and A. J. B. Zehnder. 1987b. Electophoretic mobility and hydrophobicity as a measure to predict the initial steps of bacterial adhesion. *Appl. Environ. Microbiol.* **53**:1898–1901.
Van Loosdrecht, M. C. M., J. Lyklema, W. Norde, and A. J. B. Zehnder. 1989. Bacterial adhesion: a physicochemical approach. *Microb. Ecol.* **17**:1–15.
Van Loosdrecht, M. C. M., J. Lyklema, W. Norde, and A. J. B. Zehnder. 1990a. Influence of interfaces on microbial activity. *Microbiol. Rev.* **54**:75–87.
Van Loosdrecht, M. C. M., W. Norde, J. Lyklema, and A. J. B. Zehnder. 1990b. Hydrophobic and electrostatic parameters in bacterial adhesion. *Aquat. Sci.* **52**:103–114.
Wilson, J. T., J. F. McNabb, D. L. Balwill, and W. C. Ghiorse. 1983. Enumeration and characterization of bacteria indigenous to a shallow water-table aquifer. *Ground Water* **21**:134–142.
Wood, W. W., and G. G. Ehrlich. 1978. Use of baker's yeast to trace microbial movement in ground water. *Ground Water* **16**:398–403.
Wrangstadh, M., P. L. Conway, and S. Kjelleber. 1986. The production and release of an extracellular polysaccharide during starvation of a marine *Pseudomonas* sp. and the effect thereof on adhesion. *Arch. Microbiol.* **145**:220–227.
Yao, K. M., M. T. Habibian, and C. R. O'Melia. 1971. Water and waste water filtration: concepts and applications. *Environ. Sci. Technol.* **11**:1105–1112.
Yates, M. V., and S. R. Yates. 1988. Modeling microbial fate in the subsurface environment. *Crit. Rev. Environ. Control.* **17**:307–344.

Chapter 6

Use of Models To Predict Bacterial Penetration and Movement within a Subsurface Matrix

Michael J. McInerney

Introduction 115
Experimental Systems 116
Straining and Adsorption. 118
Modeling Bacterial Transport with Fluid Flow 120
Microbial Penetration without Fluid Flow. 124
    Biological Factors 126
    Physiochemical Factors 129
Summary. 131
References. 131

The perception that the subsurface acts as a filter to remove bacteria and other particulates from groundwater is no longer valid. Bacteria can and do travel over considerable distances in aquifers and saturated soils. Coliform bacteria were transported for more than 1 km in loamy sand aquifers and up to several kilometers in fissured karstic aquifers (Crane and Moore, 1984; Gerba, 1985; Gerba and Bitton, 1984; Hagedorn, 1981; and Matthess and Pekdeger, 1985). Jack (in press) reported that a *Leuconostoc* strain injected into a well in an unconsolidated petroleum reservoir appeared in another well over a kilometer away within a day. The ability of bacteria to penetrate and propagate in deeper, less permeable petroleum reservoirs has been demonstrated (Hitzman, 1983). In a highly permeable oil reservoir (3.3 to 8.1 μm^2), the mean proliferation rate of bacteria was between 1 and 3 m/day. In addition to the above studies that show that injected bacteria can move considerable distances in the subsurface, information obtained from many studies unequivocally shows that the terrestrial subsurface contains a diverse and metabolically active microbial population, including aerobes, anaerobes, and facultative organisms (Balkwill, 1989; Ghiorse and Wilson, 1988; Jones et al., 1989; Wilson et al., 1983a; Wilson et al., 1983b). This flora exhibits a broad range of physiological

---

Michael J. McInerney • Department of Botany and Microbiology, University of Oklahoma, 770 Van Vleet Oval, Norman, Oklahoma 73019.

properties ranging from heterotrophic metabolism, including the capability of oxidizing organic pollutants, to methane production and sulfate reduction. The existence of diverse microbial populations in the subsurface raises the question as to the origin of these microorganisms. Have they been there since the original deposition of the formation or have they migrated into the formation from sources in contact with surface microorganisms? Whether it is to predict the movement of pathogens or genetically engineered bacteria in the subsurface, to develop an in situ process for the bioremediation of contaminated aquifers or for the enhancement of oil recovery, or to understand how the microbial populations colonized the subsurface, a fundamental understanding of the mechanisms of microbial movement through the subsurface is required.

The movement of microorganisms through a subsurface formation can be categorized in relation to groundwater flow rates and the mineralogy of the formation (Harvey, 1989). In aquifers with high flow rates, elevated concentrations of organic carbon, and small amounts of clays and dissolved solids, unimpeded movement of microorganisms through a portion of the reservoir may occur. This is most pronounced in aquifers with a high degree of secondary pore structure, in which substantial portions of flow occur along preferred flow paths. Slower, more discontinuous movement of microorganisms is likely to occur in aquifers with lower groundwater flow rates or with higher concentrations of clays or dissolved solids. Reversible and nonreversible interactions of the cells with the matrix material will lead to the retardation of the microorganisms relative to groundwater flow. The third mode is the directional locomotive response of many microorganisms to certain chemical stimuli. This mode is probably an important mechanism of movement in aquifers with low flow rates and provides a mechanism whereby microorganisms can propagate throughout the entire porous matrix rather than being localized in the predominant flow channels.

This classification provides the framework for the following discussion on modeling bacterial movement in the subsurface. Experimental and mathematical approaches for studying bacterial transport in consolidated and unconsolidated porous materials will be discussed in relation to the three modes of movement described above. From this, the parameters that control the penetration of bacteria through porous materials will be delineated.

EXPERIMENTAL SYSTEMS

Experimental systems are available for studying the transport and growth of microorganisms in both consolidated and unconsolidated porous materials. A variety of systems have been used to study the movement of bacteria through soils and unconsolidated aquifers (Madsen and Alexander, 1982; Smith et al., 1985; Wollum and Cassel, 1978). The transport of bacteria through soils or unconsolidated aquifers is often studied using columns filled with quartz-type sand sieved to the desired size. Mixtures of different sieve sizes, with or without the addition of clays or natural materials, are used to reconstruct native soil textures. The movement of bacteria through the soil layer is modeled by using vertical columns with continuous or periodic fluid flow. Such systems are probably not appropriate to model bacterial movement in groundwater since the movement of groundwater is perpendicular to, rather than parallel with, gravitational forces. Thus, vertically operated columns may overestimate the effect that sedimentation has on transport

(Matthess and Pekdeger, 1985). Also, groundwater velocities are very slow, usually less than a meter per day. Such slow flow rates are difficult to reproduce unless a high-quality pump is used. If native core material is to be used, it is important that it be obtained parallel with the bedding plane since directional permeability is often in this direction.

The complexity of the interconnecting network of pores present in actual subsurface porous materials is difficult to reproduce with laboratory models. Smith et al. (1985) found that 22 to 78% of the added *Escherichia coli* cells penetrated 28 cm into various soil cores and that 0.2 to 7% of the added cells penetrated the same distance when sieved and repacked cores were used, suggesting that macropores present in the original material were responsible for bacterial movement. A simple experimental system to create heterogeneity by constructing columns of porous sand with a vein of larger-diameter sand has been developed (Mills et al., in press). The columns are packed with a fine sand around a plastic soda straw which is filled with a coarse sand, whereupon the straw is removed. The vein of coarse sand creates a preferred flow path which simulates the fractures, coherent macropores, etc., found in actual porous materials. Several investigators have taken the laboratory to the field and conducted transport experiments in shallow aquifers, using natural or forced hydraulic gradients (Harvey et al., 1989; McCoy and Hagedorn, 1980).

Studies on the transport of bacteria through consolidated porous material have used Berea sandstone as model material (Jenneman et al., 1984; Jang et al., 1983; Lappin-Scott et al., 1988) because of its wide use in the petroleum industry. Berea sandstone is a relatively homogeneous porous material available in a variety of permeability classes. Since clays act as the cementing agent, adsorption may be more pronounced than with other sandstones. Fused glass bead cores which have a uniform pore size and permeability have also been used (Shaw et al., 1985; MacLeod et al., 1988). However, these cores have permeabilities (2 to 7 μm^2) that are much higher than those found in most consolidated subsurface formations.

The growth of bacteria through sandstone was studied using a growth chamber consisting of a nutrient-saturated Berea sandstone core connected to two flasks of medium (Jenneman et al., 1985; Jang et al., 1983). The system was incubated in the absence of a hydrostatic pressure gradient or other forces which would result in fluid flow. After sterilization, one of the flasks was inoculated with a bacterial suspension, and the time required for growth to occur in the uninoculated flask was measured and is defined as the penetration time. The penetration rate was obtained by dividing the penetration time by the length of the core and as such does not represent an instantaneous rate, but an average rate of penetration through the core. A similar approach was used to study the penetration of bacteria through unconsolidated sand (Reynolds et al., 1989). Glass or plastic cylinders were packed with nutrient-saturated sand by using ultrasonic vibrations to obtain a uniform and reproducible packing. One end of the core was inoculated with a suspension of bacteria, and the presence of viable cells in a nutrient-containing syringe at the distal end of the core was determined at regular intervals. This system gives a more accurate determination of penetration times since it does not include the time required for growth to occur in the flasks. Both systems allow the systematic study of a particular parameter, either biological or physiochemical, on bacterial penetration and are easily replicated for the statistical analysis of the data.

It should be obvious that different systems have been developed in response to the particular questions being studied. Simple experimental approaches are often

Table 1. Relationship between the grain size of soils and the critical pore size[a]

Soil type	Grain diam (mm)	Critical pore size (μm)
Silt		
Fine	0.002–0.006	0.72
Medium	0.006–0.02	2.4
Coarse	0.02–0.06	7.2
Sand		
Fine	0.06–0.2	24
Medium	0.2–0.6	72
Coarse	0.6–2	240
Gravel		
Fine	2–6	720
Medium	6–20	2,400
Coarse	20–63	7,200

[a] From Matthess and Pekdeger (1985) with permission of the publisher.

criticized because they do not accurately reflect actual environments and thus may not provide information that can be extrapolated to the field. However, one must accept this inherent problem of laboratory research in order to understand the factors that govern bacterial movement. Unless that information is obtained, one does not have the ability to predict. Simple experimental systems allow one variable to be studied at a time and can be replicated and reproduced so that statistically meaningful information is obtained. When the dominant factors that influence bacterial movement in the subsurface are understood, the ability to predict bacterial movement in formations where those factors prevail is greatly increased.

STRAINING AND ADSORPTION

Many factors, both biotic and abiotic, affect the movement of microorganisms through porous media. Two important physical mechanisms which retard microbial movement include straining and adsorption. The importance of straining depends on the relative size of the pore to that of the cell and is an important removal mechanism when the average cell size is greater than 5% of the grains which constitute the porous medium (Jang et al., 1983; Gruesbeck and Collins, 1982; Updegraff, 1983). Unconsolidated formations that are composed of fine sands, silt, or clays will have effective pore sizes that are within the size range of most bacteria (Table 1). This is also true for consolidated sandstones (Donaldson, 1985). Because of this, straining will be a major factor restricting the transport of bacteria. Updegraff (1983) states that the pore entrance size should be about twice the diameter of the cell to allow passage of bacteria through sandstone. Kalish et al. (1964) showed that larger cells or those that form aggregates or chains were filtered out more rapidly and caused greater permeability reductions than did smaller rods or cocci. We have made similar observations. The injection of more than 100 pore volumes of a *Pseudomonas* strain into Berea sandstone cores caused little decrease in the permeability (Jenneman et al., 1984), but the injection of a larger *Bacillus* species resulted in an almost 100% reduction in the permeability (Raiders et al., 1986). The results of these and other studies (Jang et al., 1983; Raleigh and Flock, 1965) all support the conclusion that straining is the major mechanism that retards the penetration of bacteria through consolidated porous media.

Several workers have used spores (Jang et al., 1983) or ultramicrobacteria (MacLeod et al., 1988; Lappin-Scott et al., 1988) for microbially enhanced oil recovery processes because of their smaller size. Suspensions of ultramicrobacteria were found to penetrate throughout the entire length of fused glass bead columns and sandstone cores without significantly reducing the injectivity or permeability. Thus, the use of ultramicrobacteria should prevent the accumulation of cells at the well bore and allow for greater dispersal of bacteria in oil reservoirs.

Permeabilities of less than 0.075 to 0.1 μm^2 may represent the lower limit for effective microbial transport since pore entrance size distribution would contain pores which would restrict the passage of vegetative cells. However, there have been reports of bacterial transport in cores with permeabilities of less than 0.075 μm^2 (Hart et al., 1960; Kalish et al., 1964). In one case, transport occurred in a core with a permeability of 0.1×10^{-3} μm^2 (Myers and Samiroden, 1967).

In unconsolidated porous medium, in which the grain size is often very large compared with that of the microbial cells, adsorption of microbial cells to the surface of the porous medium will govern the rate of transport of microorganisms. Adsorption is described by the Derjaguin-Landau and Verwey-Overbeek colloidal-filtration theory (Ho, 1986; Shaw, 1976), which states that the initial contact of colloidal particles with a surface depends on the relative magnitude of the attractive and repulsive forces at the interface (van der Waals and electrostatic forces). In addition to these forces, hydrophobic and steric forces can also contribute to the binding of cells to a surface (Ho, 1986). The balance of these forces may result in adhesion of particles at some distance (usually a few nanometers) from the surface. This theory has been used to describe the initial steps of adhesion of bacteria to a solid surface (Marshall et al., 1971; Fletcher, 1987). Several studies suggest that hydrophobic interactions control bacterial adhesion to surfaces (van Loosdrecht et al., 1987a, 1987b; Stenstrom, 1989; McEldowney and Fletcher, 1986). However, these studies used hydrophobic surfaces or buffers of relatively high ionic strength. At low ionic strength or with hydrophilic surfaces, electrostatic effects may control adhesion (Martin, 1990; Scholl et al., in press; Fletcher and Loeb, 1979). Scholl et al. (in press) concluded that in systems in which hydrophobicity is not a factor, as is the case in aquifers with low levels of organic carbon, electrostatic properties (surface charge) of the minerals will be a major factor controlling the initial adhesion of bacteria.

After adhesion occurs, deadsorption of cells may occur by changes in ionic strength, composition, and pH of the fluid. Irreversible binding of the cells occurs at a later time, following the synthesis of extracellular polysaccharide connecting the cells to the surface (Costerton et al., 1981; Ho, 1986). Once adsorbed, progeny cells remain attached within the extracellular polysaccharide matrix, which can then develop into a thick biofilm (Costerton et al., 1981). Most substrate decay models assume the development of a biofilm on the surface of the porous material (Costerton et al., 1981; Rittmann and McCarty, 1980a; Rittmann and Brunner, 1984; Martin, 1990). Although several studies show that most of the microbial biomass in the subsurface is associated with particles (Ghiorse and Wilson, 1988; Harvey et al.,

1984), it is not clear whether these microorganisms form uniform biofilms, as found in water treatment reactors.

Much of the work on bacterial adhesion has focused on understanding the mechanisms involved and the factors that control this process. However, these studies do not provide a parameter that is useful in predicting the deposition rates of cells in any given system. Bouwer and Martin (Bouwer, 1987; Martin, 1990) propose the use of the collision efficiency to predict deposition in groundwater aquifers. The collision efficiency is defined as the fraction of cells colliding with a surface that attach to that surface, and as such represents the probability that a collision will result in attachment. The collision efficiency depends on the interaction of various forces among the particle, the fluid, and the collector surface as the separation distance between the surfaces decreases. As an empirically determined value, it represents the net result of all of these forces. Studies on the attachment of *Pseudomonas aeruginosa* to glass surfaces showed that collision efficiencies are influenced by the ionic strength of the fluid, with the lowest-ionic-strength fluids giving the lowest collision efficiencies (Martin, 1990). In general, when the collision efficiency is high (around 1), most of the deposition of cells will occur within a short distance from the injection point. When collision efficiencies are low (0.001), greater penetration of cells into a formation will occur. Thus, in formations in which the grain size is large relative to the size of the bacterial cells, the surface charge of the porous material and the ionic strength of the groundwater will govern bacterial penetration. As the effective pore size of the porous medium decreases, straining/filtration as a mechanism of cell removal will predominate. In both cases, the heterogeneity of the porous medium itself will be an important consideration since greater depths of penetration will occur through macropores, fractures, or flow channels.

MODELING BACTERIAL TRANSPORT WITH FLUID FLOW

The transport of any component, whether bacterial cells, a substrate, or a product, in a porous material is governed by movement of the fluid (advective forces), the mixing of components due to diffusion and mechanical mixing (convective forces), and the rate of processes which lead to the loss or production of the component in the fluid phase. Mathematically, these processes are summarized in the following advection-dispersion equation written for one-dimensional horizontal linear flow through an incompressible, isotropic porous medium (Knapp et al., 1988; Zhang, 1990):

$$\partial(\phi C_i)/\partial t = -\partial(\mu_f C_i)/\partial x + \partial/\partial x \left[\phi K_D (\partial C_i/\partial x)\right] + \phi R_i \tag{1}$$

where C_i is the concentration of component i, t is time, x is length, ϕ is porosity, μ_f is the Darcy flux, K_D is the dispersion coefficient for component i in the flowing phase, and R_i is the source/sink term to describe the rate of loss or production of component i in the flowing phase. In many cases, it is assumed that the porous medium is homogeneous and thus that the Darcy flux (μ_f), porosity (ϕ), and dispersion coefficient (K_D) do not vary with length (x), which simplifies the equation to

$$\partial C/\partial t = -\mu_f(\partial C/\partial x) + K_D (\partial^2 C_i/\partial x^2) - R_i \tag{2}$$

The colloidal-filtration theory developed for water and wastewater filtration processes has been used to model the transport of bacteria through porous material (Bouwer, 1987; Martin, 1990; McDowell-Boyer et al., 1986; Yao et al., 1971). In this approach, the transport of bacterial cells is treated in a manner analogous to the transport of particles and the advection-dispersion equation is modified to account for the rate of loss of the particle, in this case bacterial cells, from the flowing fluid due to adsorption/filtration. The removal of suspended particles by a one-dimensional clean-bed filter can be described by the following equation (Yao et al., 1971):

$$N_e/N_i = e^{-(3/2)(1-\phi)\alpha\eta\,(L/d_c)} \tag{3}$$

where N_i and N_e are the influent and effluent particle concentrations, e is 2.718, L is the bed length, d_c is the grain diameter, ϕ is the porosity, η is the single collector efficiency, and α is the collision efficiency. The single collector efficiency represents the fraction of particles flowing toward a grain of the porous medium that actually collides with the grain. Particles may be brought to the grain surface as they move with the fluid (interception) or by processes which force the particle to deviate from the flow path (diffusion and sedimentation). Rajagopalan and Tien (1976) used trajectory analysis to model the single collector efficiency by formulating a force balance for a suspended particle which included the effects of gravity, fluid drag, van der Waals interactions, and increased viscous resistance to particle motion near the collector surface. This approach was modified by Tobiason (1987) to describe particle deposition in groundwater, in which fluid flow is horizontal rather than vertical and in which flow velocities are much slower. Although the single collector efficiency describes the mechanisms by which collisions will occur, it does not describe the probability that a collision will result in attachment. This is represented by the collision efficiency, which is defined as the fraction of particles colliding with the surface that attach there (Martin, 1990).

Corapcioglu and Haridas (1984, 1985) expanded the phenomenological description of bacterial transport to include terms for dispersion, convection, diffusion (Brownian), adsorption, motility, deposition, sedimentation, chemotaxis, decay, and growth:

$$\partial(\phi C_i)/\partial t = -\partial/\partial x\,(J_d + J_b + J_t + J_c + J_s) + R_{gf} + R_{df} \tag{4}$$

where ϕ is porosity, C_i is the concentration of cells, J_d is mechanical dispersion, J_b is Brownian diffusion, J_t is effective diffusivity due to tumbling, J_c is the convection term, J_s is gravitational sedimentation, and R_{gf} and R_{df} are the rates of growth and decay of bacteria in the flowing phase, respectively. This equation, a mass-conservation equation for the growth-limiting substrate, and a first-order deposition equation constitute the governing equations for bacterial transport through porous media. Although the model is extensive in its detailed description of mechanisms involved in bacterial transport, its utility is questionable since it requires estimates for a large number of parameters which may be site specific and thus difficult to obtain (Mills et al., in press).

The ability to predict how pathogenic bacteria will move in groundwater aquifers is critical in assessing the risk of waterborne diseases from damaged or improperly designed sewer lines, septic systems, and sewage infiltration beds (Gerba and Bitton, 1984). Since environmental conditions in the subsurface rarely favor the growth of enteric bacteria, survival and deposition are the two main factors determining the fate of these organisms in groundwater aquifers (Crane and Moore, 1984; Gerba, 1985). To describe the survival of the pathogen mathematically, the advection-dispersion equation is modified to include a term for the net decay rate of the organism. The net decay term includes the rate of inactivation of the organism by chemical, physical, and biological factors, offset by any growth of the organism (net decay rate = decay rate − growth rate). The rates of inactivation or decay of many pathogenic bacteria and viruses and the environmental factors that influence their survival have been studied extensively (Crane and Moore, 1984; Gerba and Bitton, 1984).

Theoretical analyses of the effects of the physical properties of the cell and the porous medium and of different fluid flow regimens on collector efficiency show that the transport of small particles (less than 1 μm in diameter) is dominated by diffusional forces and that, for particles with a diameter of more than 3 μm, gravitational forces become important (Yao et al., 1971; Martin, 1990). The ability of the colloidal-filtration theory to predict the transport of bacterial cells and three different size classes of bacterial-size microspheres in natural and forced-gradient field experiments was studied (Harvey, 1989; Harvey et al., 1989). These studies showed that the smaller-sized cells or microspheres were immobilized to a greater degree by the aquifer sediments than were larger-sized cells or spheres. This is consistent with the predictions of the colloidal-filtration theory since the rate of deposition should be larger for smaller-sized particles since particle contact is governed by diffusion, which increases with decreasing particle size. Also, this theory accurately predicted the average cell size of unattached bacteria 0.65 km down gradient of a sewage infiltration bed (Harvey, 1989).

For groundwater velocities of 1 m/day or lower and average grain diameters of 0.1 to 1 mm, very high collector efficiencies can be expected (greater than 0.1), indicating that the concentration of particles within the size range of most bacteria should decrease rapidly a short distance from the point of injection (Martin, 1990; Corapcioglu and Haridas, 1985; Germann and Douglas, 1987; Tobiason, 1987; McDowell-Boyer et al., 1986). Even when the collector efficiency is very low, the maximum theoretical penetration of bacteria in groundwater systems should be no more than 20 m (Corapcioglu and Haridas, 1985; Tobiason, 1987; McDowell-Boyer et al., 1986). However, it is well known that travel distances of 1 km or more are commonly observed in the subsurface (Gerba and Bitton, 1984; Hagedorn, 1981). This discrepancy has led several investigators to question the utility of the colloidal-filtration theory in predicting the movement of bacteria through porous material (Germann and Douglas, 1987). Several possible explanations for the discrepancy between observed travel distances and those predicted by the colloidal-filtration theory have been proposed. Martin (1990) points out that, at an ionic strength of 0.001 to 0.001 M, which is typical of a groundwater, low collision efficiencies would be expected. If the initial cell concentration and fluid velocity are high, it is probable that bacteria would be found hundreds of meters from the source (Martin, 1990).

Another possibility is that actual porous systems are not homogeneous, as is assumed in these models, but contain a considerable degree of secondary pore structure such as macropores, fractures, and solution channels. There are several reports in which the retardation factor for bacteria is substantially less than 1 indicating that the transport of microorganisms was not attenuated by the porous medium and was faster than the average groundwater velocity (Harvey, 1989

Harvey et al., 1989; Champ and Schroeter, 1988; Wood and Ehrlich, 1978). The rapid migration of bacteria through unconsolidated petroleum reservoirs has also been observed (Jack, in press). In the last study, it was estimated that the flow channel was equivalent to a 60-cm-wide pipe which was 1 km long.

In such systems, it is obvious that the colloidal-filtration theory alone will not account for the transport of bacteria. Mills et al. (in press) have developed experimental and theoretical approaches to study the transport in heterogeneous porous systems. The breakthrough of bacteria from columns containing a preferred flow path occurred with two maxima, one corresponding to transport through the preferred flow path and the other corresponding to a combination of transport through the matrix and release of bacteria from the matrix into the preferred flow path (Mills et al., in press). A dual-porosity expansion of the advection-dispersion equation (equations 5 and 6) was developed which at least qualitatively simulated the process:

$$\partial(C_{iA})/\partial t = -\mu_f (\partial C_{iA}/\partial x) + K_{DA} (\partial C_{iA}/\partial x^2) - \varepsilon(C_{iA} - C_{iB})/\phi \tag{5}$$

$$\partial(C_{iB})/\partial t = -\mu_f (\partial C_{iB}/\partial x) + K_{DB} (\partial C_{iB}/\partial x^2) - \varepsilon(C_{iB} - C_{iA})/\phi \tag{6}$$

where subscripts A and B refer to transport processes in the preferred flow path and the matrix, respectively, and ε represents the exchange coefficient describing material flow between the two porous compartments.

Microorganisms differ from inert particles in one very important property, the ability to grow. The growth of bacteria in porous material is assumed to occur as a biofilm, with the production of cells and the use of a substrate described by Monod kinetics. The accumulation of a biofilm on the surface of a pipe under turbulent flow conditions was the net result of the transport of components to the surface, the metabolic conversions within the biofilm, and the detachment and reentrainment of the biofilm (Bryers and Characklis, 1982). The equations describing substrate use, bacterial growth in the flowing phase, and biofilm accumulation are as follows:

$$Y(ds/dt) = F (S_i - S) - 1/Y(\mu VX - R_g A) \tag{7}$$

$$V(dx/dt) = F (X_i - X) + \mu VX + R_r A + R_d A \tag{8}$$

$$dB/dt = (\mu_p - K_e) B + R_d - R_d A \tag{9}$$

where S is the substrate concentration, X is the suspended biomass concentration, B is the attached biomass concentration (with the subscript values representing the initial concentrations), V is the volume of the reactor, A is the reactor surface area, F is the volumetric flow rate, μ is the specific growth rate, Y is the biomass yield coefficient, R_g is the biofilm production rate, R_d and R_r are the deposition and detachment rates, respectively, μ_p is the specific biofilm production rate, and K_e is the biofilm decay rate. Similar approaches have been used to model biofilm growth under steady-state and non-steady-state conditions (Rittmann and McCarty, 1980a, 1980b; Rittmann and Brunner, 1984). These models adequately predicted substrate utilization in laboratory-scale biofilm reactors.

The above studies assume that the biofilm is of uniform thickness and area.

While this may adequately describe the microbial populations near injection wells, infiltration beds, or other sites of nutrient or fluid input, it may not be representative of the bacterial populations within the formation. Harvey et al. (1984) reported that microcolonies containing 10 to 100 bacteria were commonly found on particles from an unconfined aquifer underlying Cape Cod, Mass. From this observation, Molz et al. (1986) developed a model to simulate microbial growth and degradation, assuming that the bulk of the microorganisms in an aquifer grow in microcolonies attached to the porous surface. In this model, both the carbon substrate and the electron acceptor (oxygen) can be limiting for growth and that growth occurs by an increase in the number of microcolonies rather than by an increase in the size of the microcolony. The transport of the substrate and oxygen is governed by advection-dispersion equations, with surface adsorption and microbial degradation as sink terms. Although the model was not verified, its predictions indicate that biodegradation can have a significant effect on contaminant transport when proper conditions for microbial growth exist.

Most bacterial transport models often exclude growth in order to simplify the computational analysis and because of the difficulties in determining in situ growth rates. Growth rates in uncontaminated aquifers appear to be several orders of magnitude lower than those observed for surface sediments (Thorn and Ventullo, 1988). However, since groundwater movement is slow, even slow growth rates can influence the extent of bacterial penetration over long periods of time. In organically contaminated aquifers, bacterial growth can be much faster, with growth rates ranging from 0.005 to 0.042/h, and approach that of some surface waters (Harvey and George, 1987). It is clear that growth must be included if an adequate description of the process is to be obtained since microbial penetration is favored under growing conditions (Craw, 1908; Jenneman et al., 1985).

MICROBIAL PENETRATION WITHOUT FLUID FLOW

The common assumption made when modeling bacterial transport in porous materials is that the bacterial cells act as particles and are carried through the formation by the advective and convective forces resulting from groundwater flow. Self-propulsion by bacteria is not thought to be an important mechanism since the velocity of the fastest recorded swimming bacterium (about 50 µm/s) (Brock, 1970) corresponds to a Reynolds number of about 10^{-4}, which is well within the range in which inertial forces are negligible to viscous forces (Martin, 1990). If it is assumed that the bacterium moves in a straight line, this would correspond to a velocity of about 4.3 m/day, which is equivalent to the average velocity of many groundwaters (1 m/day). Since bacteria frequently change direction, even in the presence of a chemotactic gradient, the actual rate of translational movement would be much less. However, this analysis does indicate that self-propulsion by bacteria may be an important mechanism for bacterial penetration in some situations, especially when groundwater velocities are low.

The movement of coliform bacteria follows the movement of nutrients from leaking septic tanks and sewer lines (Hagedorn, 1981), suggesting that conditions favorable for bacterial growth are also the conditions that favor greater bacterial penetration. Bacteria have been shown to penetrate through consolidated and unconsolidated porous media under conditions which support growth (Chang and Yen, 1984, 1985; Jang et al., 1983; Jenneman et al., 1985; Montgomery et al., 1990). In these experiments, neither a hydrostatic head nor a differential pressure was

Table 2. Penetration rates of bacteria through nutrient-saturated Berea sandstone

Bacterium	Penetration rate[a] (cm/h)	Growth rate (per h)	Permeability (μm^2)	Reference
Pseudomonas putida	3.4[b]	1.5	0.4	Chang and Yen, 1985
	0.07	1.5	0.4	Jang et al., 1983
Escherichia coli	3.3[b]		0.4	Chang and Yen, 1984
Bacillus sp. strain BCI-INS	0.47	0.94	0.18	Jenneman et al., 1985
	0.06	0.94	0.07	
Enterobacter aerogenes	0.45	1.4	0.6	Jenneman et al., 1985
Bacillus subtilis	0.16		0.4	Jang et al., 1983
Klebsiella pneumoniae	0.08	1.1	0.4	Jenneman et al., 1985
Thiobacillus denitrificans	0.016	0.014	0.5	Montgomery et al., 1990
Desulfovibrio desulfuricans	0.011	0.014	0.5	Montgomery et al., 1990

[a] Penetration was determined by the development of turbidity in the uninoculated flask except when noted.
[b] Calculated by measuring the time required to detect viable cells rather than turbidity in the distal flask.

present. Thus, the movement of the bacteria occurred without applied advective or convective forces and must have resulted from the activity of the bacteria themselves. The fact that bacteria can grow through porous materials has been known for many years. Craw (1908) showed that bacteria penetrated nutrient-saturated Doulton, Pasteur, Berkfield, Slack, and Brownlow filters and concluded that the grain size of the filter mass was the most important factor governing the growth of bacteria through filters. More recently, growth through filters has been used as a method to enrich selectively for filamentous actinomycetes (Hirsch and Christensen, 1983).

The penetration rates of several bacteria through nutrient-saturated Berea sandstone cores are shown in Table 2. The penetration rate of a bacterium to be used in enhanced oil recovery, *Bacillus* sp. strain BCI-INS, through cores with permeabilities of greater than 0.1 μm^2 was close to the velocities of most waterfloods (0.6 to 1.3 cm/h). Penetration of BCI-INS through the core was not observed when the carbon source was deleted from the medium. Even a nonmotile bacterium, *Klebsiella pneumoniae*, grew through Berea sandstone under static conditions. The penetration rates given in Table 2 include the time required to observe turbid growth in the uninoculated flask (Jenneman et al., 1985). When the penetration time of *Pseudomonas putida* was measured by determining the time required for viable cells rather than turbidity to be detected in the uninoculated flask (Chang and Yen, 1985) (Table 2), the penetration time was much faster, suggesting that the penetration rates of many of the bacteria listed in Table 2 may be underestimated. Whether these rates are underestimated or not remains to be determined. However, these data do show that, under favorable nutrient conditions, bacteria can penetrate consolidated rock at a rate which is within an order of magnitude of observed flow rates for natural groundwater aquifers and waterfloods associated with oil recovery. How bacteria penetrate porous material in the absence of fluid flow will now be discussed.

Table 3. Importance of chemotaxis, motility, and gas production in the penetration rate of E. coli mutants through nutrient-saturated sand-packed cores[a]

Strain	Phenotype	Growth rate (per h)	Penetration rate (cm/h) With galactose	Penetration rate (cm/h) Without galactose
Motile				
RP437	Chemotactic wild type	0.27	0.073	<0.01
RP5232	Nontumbling	0.50	0.28	0.31
RP1616	Tumbly	0.35	0.16	0.24
RW262	Gas producer	0.66	0.42	0.50
FM2629	Non-gas producer	0.57	0.38	0.50
Nonmotile				
RP2912	Flagellated	0.37	0.09	<0.01
MC4100	Gas producer	0.58	0.085	0.031
FM909	Non-gas producer	0.46	0.015	<0.01

[a] Modified from Reynolds et al. (1989).

Biological Factors

The mechanisms by which bacteria penetrate porous material under static conditions could include growth, motility, chemotaxis, or a combination of these modes of locomotion. The importance of motility in the penetration of bacteria through nutrient-saturated Berea sandstone was implicated since the penetration rate of the motile bacterium *Enterobacter aerogenes* was three to eight times faster than that of the nonmotile bacterium *K. pneumoniae* (Jenneman et al., 1985). However, since these bacteria do not have identical physiologies, other factors may have contributed to the observed differences in penetration rates. To avoid this problem, the penetration rates of motile and nonmotile strains of *E. coli* through unconsolidated sand-packed cores were compared (Reynolds et al., 1989). In general, motile strains of *E. coli* penetrated four times faster than did nonmotile mutants. Motile, nonchemotactic strains (RP5232 and RP1616) had faster penetration rates than did the parental chemotactic strain (RP437) (Table 3). This plus the fact that the penetration of a mutant which is capable of chemotaxis toward, but not the metabolism of, galactose was the same whether or not a galactose gradient was present in the core suggests that chemotaxis may not be required for bacterial penetration through unconsolidated porous material.

The reason that chemotactic strains had slower penetration times than did nonchemotactic strains may be related to the mode of growth through the unconsolidated material. Adler and Dahl (1967) showed that the movement of nonchemotactic bacteria through capillary tubes can be described by an equation for diffusion of a thin layer of cells in a column of liquid, for which the plot of time versus the square of the distance that the fastest cells have moved from the origin gives a straight line with a y intercept of zero. To determine whether the cells penetrated in a diffuse manner, the concentration of viable cells in each section of sand-packed cores was determined with time for the motile chemotactic strain RP437 and the isogenic nonchemotactic mutant RP5232 (P. Sharma and M. J. McInerney, unpublished data). RP437 progressively grew through the core, with large numbers of cells (more than 10^6/ml) accumulating in one section of the core before cells were detected in the next section. On the other hand, RP5232 grew in a more diffuse manner, with low concentrations of cells (10^2 to 10^3/ml) detected in distal sections before high cell concentrations were found in the proximal sections. Plots of time versus the square of the distance that the fastest cells had moved were linear for the nonchemotactic strain, but were curve linear for the wild-type strain, indicating that the penetration of the nonchemotactic strain resembled a diffusion process while the wild-type strain penetrated in an ordered, bandlike fashion. This suggests that the chemotactic strains which can sense a nutrient-rich environment in all directions grew throughout the entire pore volume of the core, while for nonchemotactic strains which cannot sense a nutrient gradient, the route taken by the fastest cells may have been less tortuous.

In addition to motility, the growth rate of the strain is also an important factor for bacterial penetration (Reynolds et al., 1989). Table 3 shows that the strains with the fastest growth rates had the fastest penetration times. The relationship between the penetration rate and the growth rate under nutrient-rich conditions for motile strains of *E. coli* is shown in Fig. 1 and can be described by the following sigmoidal relationship:

$$t_p = e^{\sigma\mu}/(1 + e^{\sigma\mu} + \tau\mu^{\bar{\omega}}) \qquad (10)$$

where t_p is the penetration time (in hours), μ is the specific growth rate (per hour), σ is the upper limit of the sigmoidal relationship (estimated from Fig. 1 as 0.5), τ represents the midpoint of the sigmoidal relationship, and $\bar{\omega}$ is an empirical constant (estimated from Fig. 1 as -2.95). A sigmoidal relationship is suggested since one would expect that other factors would limit the penetration rates of even very fast-growing organisms and that some penetration would occur, albeit very slowly, even at very slow growth rates. The penetration rate of *Desulfovibrio desulfuricans* increased when a medium with a higher lactate concentration, which supported a faster growth rate, was used (M. Rozmin and M. J. McInerney, unpublished results). This suggests that the nutritional status of the formation is an important factor controlling the rate of bacterial penetration; i.e., conditions which support faster growth rates would support faster penetration rates. Although this hypothesis is based on a limited number of observations, it is an attractive one since it may be possible to predict the rate of bacterial penetration, at least qualitatively, from easily obtained parameters such as the total organic carbon content of aquifer material.

If motility is important for penetration, the question of how nonmotile organisms penetrate porous material is a puzzling one. For the nonmotile strain RP2912, penetration of sand-packed cores occurred only when galactose was present (Table 3). This is also the nutrient condition that results in the production of a large amount of gas inside cores. The importance of gas production for penetration of nonmotile strains was confirmed when penetration rates of motile (RW262) and nonmotile (MC4100) gas-producing strains were compared with those of their isogenic, non-gas-producing strains (FM2629 and FM909), which lack formate dehydrogenase (Table 3). The penetration of the two motile strains was similar whether or not galactose was present. When galactose was present, the penetration rate of MC4100 was about five times faster than that of FM909. In the absence of galactose, when gas production would be minimal, the penetration of FM909 was not observed.

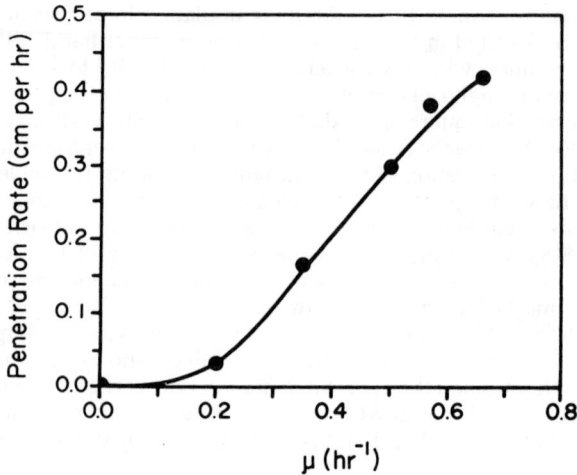

Figure 1. Relationship between the growth rate and the penetration rate for motile strains of *E. coli* through nutrient-saturated sand-packed cores (from Reynolds et al., 1989).

One possible mode of penetration of nonmotile bacteria through porous media would be if growth was perceived to occur in a filamentous fashion. Thus, the penetration rate would be a function of the length of the cell and of the core. The actual penetration rates were much slower than that predicted by this relationship, suggesting that the actual path taken by the cells is more tortuous than a straight line (Jenneman et al., 1985; Reynolds et al., 1989). If the penetration time is a function not only of the core length but also of the pore volume and the cell volume, then an equation that describes the penetration of nonmotile cells without gas production would be as follows:

$$t_p = t_d l_x [\ln(V_x/V_c)/\ln 2] \tag{11}$$

where t_p is the penetration time (in hours), t_d is the doubling time (in hours) of the bacterium, l_x is the length of the core (in centimeters), and V_x and V_c are the liquid volume of the core per centimeter and the average cellular volume, respectively. This equation predicts a penetration time of 511 h for a single cell with a volume of 1.2 μm³ and a doubling time of 2.7 h to transverse an 8-cm-long core with a pore volume of 7 ml (Reynolds et al., 1989). The actual penetration time was 513 h. With equation 11, the calculated penetration time of the nonmotile bacterium *K. pneumoniae* through a 4.8-cm-long core of Berea sandstone (pore volume, 0.43 ml) is about 73 h, which is in the range of the observed penetration times, 48 to 83 h (Jenneman et al., 1985). These analyses indicate that nonmotile cells penetrate the core by growing throughout the available pore volume. This conclusion is supported by studies that show that the growth of both motile and nonmotile strains inside the cores proceeds in an ordered or bandlike manner from the point of inoculation (Sharma and McInerney, unpublished data). Gas production then acts to increase the rate of penetration.

Physiochemical Factors

Permeability is a measure of the ability of the porous material to transmit fluids, which is related to the median pore entrance size distribution of the rock. For cores of equal porosity, low-permeability cores will have median pore entrance size distributions skewed to the smaller pore entrance sizes, while the reverse will be true for high-permeability cores. However, the relationship between permeability and pore entrance size is a complex one. Consolidated rock with a large proportion of small pores can have a high permeability if there are a few very large pores or fissures. For a relatively homogeneous rock such as Berea sandstone, the median pore entrance size generally decreases with increasing permeability. Kalish et al. (1964), using different permeability classes of Berea sandstone (270 to 400 μm^2, high; 130 to 160 μm^2, medium; and 17 to 48 μm^2, low), found that the median pore entrance size distribution was 3.5 to 4.0 for the low-permeability range, 4.5 to 5.0 for the medium-permeability range, and 5.5 to 6.0 for the high-permeability range. Because bacteria have sizes within the median range of the pore entrance sizes, the permeability of the rock should be an important factor controlling bacterial penetration.

To test this hypothesis, the penetration rate of a bacterium to be used in enhanced oil recovery was studied as a function of the permeability of Berea sandstone (Fig. 2) (Jenneman et al., 1985). The penetration rate of *Bacillus* sp. strain BCI-INS was fastest and generally independent of permeability above 0.1 to 0.2 μm^2 and decreased rapidly for cores with permeabilities below these values. Multiple regression analysis showed that 80% of the variance in the penetration rate could be explained by differences in the core length and the permeability class. The rate of penetration was independent of permeability above 0.1 μm^2 and was zero order with respect to length for cores with permeabilities above 0.1 μm^2 and first order with respect to length for cores below 0.1 μm^2. Thus, the penetration of *Bacillus* sp. strain BCI-INS can be described by two simple equations, one for cores with permeabilities above 0.1 μm^2 (equation 12) and one for cores with permeabilities below 0.1 μm^2 (equation 13):

$$t_p = 2.5L + 2.0 \tag{12}$$

$$t_p = 1.4L^{1.8} \tag{13}$$

where t_p is the penetration time in hours and L is the length of the core (in centimeters). A good conceptual model of a rock is that it is made up of a series of capillaries. If one assumes that in general as permeability decreases the pore diameter also decreases, then it follows that as permeability decreases and length increases the probability becomes greater that a larger number of these capillaries will have a restriction that is inaccessible to bacterial penetration. Thus, the apparent decrease in penetration rate with core length for low-permeability cores probably does not reflect a slowing down of the bacteria as they penetrate longer lengths, but may simply indicate that fewer and fewer cells penetrate further into the rock. The fact that, in high-permeability cores, the penetration rate is independent of permeability implies that neither pore size nor permeability is restrictive to bacterial penetration.

These equations were derived by using a single bacterial strain and a single

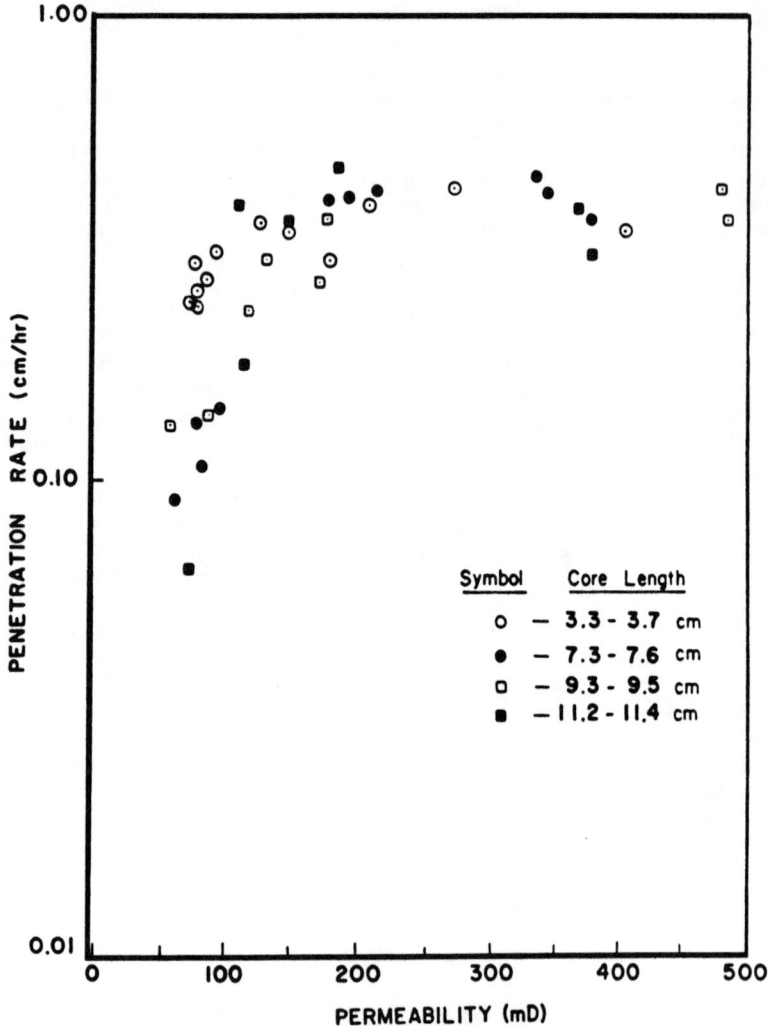

Figure 2. Effect of permeability on the penetration rate of *Bacillus* sp. strain BCI-INS through nutrient-saturated Berea sandstone cores (from Jenneman et al., 1985). mD, MilliDarcy (10^{-3} μm^2).

porous medium. It is important to determine whether these relationships (equations 12 and 13) hold for other bacteria and with other porous media since they offer a simple approach for estimating microbial penetration through subsurface formations. The penetration time of motile strains of *E. coli* through unconsolidated sand-packed cores increased linearly with core volume (length), showing that the penetration rate was independent of permeability and zero order with respect to length, with an average penetration rate of 0.26 cm/h (Reynolds et al., 1989). Thus, in unconsolidated porous material a simple relationship can be derived to predict penetration rates of bacteria.

Surface properties of the porous material are also an important factor controlling the penetration of the bacteria under static conditions. Jenneman et al. (1986) found that the penetration rate of *Bacillus* sp. strain BCI-INS increased when Berea sandstone was sterilized by autoclaving rather than by use of dry heat. Changes in the permeability, porosity, and pore entrance size distribution of the rock as a result of autoclaving were not of sufficient magnitude to explain the differences in penetration time. Electron dispersion spectroscopy showed that the surface of dry-heat-sterilized Berea sandstone had higher silica, aluminum, and potassium contents and a much lower chloride content than did autoclaved sandstone. Scanning electron microscopy showed that clay morphology changed as a result of autoclaving. These data suggest that the changes in penetration rates caused by autoclaving were probably the result of a change in the surface charge of the pores and a reduction in the surface area of the clays available for adhesion. Thus, as is the case for bacterial penetration during advective processes, adsorption also plays an important role in controlling bacterial penetration in the absence of fluid flow.

SUMMARY

The transport of bacterial cells is governed by groundwater flow, the mixing of components due to diffusion and mechanical factors, and the rate of processes that result in the loss (adsorption, straining, and death) or production (growth) of bacteria in the flowing phase. In formations in which the grain size is large, the surface charge of the porous medium and the ionic strength of the groundwater will govern bacterial penetration. As the effective pore size of the porous medium decreases, straining/filtration will become an important factor. The collision efficiency represents the net result of the various forces involved in bacterial adsorption and as such is a useful parameter in predicting deposition rates of cells in any given porous medium.

Models based on the colloidal-filtration theory have been shown to adequately predict the transport of cells through porous media in laboratory experiments and in an unconfined contaminated aquifer. Verification of these models has often been hindered by a lack of precise estimates for important parameters such as dispersion coefficients and collection efficiencies for bacterial cells. Greater depths of bacterial penetration will occur in heterogeneous formations that contain macropores, fractures, or flow channels. In the absence of fluid flow, bacterial penetration does occur under growing conditions, with the growth rate of the organism and the permeability of the porous medium being important factors controlling the rate of penetration.

I thank A. L. Mills, E. J. Bouwer, and R. E. Martin for providing information prior to publication and R. M. Knapp for helpful discussions.

REFERENCES

Adler, J., and M. M. Dahl. 1967. A method for measuring the motility of bacteria and for comparing random and non-random motility. *J. Gen. Microbiol.* **46**:161–173.
Balkwill, D. L. 1989. Numbers, diversity, and morphological characteristics of aerobic,

chemoheterotrophic bacteria in deep subsurface sediments from a site in South Carolina. *Geomicrobiol. J.* **7**:33–52.
Bouwer, E. 1987. Theoretical investigation of particle deposition in biofilm systems. *Water Res.* **21**:1489–1498.
Brock, T. D. 1970. *Biology of Microorganisms*. Prentice-Hall, Inc., Englewood Cliffs, N.J.
Bryers, J. D., and W. G. Characklis. 1982. Processes governing primary biofilm formation. *Biotechnol. Bioeng.* **24**:2451–2476.
Champ, D. R., and J. Schroeter. 1988. Bacterial transport in fractured rock—a field scale tracer test at Chalk River Nuclear Laboratory, p. 14:1–14:7. *In* Proceedings of the International Conference on Water and Wastewater Microbiology, Newport Beach, Calif., 8–11 February 1988.
Chang, P. L., and T. F. Yen. 1984. Interaction of *Escherichia coli* B and B/4 and bacteriophage T4D with Berea sandstone rock in relation to enhanced oil recovery. *Appl. Environ. Microbiol.* **47**:544–550.
Chang, P. L., and T. F. Yen. 1985. Interaction of *Pseudomonas putida* ATCC 12633 and bacteriophage gh-1 in Berea sandstone rock. *Appl. Environ. Microbiol.* **50**:1545–1547.
Corapcioglu, M. Y., and A. Haridas. 1984. Transport and fate of microorganisms in porous media: a theoretical investigation. *J. Hydrol.* (Amsterdam) **72**:149–169.
Corapcioglu, M. Y., and A. Haridas. 1985. Microbial transport in soils and ground water: a numerical model. *Adv. Water Res.* **8**:188–200.
Costerton, J. W., R. T. Irvin, and K.-J. Cheng. 1981. The bacterial glycocalyx in nature and disease. *Annu. Rev. Microbiol.* **35**:299–324.
Crane, S. R., and J. A. Moore. 1984. Bacterial pollution of groundwater: a review. *Water Air Soil Pollut.* **22**:67–83.
Craw, J. A. 1908. On the grain of filters and the growth of bacteria through them. *J. Hyg.* **8**:70–73.
Donaldson, E. C. 1985. Use of capillary pressure curves for analysis of production formation damage. Paper SPE 13809, Society of Petroleum Engineers 1985 Production Operations Symposium, Oklahoma City, Okla.
Fletcher, M. 1987. How do bacteria attach to solid surfaces? *Microbiol. Sci.* **4**:133–135.
Fletcher, M., and G. I. Loeb. 1979. Influence of substratum characteristics on the attachment of a marine pseudomonad to solid surfaces. *Appl. Environ. Microbiol.* **37**:67–72.
Gerba, C. P. 1985. Microbial contamination of the subsurface, p. 53–67. *In* C. H. Ward, W. Giger, and P. L. McCarty (ed.), *Ground Water Quality*. John Wiley & Sons, Inc., New York.
Gerba, C. P., and G. Bitton. 1984. Microbial pollutants: their survival and transport pattern to groundwater, p. 65–88. *In* G. Bitton and C. P. Gerba (ed.), *Groundwater Pollution Microbiology*. Wiley Interscience, New York.
Germann, P. F., and L. A. Douglas. 1987. Comments on "particle transport through porous media." *Water Resour. Res.* **23**:1697–1698.
Ghiorse, W. C., and J. T. Wilson. 1988. Microbial ecology of the terrestrial subsurface. *Adv. Appl. Microbiol.* **33**:107–173.
Gruesbeck, C., and R. E. Collins. 1982. Entrainment and deposition of fine particles in porous media. *Soc. Petrol. Eng. J.* **22**:847–856.
Hagedorn, C. 1981. Transport and fate: bacterial pathogens in groundwater, p. 153–172. *In* L. W. Canter, E. W. Akin, J. F. Kreissel, and J. F. McNabb (ed.), *Microbial Health Considerations of Soil Disposal of Domestic Wastewaters*. Publication no. EPA-600/9-83-017. U.S. Environmental Protection Agency, Washington, D.C.
Hart, R. T., T. Fekete, and D. L. Flock. 1960. The plugging effect of bacteria in sandstone systems. *Can. Mining Metallurgical Bull.* **July**:495–501.
Harvey, R. W. 1989. Transport of bacteria in a contaminated aquifer, p. 183–188. *In* G. E. Mallard and S. E. Ragone (ed.), *U.S. Geological Survey Toxic Substances Hydrology Program—Proceedings of the Technical Meeting, Phoenix, Ariz., 26–30 September 1988*. U.S. Geological Survey, Reston, Va.
Harvey, R. W., and L. H. George. 1987. Growth determinations for unattached bacteria in a contaminated aquifer. *Appl. Environ. Microbiol.* **53**:2992–2996.
Harvey, R. W., L. H. George, R. L. Smith, and D. R. LeBlanc. 1989. Transport of

microspheres and indigenous bacteria through a sandy aquifer: results of natural and forced-gradient tracer experiments. *Environ. Sci. Technol.* **23:**51–56.

Harvey, R. W., R. L. Smith, and L. George. 1984. Effect of organic contamination upon microbial distributions and heterotrophic uptake in a Cape Cod, Mass., aquifer. *Appl. Environ. Microbiol.* **48:**1197–1202.

Hirsch, C. F., and D. L. Christensen. 1983. Novel method for selective isolation of actinomycetes. *Appl. Environ. Microbiol.* **46:**925–929.

Hitzman, D. O. 1983. Petroleum microbiology and the history of its role in enhanced oil recovery, p. 162–218. *In* E. C. Donaldson and J. B. Clark (ed.), *Proceedings of the 1982 International Conference on Microbial Enhancement of Oil Recovery*. Department of Energy, Bartlesville, Okla. (Available through National Technical Information Service [CONF 8205140].)

Ho, C. S. 1986. An understanding of the forces in the adhesion of microorganisms to surfaces. *Process Biochem.* **21:**148–152.

Jack, T. Field and laboratory results for a bacterial selective plugging system. *In* E. C. Donaldson (ed.), *Proceedings of the 1990 MEOR International Conference*, in press. Elsevier Science Publishers, B. V., Amsterdam.

Jang, L. K., P. W. Chang, J. E. Findley, and T. F. Yen. 1983. Selection of bacteria with favorable transport properties through porous rock for the application of microbial-enhanced oil recovery. *Appl. Environ. Microbiol.* **46:**1066–1072.

Jenneman, G. E., R. M. Knapp, M. J. McInerney, D. E. Menzie, and D. E. Revus. 1984. Experimental studies of in situ microbial enhanced oil recovery. *Soc. Petrol. Eng. J.* **24:**33–37.

Jenneman, G. E., M. J. McInerney, M. E. Crocker, and R. M. Knapp. 1986. Effect of sterilization by dry heat or autoclaving on bacterial penetration through Berea sandstone. *Appl. Environ. Microbiol.* **51:**39–43.

Jenneman, G. E., M. J. McInerney, and R. M. Knapp. 1985. Microbial penetration through nutrient-saturated Berea sandstone. *Appl. Environ. Microbiol.* **50:**383–391.

Jones, R. E., R. E. Beeman, and J. M. Suflita. 1989. Anaerobic metabolic processes in deep terrestrial subsurface. *Geomicrobiol. J.* **7:**117–130.

Kalish, O. J., J. A. Stewart, W. F. Rogers, and E. O. Bennett. 1964. The effect of bacteria on sandstone permeability. *J. Petrol. Technol.* **16:**805–814.

Knapp, R. M., F. Civan, and M. J. McInerney. 1988. Modeling growth and transport of microorganisms in porous formations, p. 676–679. *In* R. Vichnevetsky, P. Borne, and J. Vignes (ed.), *IMACS 1988 Proceedings of the 12th World Congress on Scientific Computation*, vol. 3. International Association for Mathematics and Computers in Simulation, Paris.

Lappin-Scott, H. M., F. Cusack, and J. W. Costerton. 1988. Nutrient resuscitation and growth of starved cells in sandstone cores: a novel approach to enhanced oil recovery. *Appl. Environ. Microbiol.* **54:**1373–1382.

MacLeod, F. A., H. M. Lappin-Scott, and J. W. Costerton. 1988. Plugging of a model rock system by using starved bacteria. *Appl. Environ. Microbiol.* **54:**1365–1372.

Madsen, E. I., and M. Alexander. 1982. Transport of *Rhizobium* and *Pseudomonas* through soil. *Soil Sci. Soc. Am. J.* **46:**557–560.

Marshall, K. C., R. Stout, and R. Mitchell. 1971. Mechanism of the initial events in the sorption of marine bacteria to surfaces. *J. Gen. Microbiol.* **68:**337–348.

Martin, R. E. 1990. Quantitative description of bacterial deposition and initial biofilm development in porous media. Ph.D. dissertation. Johns Hopkins University, Baltimore, Md.

Matthess, G., and A. Pekdeger. 1985. Survival and transport of pathogenic bacteria and viruses in ground water, p. 472–482. *In* C. H. Ward, W. Giger, and P. L. McCarty (ed.), *Ground Water Quality*. John Wiley & Sons, Inc., New York.

McCoy, E. L., and C. Hagedorn. 1980. Transport of resistance-labeled *Escherichia coli* through a transition between two soils in a topographic sequence. *J. Environ. Qual.* **9:**686–691.

McDowell-Boyer, L. M., J. R. Hunt, and N. Sitar. 1986. Particle transport through porous media. *Water Resour. Res.* **22:**1901–1921.

McEldowney, S., and M. Fletcher. 1986. Effect of growth conditions and surface characteristics of aquatic bacteria on their attachment to solid surfaces. *J. Gen. Microbiol.* **132:**513–523.

Mills, A. L., G. M. Hornberger, J. S. Herman, J. E. Saiers, and D. E. Fontes. Bacterial transport in heterogeneous porous media. In C. B. Fliermans and T. C. Hazen (ed.), *Proceedings of the International Symposium on Deep Subsurface Microbiology*, in press. U.S. Department of Energy, Washington, D.C.

Molz, F. J., M. A. Widdowson, and L. D. Benefield. 1986. Simulation of microbial growth dynamics coupled to nutrient and oxygen transport in porous media. *Water Resour. Res.* **22**:1207–1216.

Montgomery, A. D., M. J. McInerney, and K. L. Sublette. 1990. Microbial control of the production of hydrogen sulfide by sulfate reducing bacteria. *Biotechnol. Bioeng.* **35**:533–539.

Myers, G. E., and W. D. Samiroden. 1967. Bacterial penetration in petroliferous rock. *Producers Monthly* **31**:22–25.

Raiders, R. A., M. J. McInerney, D. E. Revus, H. Torbati, R. M. Knapp, and G. E. Jenneman. 1986. Selectivity and depth of microbial plugging in Berea sandstone cores. *J. Ind. Microbiol.* **1**:195–203.

Rajagopalan, R., and C. Tien. 1976. Trajectory analysis of deep-bed filtration with sphere-in-cell porous media model. *Am. Inst. Chem. Eng. J.* **22**:523–533.

Raleigh, J. T., and D. L. Flock. 1965. A study of formation plugging with bacteria. *J. Petrol. Technol.* **17**:201–206.

Reynolds, P. J., P. Sharma, G. E. Jenneman, and M. J. McInerney. 1989. Mechanisms of microbial movement in subsurface materials. *Appl. Environ. Microbiol.* **55**:2280–2286.

Rittmann, B. E., and C. W. Brunner. 1984. The non-steady-state biofilm process for advanced organics removal. *J. Water Pollut. Control Fed.* **56**:874–880.

Rittmann, B. E., and P. L. McCarty. 1980a. Model of steady-state biofilm kinetics. *Biotechnol. Bioeng.* **22**:2343–2357.

Rittmann, B. E., and P. L. McCarty. 1980b. Evaluation of steady-state biofilm kinetics. *Biotechnol. Bioeng.* **22**:2359–2373.

Scholl, M. A., A. L. Mills, J. S. Herman, and G. M. Hornberger. The influence of minerology and solution chemistry on the attachment of bacteria to representative aquifer materials. *J. Contam. Hydrol.*, in press.

Shaw, D. J. 1976. *Introduction to Colloid and Surface Chemistry*, 2nd ed., p. 168–177. Butterworth Inc., Boston.

Shaw, J. C., B. Bramhill, N. C. Wardlaw, and J. W. Costerton. 1985. Bacterial fouling in a model core system. *Appl. Environ. Microbiol.* **49**:693–701.

Smith, M. S., G. W. Thomas, R. E. White, and D. Ritonga. 1985. Transport of *Escherichia coli* through intact and disturbed columns. *J. Environ. Qual.* **14**:87–91.

Stenstrom, T. A. 1989. Bacterial hydrophobicity, an overall parameter for the measurement of adhesion potential to soil particles. *Appl. Environ. Microbiol.* **55**:142–147.

Thorn, P. M., and R. M. Ventullo. 1988. Measurement of bacterial growth rates in subsurface sediments using the incorporation of tritiated thymidine into DNA. *Microb. Ecol.* **16**:3–16.

Tobiason, J. C. 1987. Physiochemical aspects of particle deposition in porous media. Ph.D. dissertation. Johns Hopkins University, Baltimore, Md.

Updegraff, D. M. 1983. Plugging and penetration of petroleum reservoir rock by microorganisms, p. 80–85. In E. C. Donaldson and J. B. Clark (ed.), *Proceedings of the 1982 International Conference on Microbial Enhancement of Oil Recovery*. Department of Energy, Bartlesville Technology Center, Bartlesville, Okla. (Available through National Technical Information Service [CONF 8205140].)

van Loosdrecht, M. C. M., J. Lyklema, W. Norde, G. Schraa, and A. J. B. Zehnder. 1987a. The role of bacterial cell wall hydrophobicity in adhesion. *Appl. Environ. Microbiol.* **53**:1893–1897.

van Loosdrecht, M. C. M., J. Lyklema, W. Norde, G. Schraa, and A. J. B. Zehnder. 1987b. Electrophoretic mobility and hydrophobicity as a measure to predict the initial steps of bacterial adhesion. *Appl. Environ. Microbiol.* **53**:1898–1901.

Wilson, J. T., J. F. McNabb, D. L. Balkwill, and W. C. Ghiorse. 1983a. Enumeration and characterization of bacteria indigenous to a shallow water-table aquifer. *Ground Water* **21**:134–142.

Wilson, J. T., J. F. McNabb, B. S. Wilson, and M. S. Noonan. 1983b. Biotransformation of selected organic pollutants in ground water. *Dev. Ind. Microbiol.* **24**:225–233.

Wollum, A. G., and D. K. Cassel. 1978. Transport of microorganisms in sand columns. *Soil Sci. Soc. Am. J.* **42**:72–76.

Wood, W. W., and G. G. Ehrlich. 1978. Use of baker's yeast to trace microbial movement in ground water. *Ground Water* **16**:398–403.

Yao, K.-M., M. T. Habibian, and C. R. O'Melia. 1971. Water and waste water filtration: concepts and applications. *Environ. Sci. Technol.* **5**:1105–1112.

Zhang, X. 1990. Mathematical modeling of microbially enhanced oil recovery. M.S. thesis. University of Oklahoma, Norman.

III. FATE IN THE SURFACE WORLD

Chapter 7

Using Linear and Polynomial Models To Examine the Environmental Stability of Viruses

Christon J. Hurst

Introduction	137
Experimental Design	138
Statistical Design	139
Linear Regression	139
Multiple Linear Regression	143
Polynomial Regression	156
Summary	157
References	159

Science is, of course, based upon careful experimentation and observation of produced results. Model equations can be of great value by helping to understand the interactions which led to the observed results and to provide a basis for anticipating the future outcome of similar situations. The most important thing to remember in developing models is that they are only estimates and have validity within specific ranges. Other important factors which need to be considered include the possibility that random variability which exists in the experimental data can interfere with the modeling effort, as well as the fact that interrelations between the independent variables either should be eliminated from the models or else should be recognized and their influence understood within the general framework of the developed models.

This chapter demonstrates the use of both linear and polynomial modeling in an effort to describe the influence of environmental factors and material characteristics upon the stability of viral infectivity in soils, fresh waters obtained from surface sources, and wastewater sludges examined both during the course of digestion treatment and following their disposal onto land surfaces. Virus populations undergo changes in titer during the time that they are suspended in these

---

Christon J. Hurst • U.S. Environmental Protection Agency, 26 Martin Luther King Drive West, Cincinnati, Ohio 45268.

materials. Those changes presented in this chapter generally reflect decreases in viral titer, since the experimental conditions used in developing the data were not conducive to the support of viral replication. In the case of the human enteric viruses, there was a lack of proper host cells within the sample materials that could serve to support viral replication. Conditions in the samples presumably were inadequate for maintaining a sufficiently high concentration of metabolically active host cells to support continuous replication of the bacterial viruses. Change in virus titer is used as the dependent variable for these models. The independent variables variously consist of experimental parameters such as temperature, physical or chemical characteristics inherent to the environmental materials, and loss of moisture in unsaturated environmental materials such as wastewater sludge and soil.

EXPERIMENTAL DESIGN

Variability within the data set being modeled is a crucial factor which can cause even the best-intended analysis to produce spurious or erroneous results. Some sources of variability can be eliminated by the use of appropriate techniques in designing and performing the experiments. In this regard, the design stage of a survival experiment is probably the most important. Techniques for eliminating unwanted error include following the fate of the viruses simultaneously in each of the sample materials being examined and under all of the possible combinations of the test conditions. This not only helps to eliminate the variation which would naturally occur if the combinations of conditions were followed at independent times or in different places, but also provides a factorial design which assures the acquisition of a complete set of data upon which to base statistical models.

Another important means of eliminating error at the design stage of survival experiments consists of reducing the amount of contaminating materials, such as deleterious chemicals or potential microbial nutrients, which may be introduced into the experimental system along with the seed viruses. In addition to general cleanliness, techniques that can be used for purification of the virus preparations themselves include gradient centrifugation and dialysis. Diluting the viruses in an appropriate suspending medium prior to placing them in the experimental system provides an additional means of reducing the possible influence of contaminants upon the outcome of the experiments. All samples must be treated in an identical manner, and this includes the performance of any dilutions or special handling procedures.

Many sources of variability cannot be eliminated, although the variation which they contribute to the body of data may be controlled to some extent through randomization. One such source of variability is sampling error. The material being studied should be thoroughly mixed before a sample is removed. Alternatively, if the material has been aliquoted prior to beginning the survival study, or if the quantity of material being studied is very large, then the samples must be collected in a randomized manner. In a large study, daily variability in the sensitivity of viral assay techniques can contribute a substantial amount of variation to the body of data. For this reason the order in which related samples are assayed may become critical. Viral samples should be assayed either on a randomized schedule or on a schedule in which all samples representing the same sampling date are assayed simultaneously.

STATISTICAL DESIGN

This chapter focuses on the use of linear regression to develop statistical models for describing changes in virus titer as a function of experimental parameters or sample characteristics. Statistical models can only approximate the relationship between the dependent variable and independent variables. Linear regression is a concept which is simple to understand and produces results that can be visualized readily and are easily compared. It must be understood that using a linear approach to develop models places rigid constraints upon those models, and also that linear models may not provide the best possible fit in terms of describing any particular set of data. The most important constraint involved in using linear models to describe microbial inactivation is the necessity of performing a logarithmic transformation on one of the variables, most commonly that part of the dependent (Y) variable which represents microbial titer. The survival models presented in this chapter employ $\log_{10}$ transformation of either the actual microbial titer itself (for equations 21 and 22 and Fig. 11) or the titer ratio N_t/N_0. N_t represents the titer at elapsed time t, and N_0 represents the titer at time 0, the outset of the study.

Linear Regression

This section presents use of the basic linear regression format for modeling the influence of a single independent variable upon the fate of viruses. The independent variable can be either an individually definable factor, such as temperature, or a combined set of factors which cannot be separated from one another, such as the effects of field drying. Equation 1 presents the format of the linear regression equations:

$$Y = \beta_0 + \beta_1 X \tag{1}$$

where Y represents the dependent value or response, β_0 represents the y-axis intercept, X represents the independent variable, and β_1 is the slope. While all of the models presented in this chapter are empirical in nature, it still must be recognized that there are questionable assumptions involved if statements of cause-and-effect relationships are made on the basis of these models.

Survival in wastewater sludge

Figure 1 presents the use of linear regression to describe the loss of titer for indigenous viruses contained in aerobically digested wastewater sludge following land surface disposal. The individual points shown on the graph represent the average value of two independent trials each for days 1, 2, and 7 and a single value for day 5. This plot demonstrates that the relationship between days of field drying and change in virus titer can be approximated reasonably well by using linear regression. In this example the independent variable, days drying in the field, represents a combined effect of several different factors which may include temperature and possible deleterious microbial or chemical influences in addition to the loss of sludge moisture content.

Tables 1 and 2 present the use of linear regression to compare two sets of data. Table 1 compares the data set generated by the study presented in Fig. 1 (Hurst et

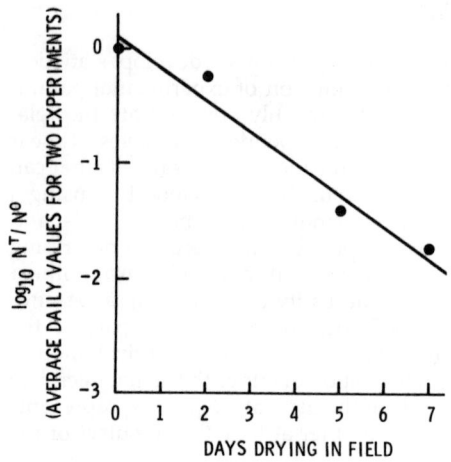

Figure 1. Loss of recoverable indigenous viruses during the field drying of aerobically digested wastewater sludge solids. Values presented are the averages of two studies (from Hurst et al., 1978).

al., 1978) with a data set generated by Farrah and co-workers (Farrah et al., 1981). Both of these studies examined the fate of indigenous viruses in land surface-disposed, aerobically digested sludge solids during the same season of the year in geographical areas that have roughly similar environmental conditions. The first two regressions in this table represent the use of changes in titer for the indigenous virus populations as the dependent variable (expressed as $\log_{10}$ transformed titer ratio values, $\log_{10} N_t/N_0$), against days of drying in the field as the independent variable for the individual data sets from the studies by Farrah et al. (first regression) and Hurst et al. (second regression). The last (third) regression listed in Table 1 shows the result obtained by pooling the sets of viral titer data from these two different studies and then regressing the log transformed titer ratio values in that pool as the dependent variable against time of drying in the field. This table also presents the respective Pearson correlation coefficients (r values) and statements of probability (P values) that the regression values expressed are different

Table 1. Regression equations describing data for inactivation of indigenous viruses following land surface spreading of sludge[a]

Description	Regression equation		r value for the regression	P value for the regression
	Slope	y Intercept		
Data from study by Farrah et al. (1981)	0.24325	0.20603	0.984	0.002
Data from study by Hurst et al. (1978)	0.28769	−0.10325	0.950	0.001
Both of the above, analyzed as a single group	0.26601	0.03114	0.959	<0.00001

[a] Data used in these analyses were presented in Table 20 of Hurst (1989). Linear regression analysis was performed with $\log_{10}$ reduction in titer since day of spreading as the dependent variable versus number of days since sludge was placed on land as the independent variable.

Table 2. Regression equations describing inactivation of viruses in anaerobically digesting sludge[a]

Description	Regression equation		r value for the regression	P value for the regression
	Slope	y Intercept		
Viruses suspended directly in anaerobically digesting sludge	0.420	−13.623	0.944	0.00004
Viruses adsorbed to filters immersed in anaerobically digesting sludge	4.076	−145.027	0.575	NS[b]
Both of the above, analyzed as a single group	14.617	−516.102	0.520	0.027

[a] Linear regression analyses were performed with rate of viral inactivation as the dependent variable versus sludge digestion temperature as the independent variable. The data used in these regressions were presented in Tables 9 and 10 of Hurst (1989).
[b] NS, Not significant ($P > 0.05$).

from zero. It can be seen that regression of the pooled data generates values for both slope and y intercept that are intermediate between the corresponding values yielded when the two sets of data are regressed individually. In addition, it can be seen that regression on the combined set of data still generates a strong r value and produces a better (lower) P value than does regression of the two sets of data independently. Thus, these two sets of data appear to be compatible.

The regressions presented in Table 2 demonstrate incompatibility of data sets. The first regression presented in Table 2 represents data developed for the influence of temperature upon the inactivation of viruses freely suspended in anaerobically digesting wastewater sludge. The second regression in Table 2 represents data developed for the influence of temperature upon viruses which are exposed to anaerobically digesting wastewater sludge while the viruses are adsorbed to a filter. Incompatibility between these two data sets is evidenced by the fact that when the two data sets are pooled and regressed together against temperature (third regression presented in Table 2), the resulting slope and y-intercept values are outside the range of values established by regressing the two sets of data independently. Also, despite the fact that the P value generated by regression of the pooled data is statistically significant (<0.05), it is far worse (higher) than that yielded by the first regression (viruses suspended directly in digesting sludge).

Survival in soil

Figure 2 outlines the basic approach used for examining the fate of seed viruses in different types of soil under a variety of environmental conditions. This illustration represents the study described by Hurst et al. (1980). Virus survival appears to be influenced by soil moisture level. For this reason, perhaps the most important aspect in the design of that study was for the experiments to be performed with the soil maintained under conditions where loss of soil moisture (drying) would be minimized. Minimizing moisture loss was accomplished by incubating the vials of seeded soil either in sealed GasPak anaerobic jars containing

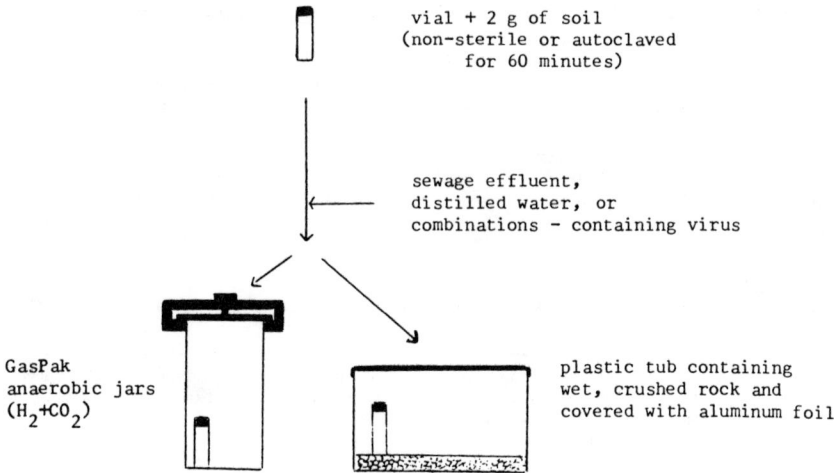

Figure 2. Procedure for following the survival of seeded viruses in soil.

wetted paper towels or in plastic tubs which were tightly closed and contained wet crushed rock.

Figure 3 describes the basis for a two-step linear regression analysis of survival data. By this approach, $\log_{10}$ transformed titer ratio values are first regressed against incubation time in days. The resulting slope values are then used as a dependent variable and regressed against either a single independent variable, such as soil moisture level or incubation temperature, or against a set of independent variables in a multiple regression. Use of the multiple regression approach will be described later in this chapter. The advantage of using this two-step regression technique is illustrated in Fig. 4 and 5. Figure 4 shows a plot of $\log_{10}$ titer ratio values versus incubation time in days for the survival of seed viruses at different moisture levels in a single soil type without any statistical analysis of the data. It is impossible to look at the data as presented in this figure and determine whether or not soil moisture level had any influence upon viral survival. Use of the two-step regression technique to analyze this same data yielded the graph shown in Fig. 5. From this latter plot, it can be seen that viral persistence (presented as survival slope values, which indicate the rate of $\log_{10}$ change in viral titer per day) seemed to be worse at a soil moisture level of 15% and that viral survival was better at moisture levels which were further to either side of this particular moisture level. This type of relationship between rate of change in viral titer and soil moisture content is not linear, and so it would not have been evident from simply examining the results from a linear regression of these survival slope values versus soil moisture level. However, the nature of the relationship immediately becomes apparent upon examining the scatterplot of the values (Fig. 5). The finding that viral survival was worst at a soil moisture level of 15% has potential importance in light of the fact that 15% is the approximate field saturation level for this particular soil. The two-step regression approach was also used when calculating the results shown in Table 2, which describe the inactivation of viruses in anaerobically digesting sludge. In the latter case, the product of the first step of the regression

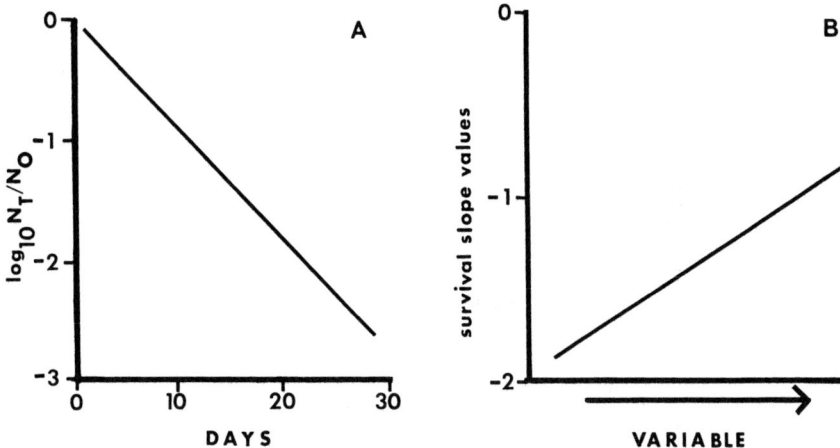

Figure 3. Format used for two-step regression analysis. (A) Survival slope determination: survival slope values expressed as the rate of $\log_{10}$ change in virus titer per day. (B) Survival slope values versus independent variable: the independent variables used in this stage of the analysis represent experimental conditions and may include temperature, in some instances, and soil or water characteristics. The example shown in panel B indicates a positive correlation between the variable and virus survival, with survival slope values decreasing as the value of the variable increases. Alternatively, the values resulting from the first step of the regression can be expressed as viral inactivation rates (expressed in terms of $\log_{10}$ units per day) by inverting the signs of the slope value and y-axis intercept. The inactivation rates would then be regressed against experimental conditions or soil or water characteristics in a similar manner as shown in panel B. (From Hurst et al., 1980.)

analysis was expressed in terms of viral inactivation rates (rate of $\log_{10}$ decrease in viral titer per day). Both survival slope values and viral inactivation rate values are expressed in the same units. The only difference between these values is the sign of their slope and y-axis intercept values. As such, these values are directly interconvertible. The only advantage of expressing these values as viral inactivation rates is that doing so eliminates the presence of "negative" signs, which would usually be needed when presenting survival slope values (due to the fact that the viruses do not multiply in the environmental samples but rather are inactivated). In choosing between these two forms of presentation, the viral inactivation rate format may be understood more readily by people who have limited previous experience with statistical analysis.

Multiple Linear Regression

The above section provided examples of how the second step of the two-step regression technique presented in Fig. 3 is useful for evaluating survival as a function of a single independent variable. This same approach can be expanded to incorporate more than one independent variable into the second step of the regression. When doing so, either survival slope value or inactivation rate value

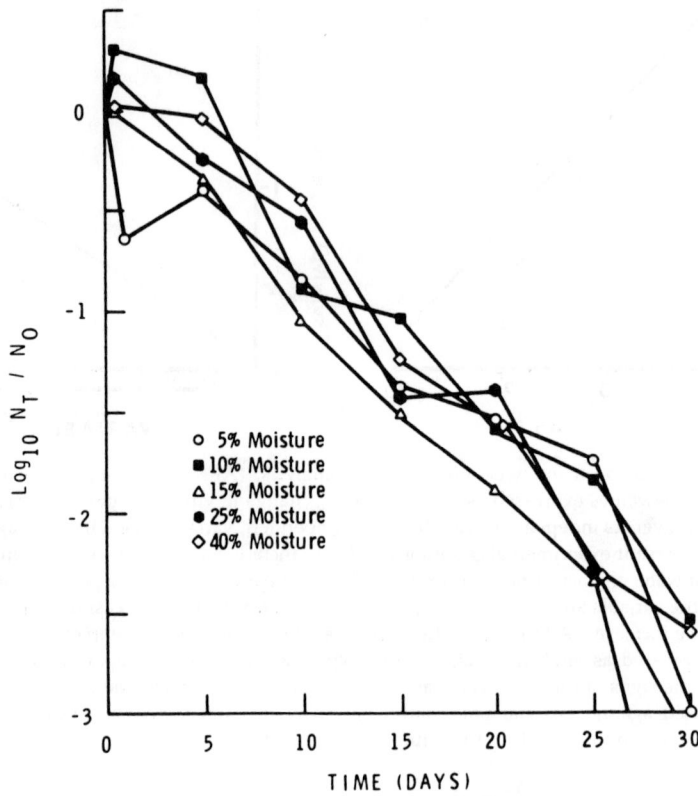

Figure 4. Plot showing daily changes in virus titer which occurred during storage of seeded poliovirus 1 in Flushing Meadows soil at 23°C with different levels of soil moisture content.

(whichever form of presentation is desired) is used as the dependent variable and regressed against a multiple number of independent variables that represent either experimental conditions or characteristics of the environmental material that is being studied. The format for multiple linear first-order regression is presented in equation 2.

$$Y = \beta_0 + \beta_1 X_1 + \beta_2 X_2 + \ldots + \beta_n X_n \qquad (2)$$

In such equations Y is the dependent, or response, variable and has the form of either survival slope values or viral inactivation rate values; β_0 represents the y-axis intercept; X_1 through X_n represent the different independent variables; and β_1 through β_n represent the slope coefficients assigned to those different independent variables.

The multiple regression equation is easiest to interpret if the different inde-

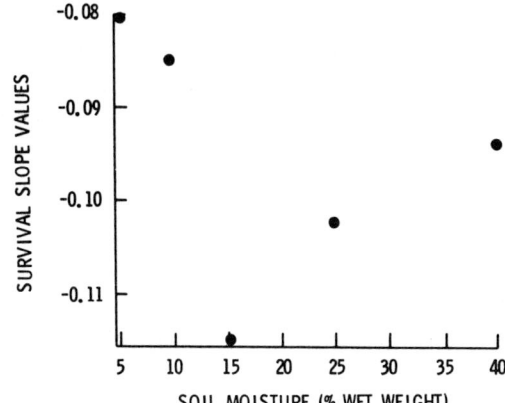

Figure 5. Scatter diagram of the survival slope values calculated from the data presented in Fig. 4 versus soil moisture content; $r = -0.31$, $P = 0.31$. (From Hurst et al., 1980.)

pendent variables are unrelated. However, in practice, this rarely happens. For this reason, when selecting candidate independent variables for inclusion into a multiple linear regression model, it is important to first examine the set of all possible independent variables for the presence of any strong cross correlations between them. This examination can be done quite easily by using Pearson's test. If two or more of the candidate variables correlate with one another in a highly significant manner, then it would probably be better to allow only one of those variables to be included in a multiple regression model. Having first examined the candidate independent variables for correlation with one another, and then having eliminated the potential problem of including highly cross-correlated variables in a model, it becomes time to select which of the remaining candidate variables will be incorporated into the actual model equation. All of the multiple linear regression models presented in this chapter were prepared by using a stepwise technique beginning with that candidate independent variable which demonstrated the strongest correlation with the dependent variable (either survival slope value or inactivation rate value). Additional individual independent variables were incorporated in a successive manner to improve the model fit. When developing such models it is helpful to establish criteria for deciding when to stop adding additional independent variables to the equation. The criterion used for the models presented in this chapter was to include at least one independent variable and to stop adding further variables when the next to be incorporated in sequence either would increase the P value for the overall model to above 0.05, would produce no improvement in the overall P value for the model, or would contribute an amount of less than 0.001 to the overall r value for the model.

Survival in soil

This section summarizes results obtained by modeling the survival of seed viruses incubated at a single temperature (23°C) in nine different characterized soils (Hurst et al., 1980). Table 3 summarizes the experimental design factors used for developing the data base that was modeled. The three different virus types listed in Table 3 (human poliovirus 1, bacteriophage MS-2, and bacteriophage T2) were inoculated independently into different samples of each soil. Viral survival was

Table 3. Experimental design factors for stability of seed viruses in soil[a]

Soil type	Virus type	Temp (°C)
Vernon	Poliovirus 1	23
Anthony	Bacteriophage MS-2	
Rubicond	Bacteriophage T2	
Windhorst		
Chigley		
Flushing Meadows		
Pomello		
Clarita		
Unclassified sample		

[a] Design factors for study described in Hurst et al. (1980).

then followed simultaneously for each of the possible virus-soil combinations under aerobic, nonsterile conditions. The soil moisture level was 15%, provided by adding a 50% solution of sewage effluent in distilled water to air-dried soil. The data for each of the possible combinations of virus type and soil type were thus obtained individually, and their survival slope values were derived independently (27 survival slope values in total). These survival slope values were then pooled to form a single data set for use in the modeling analysis. Figure 6 is an example of what the $\log_{10}$ transformed viral titer ratio values looked like when plotted versus incubation time in days without the benefit of statistical analysis. The information presented in Fig. 6 is for only one of the soil types, Flushing Meadows. This soil was of particular interest in the described study because it is of the same type employed at a municipal site near Phoenix, Ariz., used for the land surface discharge of wastewater effluent by rapid infiltration. The survival of four additional virus types was examined for this one soil although, to maintain consistency, only those data which represented either human poliovirus 1, bacteriophage MS-2, or bacteriophage T2 were included in the modeling effort. Table 4 lists the different chemical and physical soil characteristics that were considered for inclusion as independent variables in the multiple linear regression model developed to describe the survival of these three different virus types in the nine soils under the stated experimental conditions. The best stepwise regression fit for modeling virus survival based on these soil characteristics is presented in Table 5, with the actual equation presented as equation 3:

$$Y = 0.1005 + (0.0025)(\text{viral adsorption to soil}) \quad (3)$$
$$- (0.0008)(\text{extractable phosphorus})$$
$$- (0.0007)(\text{exchangeable aluminum})$$
$$- (0.0510)(\text{saturation pH})$$

This model includes four independent variables: the extent of viral adsorption to the particles of a test soil and the levels of resin-extractable phosphorus, exchangeable aluminum, and saturation pH for each of the soils. When looking at the model presented in Table 5, it is important to notice that stepwise regression analysis included exchangeable aluminum as an independent variable and entered it as th

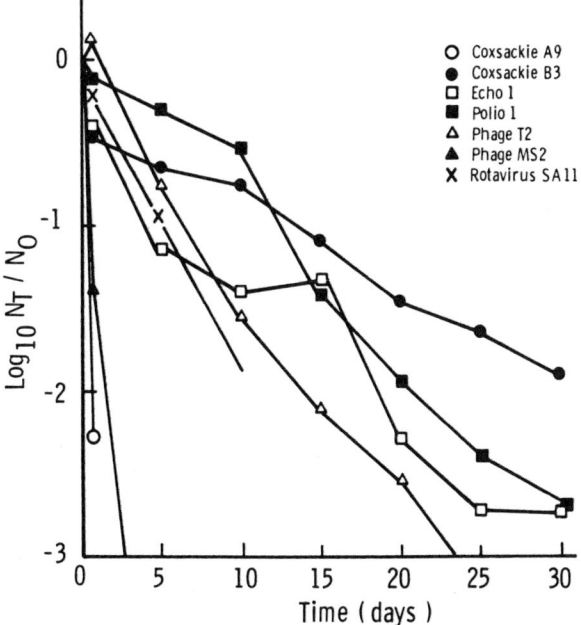

Figure 6. Comparative survival of different seeded viruses in Flushing Meadows soil. (From Hurst et al., 1980.)

third out of four independent variables. The individual P value for exchangeable aluminum when other independent variables are considered in the model is 0.257 and is therefore statistically nonsignificant (>0.05). Yet, the individual P value of the fourth variable entered in the model, saturation pH, is statistically significant (0.026). Eliminating either exchangeable aluminum alone, or both resin-extractable phosphorus and exchangeable aluminum, as candidate independent variables for inclusion in the model resulted in the variable of saturation pH becoming statistically nonsignificant in terms of describing viral survival. The explanation derived from this finding was that the apparent effect of pH upon viral survival resulted from pH-dependent changes in the availability of aluminum and phosphorus within the soil

Figures 7 through 10 are scatterplots of the pooled viral survival slope values representing the fate of the three test viruses in the nine different soils ($n = 27$) versus the four individual soil characteristics included in the multiple regression model for viral survival in soil (Table 5). The scatterplot of the pooled survival slope values as the dependent variable versus viral percent adsorption to soil as the independent variable (Fig. 7) shows a strong upward trend with increasing level of virus adsorption. This corresponds with the fact that the r value from a linear regression of these two variables using the same relationship (survival slope as dependent variable, adsorption to soil as independent variable) has a positive value (see legend of Fig. 7) and suggests that adsorption of viruses to the soil particles

Table 4. Soil characteristics evaluated as potential correlates with viral stability in soil[a]

Chemical	Physical
Organic matter (%)	Surface area (m^2/g)
Saturation pH	Clay (%)
Conductivity (μmhos/cm)	Silt (%)
Total phosphorus (ppm)	Sand (%)
Resin extractable phosphorus (ppm)	Extent of viral adsorption to the test soil
Total iron (%)	
Exchangeable iron (ppm)	
Total aluminum (%)	
Exchangeable aluminum (ppm)	
Total calcium (%)	
Exchangeable calcium (ppm)	
Total magnesium (%)	
Exchangeable magnesium (ppm)	
Cation exchange capacity (meq/100 g)	

[a] Soil characteristics evaluated for study described in Hurst et al. (1980).

was beneficial to viral survival. The scatterplot of survival slope values as dependent variable versus soil extractable phosphorus as the independent variable (Fig. 8) shows a slight downward trend with increasing level of extractable phosphorus. This corresponds with the fact that the r value derived by linear regression of these two variables in the same relationship has a negative value (see legend of Fig. 8) and suggests that extractable phosphorus is detrimental to viral survival. The scatterplot of survival slope values as the dependent variable versus soil exchangeable aluminum as the independent variable (Fig. 9) shows a slight upward trend with increasing level of exchangeable aluminum. This corresponds with the fact that the r value derived by linear regression of these two variables in the same relationship has a positive value (see legend of Fig. 9) and suggests that exchangeable aluminum is beneficial to viral survival. The scatterplot of survival

Table 5. Best stepwise regression fit for modeling virus survival versus soil characteristics[a]

Variable entered into model	Individual P value for variable	Change in overall r^2 achieved by adding this variable	Overall r^2 for model
X_1 = adsorption	0.010	0.235	0.235
X_2 = resin-extractable phosphorus	0.056	0.110	0.345
X_3 = exchangeable aluminum	0.257	0.036	0.381
X_4 = saturation pH	0.026	0.127	0.508

[a] The significance of this equation to describe the average survival for poliovirus 1, bacteriophage T2, and bacteriophage MS-2 in the examined soils at the study temperature (23°C) was $P = 0.003$. Further addition of variables did not improve (decrease) the P value associated with the overall model. The dependent variable for this model was survival slope, representing rate of change in virus titer expressed in terms of $\log_{10}$ units per day.

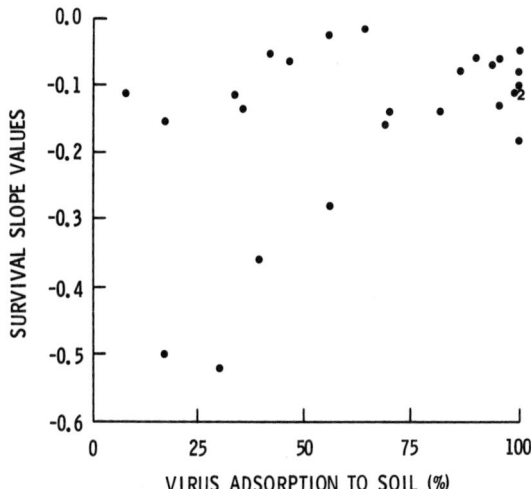

Figure 7. Scatter diagram of the survival slope values for survival of poliovirus 1 and the bacteriophages MS-2 and T2 at 23°C in nine different soils versus the level of virus adsorption to soil; $r = 0.48$, $P = 0.0052$. (From Hurst et al., 1980.) The number appearing within the field of data points indicates that more than one value plotted in that particular position.

slope values as the dependent variable versus soil saturation pH as the independent variable (Fig. 10) shows a strong downward trend with increasing level of pH. This corresponds with the fact that the r value from a linear regression of these two variables in the same relationship has a negative value and suggests that increased soil pH is associated with reduced viral survival.

It must be remembered that only one incubation temperature was used for this particular effort at modeling virus survival. Temperature is a crucial factor in viral survival, and the models presented in the next section will address the question of assessing models with respect to incubation temperature.

Survival in water

This section summarizes results obtained by modeling the survival of seeded viruses incubated independently at different temperatures (-20, 1, and 22°C) under aerobic, nonsterile conditions in samples of characterized fresh waters collected from different surface sources (Hurst et al., 1989a). Table 6 summarizes the experimental design factors used for developing the data base that was modeled. The three different virus types listed in Table 6 (human coxsackievirus B3, human echovirus 7, and human poliovirus 1) were inoculated independently into different samples of each water. The data for the three different viruses were thus obtained individually. The survival data from this study were modeled as viral inactivation rate values. There were 15 independently developed rate values for each incubation

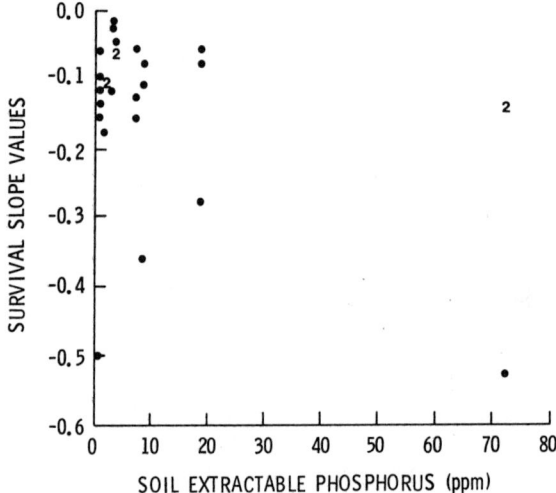

Figure 8. Scatter diagram of the survival slope values for survival of poliovirus 1 and the bacteriophages MS-2 and T2 at 23°C in nine different soils versus the level of soil resin-extractable phosphorus; $r = -0.34$, $P = 0.040$. (From Hurst et al., 1980.) Numbers appearing within the field of data points indicate that more than one value plotted in those particular positions.

temperature (water from 5 sources × 3 virus types), adding up to a pool of 45 rate values developed for the entire study (15 rate values per temperature × 3 temperatures). Pooling of the inactivation rate values into a single data set for the purpose of modeling is permissible since, again, virus survival was followed simultaneously for all possible combinations of incubation temperature, virus type, and suspending material (in this case, water). A list of the water characteristics evaluated as potential correlates with viral stability for the modeling of these data is presented in Table 7. Five of the characteristics listed in Table 7 correlated with viral inactivation rates. These characteristics were hardness, conductivity, turbidity, suspended solids, and the capacity of indigenous nutrients present in separate portions of the different water samples to support growth of seed bacteria following elimination of the indigenous bacterial population in the water by either pasteurization or sterile filtration. Of these five characteristics, hardness and conductivity so strongly correlated with one another that only one of the two would be allowed to occur in any given model. Likewise, turbidity and suspended solids so strongly correlated with one another that only one of these two would be allowed to occur in any given model.

Table 8 presents the regression which had the best fit for describing virus survival during incubation at 22°C. Two independent variables were included in this model, hardness and the mean of the number of generations of growth which the water could support, for three different bacterial species following their being independently seeded into sterilized or pasteurized samples of water from the

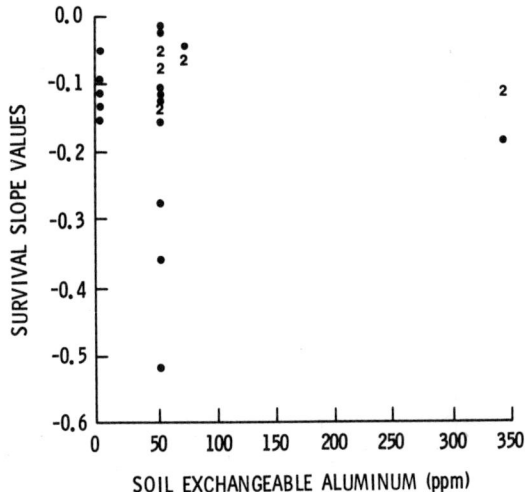

Figure 9. Scatter diagram of the survival slope values for survival of poliovirus 1 and the bacteriophages MS-2 and T2 at 23°C in nine different soils versus the level of soil exchangeable aluminum; $r = 0.08$, $P = 0.34$. (From Hurst et al., 1980.) Numbers appearing within the field of data points indicate that more than one value plotted in those particular positions.

different sources. This second parameter has been abbreviated as "Mean Gen PEK." The bacteria species used were *Pseudomonas fluorescens*, *Enterobacter cloacae*, and *Klebsiella oxytoca*. All three of these species are heterotrophic organisms common to water samples. Equation 4 represents the application of the regression model listed in Table 8 to describe viral inactivation in water incubated at 22°C. Equations 5 through 7, respectively, represent the application of this same regression to virus survival at 1°C, −20°C, and all three experimental incubation temperatures taken together as a single group.

$$Y = 2.04 \times 10^{-2} + (1.37 \times 10^{-3})(\text{hardness})$$
$$- (2.93 \times 10^{-3})(\text{Mean Gen PEK}) \quad [22°C] \quad (4)$$

$$Y = 3.50 \times 10^{-2} + (1.44 \times 10^{-5})(\text{hardness})$$
$$+ (1.10 \times 10^{-3})(\text{Mean Gen PEK}) \quad [1°C] \quad (5)$$

$$Y = 1.14 \times 10^{-2} + (5.50 \times 10^{-6})(\text{hardness})$$
$$- (6.52 \times 10^{-4})(\text{Mean Gen PEK}) \quad [-20°C] \quad (6)$$

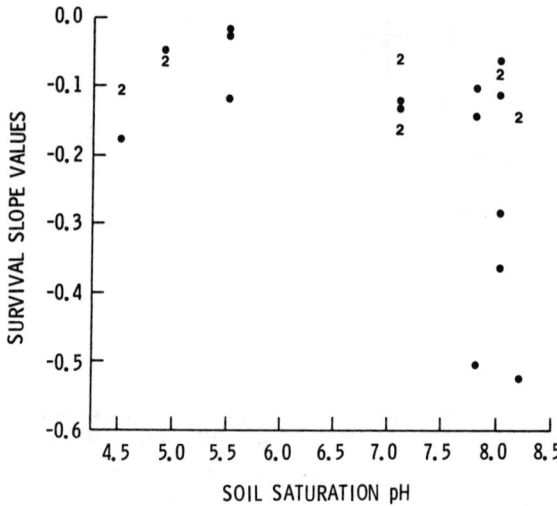

Figure 10. Scatter diagram of the survival slope values for survival of poliovirus 1 and the bacteriophages MS-2 and T2 at 23°C in nine different soils versus the level of soil saturation pH; $r = -0.39$, $P = 0.022$. (From Hurst et al., 1980.) Numbers appearing within the field of data points indicate that more than a single value plotted in those particular positions.

$$Y = 2.23 \times 10^{-2} + (4.64 \times 10^{-4})(\text{hardness})$$
$$- (8.27 \times 10^{-4})(\text{Mean Gen PEK}) \quad \text{[all three temperatures]} \quad (7)$$

As mentioned earlier in this chapter, temperature is a crucial factor in viral survival. The rate at which viral inactivation occurs increases with temperature. This fact provides a means to consider the potential appropriateness of a particular model to describe viral persistence in the different waters that were studied. When examining the equations in which a given model is applied to data from successively lower temperatures, it should be expected either that the corresponding

Table 6. Experimental design factors for stability of seed viruses in surface fresh waters[a]

Water source	Virus type	Temp (°C)
Burnet Pond (1.3 hectares)	Coxsackievirus B3	22
Winton Lake (1.5×10^2 hectares)	Echovirus 7	1
Nagel Creek (1.0×10^2 m³/day)	Poliovirus 1	−20
Little Miami River (6.9×10^4 m³/day)		
Ohio River (1.1×10^7 m³/day)		

[a] Design factors for study described by Hurst et al. (1989a, 1989b).

Table 7. Water characteristics evaluated as potential correlates with viral stability in surface fresh waters[a]

Chemical	Bacteriological	Physical
pH	Enterococci	Total solids
Chloride	Total coliform count	Suspended solids
Hardness (Ca+Mg)	Heterotrophic plate count	Dissolved solids
Turbidity		
Conductivity		
Total organic carbon		
Capability of supporting bacterial growth		

[a] Water characteristics evaluated for study described by Hurst et al. (1989a, 1989b).

values for β_0 (as defined in equation 2) will be numerically close to one another, indicating that the equations have a similar y-axis intercept, or that these values will decrease with the level of temperature that is represented. Also, as suggested by the regressions presented earlier for wastewater sludge, the β_0 value for a regression equation developed by applying a given model to the pooled data from all three incubation temperatures typically should lie within the range of values established by equations representing application of that same model to the three individual temperatures. Corresponding estimates predicted for any one of the other "betas" (i.e., β_1, β_2, or β_3) in equations which represent the application of a given model to data for successively lower temperatures should likewise be expected to show a decreasing trend in their numerical value, and the estimated value for that particular beta which represents application of the same model to pooled data for all three temperatures should be expected to lie within the range of values established by equations representing the individual temperatures.

The corresponding β_0 values for equations 4 through 6 are numerically close to one another, and the β_0 value for equation 7 is within the range established by

Table 8. Best stepwise regression fit for modeling virus survival versus water characteristics during incubation at 22°C[a,b]

Variable entered into model	Individual P value for variable	Change in overall r^2 achieved by adding this variable	Overall r^2 for model
X_1 = hardness	0.010	0.497	0.497
X_2 = mean generations of growth supported for P. fluorescens, E. cloacae, and K. oxytoca	0.041	0.002	0.499

[a] The significance (overall P value) of this equation to describe the average survival for coxsackievirus B3, echovirus 7, and poliovirus 1 in the examined waters at a temperature of 22°C was $P = 0.045$. Addition of any further variables made the value for overall $P > 0.05$. Values for performance of this same regression to represent incubation at other temperatures yielded: for 1°C, overall $r^2 = 0.102$ and overall $P = 0.616$; for −20°C, overall $r^2 = 0.124$ and overall $P = 0.550$; for all three temperatures in combination, overall $r^2 = 0.047$ and overall $P = 0.455$.
[b] The dependent variable for these models was the "inactivation rate" for the viruses, expressed in terms of $\log_{10}$ units per day.

Table 9. Best stepwise regression fit for modeling virus survival versus water characteristics during incubation at 1°C[a,b]

Variable entered into model	Individual P value for variable	Change in overall r^2 achieved by adding this variable	Overall r^2 for model
X_1 = turbidity	0.023	0.419	0.419
X_2 = generations of growth supported for K. oxytoca	0.232	0.090	0.509

[a] The significance (overall P value) of this equation to describe the average survival for coxsackievirus B3, echovirus 7, and poliovirus 1 in the examined waters at a temperature of 1°C was $P = 0.041$. Addition of any further variables made the value for overall $P > 0.05$. Values for performance of this same regression to represent incubation at other temperatures yielded: for 22°C, $r^2 = 0.432$ and overall $P = 0.079$; for −20°C, $r^2 = 0.350$ and overall $P = 0.144$; for all three temperatures in combination, $r^2 = 0.039$ and overall $P = 0.514$.
[b] See Table 8, footnote b.

equations 4 through 6. Similarly, it can be seen that the β_1 values (coefficients for hardness) of equations 4, 5, and 6 show a successive decrease in absolute numerical value as the incubation temperature which those equations represent decreases, and that the β_1 value from equation 7 lies within the range of β_1 values established by equations 4 through 6. The β_2 values (coefficients for Mean Gen PEK) of equations 4, 5, and 6 also show a successive decrease in absolute numerical value as the incubation temperature which those equations represent decreases, and the β_2 value from equation 7 lies within the range of β_2 values established by equations 4 through 6. These findings, derived by comparing the beta values for equation 7, which represent the pool of viral inactivation rate values from all three inactivation temperatures, with their corresponding values from equations 4 through 6, which represent the three successively lower individual incubation temperatures, improve confidence in the possible appropriateness of the model presented in Table 8 for describing viral survival in the examined waters.

The best regression fit for viruses incubated in water samples at 1°C is presented in Table 9. Two independent variables were included in this model, the first being turbidity measured as nephelometric turbidity units (NTU) and the second being the number of generations of growth observed for K. oxytoca seeded into sterilized water samples from the different sources. This latter parameter has been abbreviated as "Gen K." Equations 8 through 11 represent the application of the regression model listed in Table 9 to describe inactivation of viruses in water incubated at 22°C, 1°C, −20°C, and all three incubation temperatures taken together as a single group.

$$Y = 7.44 \times 10^{-5} + (2.76 \times 10^{-3})(NTU) + (1.75 \times 10^{-2})(Gen\ K) \quad [22°C] \quad (8)$$

$$Y = 4.99 \times 10^{-2} - (2.09 \times 10^{-4})(NTU) + (6.87 \times 10^{-4})(Gen\ K) \quad [1°C] \quad (9)$$

$$Y = 5.47 \times 10^{-3} + (4.79 \times 10^{-5})(NTU) - (2.28 \times 10^{-4})(Gen\ K) \quad [-20°C] \quad (10)$$

Table 10. Best stepwise regression fit for modeling virus survival versus water characteristics during incubation at $-20°C^{a,b}$

Variable entered into model	Individual P value for variable	Change in overall r^2 achieved by adding this variable	Overall r^2 for model
X_1 = turbidity	0.092	0.258	0.258

[a] The significance (overall P value) of this equation to describe the average survival for coxsackievirus B3, echovirus 7, and poliovirus 1 in the examined waters at a temperature of $-20°C$ was $P = 0.092$. Addition of any further variables increased the value for overall P. No regressions demonstrated overall significance of $P \leq 0.05$. Values for performance of this same regression to represent incubation at other temperatures yielded: for 1°C, $r^2 = 0.419$ and overall $P = 0.023$; for 22°C, $r^2 = 0.021$ and overall $P = 0.649$; for all three temperatures in combination, $r^2 = 0.001$ and overall $P = 0.854$.
[b] See Table 8, footnote b.

$$Y = 1.85 \times 10^{-2} + (8.65 \times 10^{-4})(NTU) + (6.00 \times 10^{-3}) \text{ (Gen K)}$$
[all three temperatures] (11)

The corresponding β_0 values for equations 8 through 10 do not show the expected temperature-related trend, since the β_0 value for 22°C (7×10^{-5}, equation 8) is smaller by almost two orders of magnitude than the lesser of the β_0 values for the lower incubation temperatures (5×10^{-3} for $-20°C$, equation 10). This gives a possible reason for preferring the model presented in Table 8 over that presented in Table 9 for describing viral survival in the examined waters.

Table 10 represents the best regression fit for viruses incubated in water samples at $-20°C$. Only one independent variable was included in this model, turbidity measured as NTU. Equations 12 through 15 represent the application of the regression model listed in Table 10 to describe inactivation of viruses in water incubated at 22°C, 1°C, $-20°C$, and all three incubation temperatures taken together as a single group.

$$Y = 1.69 \times 10^{-1} + (7.75 \times 10^{-4})(NTU) \quad [22°C] \quad (12)$$

$$Y = 5.65 \times 10^{-2} - (2.87 \times 10^{-4})(NTU) \quad [1°C] \quad (13)$$

$$Y = 3.27 \times 10^{-3} + (7.37 \times 10^{-5})(NTU) \quad [-20°C] \quad (14)$$

$$Y = 7.61 \times 10^{-2} + (1.87 \times 10^{-4})(NTU) \quad [\text{all three temperatures}] \quad (15)$$

The corresponding β_0 values, and also the β_1 values, for equations 12 through 14 do show the expected temperature-related trends. This, plus the fact that the β_0 and β_1 values from equation 15 fall within the ranges established by their corresponding values in equations 12 through 14, gives confidence in the appropriateness of the model presented in Table 10 for describing viral survival in the examined waters.

Table 11 represents the best regression fit for viruses in water samples representing all three incubation temperatures. Two independent variables are

Table 11. Best stepwise regression fit for modeling virus survival versus water characteristics and incubation temperature (data for all three temperatures combined)[a,b]

Variable entered into model	Individual P value for variable	Change in overall r^2 achieved by adding this variable	Overall r^2 for model
X_1 = incubation temperature	<1.0 × 10^{-8}	0.670	0.670
X_2 = hardness	0.206	0.046	0.716

[a] The significance (overall P value) of this equation to describe the average survival for coxsackievirus B3, echovirus 7, and poliovirus 1 in the examined waters, across the range of incubation temperatures −20 to 22°C, was P = <1.0 × 10^{-8}. Addition of any further variables made a negligible increase in r^2 (<0.001).
[b] See Table 8, footnote b.

included in this model, the first being incubation temperature in degrees Celsius, abbreviated as "T" in the equations for this study, and the second independent variable being hardness. The application of this regression model to describe inactivation of viruses in water over the entire experimental range of −20 through 22°C is represented by equation 16. In recognition of the overwhelming importance of temperature in describing viral inactivation rates, equation 17 represents the regression of inactivation rate versus incubation temperature alone over the full experimental range of −20 through 22°C.

$$Y = 1.22 \times 10^{-2} + (4.37 \times 10^{-3})(T) + (4.42 \times 10^{-4})(\text{hardness}) \quad (16)$$

$$Y = 7.66 \times 10^{-2} + (4.37 \times 10^{-3})(T) \quad (17)$$

Polynomial Regression

The above sections of this chapter have addressed the application of models based upon linear regression analysis to describe the fate of viruses in wastewater sludges, soils, and water. There are instances when polynomial regression models may be more appropriate for describing a particular effect. Two examples of this are presented below.

An example of the use of a quadratic (second-order polynomial) equation to describe the fate of viruses has been presented by Hurst and Brashear (1987), who attempted to describe the leaching of indigenous viruses from wastewater sludge solids. The basic form of a second-order polynomial equation is shown in equation 18.

$$Y = \beta_0 + \beta_1 X + \beta_2 X^2 \quad (18)$$

Hurst and Brashear used as their dependent variable the ratio of the level of viruses in nascent leachate fluid collected from wastewater sludges by vacuum filtration to the level of viruses that could be recovered by subsequently passing volumes of test fluids through the same sludge samples. The volumes of test fluid were likewise collected by vacuum filtration following passage through the sludge. The independent variable used in the equation of Hurst and Brashear was the percent solids content of those sludge samples that were subjected to the experimental procedure. The equation developed by Hurst and Brashear is presented as equation 19:

$$Y = 43.00 - (30.15)(X) + (5.76)(X^2) \tag{19}$$

($P = 0.0001$; no r value available) and was shown to be better based on the P value and an examination of residuals than was a linear equation evaluated for describing the same effect (Hurst and Brashear, 1987).

R. Sullivan (personal communication) has compared the fit of both a linear regression equation (equation 1) and a cubic (third-order polynomial) regression equation (equation 20):

$$Y = \beta_0 + \beta_1 X + \beta_2 X^2 + \beta_3 X^3 \tag{20}$$

to describe the inactivation of seeded *Escherichia coli* organisms in primary sewage effluent. The regression equations developed for describing one such set of data are presented as equation 21 (linear) and equation 22 (cubic).

$$Y = 3.38 - (2.70 \times 10^{-2})(X) \tag{21}$$

($r = 0.868$; no P value available)

$$Y = 4.11 - (1.23 \times 10^{-1})(X) + (1.96 \times 10^{-3})(X^2) - (1.10 \times 10^{-5})(X^3) \tag{22}$$

($r = 0.998$; no P value available). For equations 21 and 22, the dependent variable is the bacterial level in the sample being disinfected, expressed as colony-forming units per milliliter. The independent variable for these two equations is the length of time, expressed in minutes, during which the bacterial organisms were suspended in the disinfectant solution (monochloramine formed in situ). Figure 11 presents a visual comparison of the outcome from this modeling effort and suggests that a polynomial regression equation may come closer to fitting such data than does a first-order equation. Use of a third-order polynomial may be overfitting this particular data set, due to the limited sample size that is involved. Unfortunately, a quadratic model was not evaluated for these data, and the appropriateness of using a cubic model cannot be determined accurately without knowledge of how well a quadratic model would fit the same data.

SUMMARY

This chapter has presented the development of model equations for describing the fate of viral infectivity in environmental samples. Most of the presented models were based upon the use of a two-step linear regression approach. The first step employs regression of $\log_{10}$ transformed viral titer ratios from various sampling dates as the dependent variable, versus the length of time that the viruses were incubated in the test material as the independent variable. The slope values derived from this first step of the regression technique are then used as the dependent variable in the second step of the analysis, when they are linearly regressed against either a single independent variable, such as soil moisture level or incubation

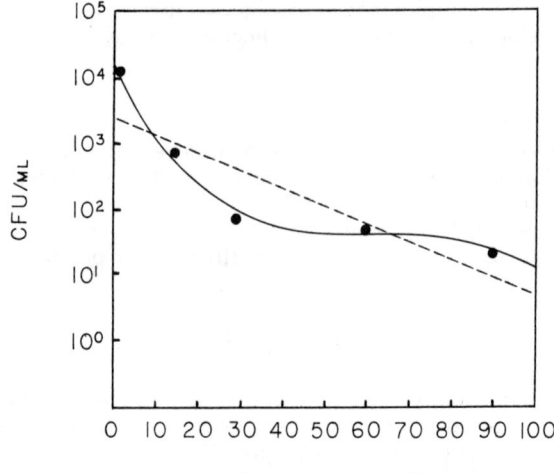

Figure 11. Comparison of linear regression (as per equation 1) versus cubic regression using a third-order polynomial (as per equation 20) to describe the timewise loss in viability of seeded *E. coli* at 10 to 15°C in primary sewage effluent containing monochloramine disinfectant. The dashed line represents the linear regression (equation 21) with $r = 0.868$ and no P value available; the solid line represents cubic regression (equation 22) with $r = 0.998$ and no P value available (Sullivan, personal communication).

temperature, or against a set of independent variables in a multiple regression. A variety of examples based upon experimental data were used to demonstrate the application and benefits of this two-step regression technique.

A method was demonstrated for determining if two or more sets of data (representing the same dependent variable) were compatible with one another. This method was based upon individually regressing the sets of data against their common independent variable and then pooling those sets of data and regressing that entire data pool against the independent variable. For those demonstrations of this technique that were presented in this chapter, the relationship between the dependent and independent variables was assumed to be linear in nature. Most of the models presented in this chapter were of first order, although examples were also provided of instances when higher-order polynomial regression equations proved superior to first-order equations.

It must be understood that linear regression imposes constraints upon the developed model equations and will not always provide the best fit in terms of describing the relationship between the dependent and independent variables. Nevertheless, linear regression can be a very valuable tool in the study of microbial activity. The subject of nonlinear models is addressed in several of the other chapters in this book.

REFERENCES

Farrah, S. R., G. Bitton, E. M. Hoffmann, O. Lanni, O. C. Pancorbo, M. C. Lutrick, and J. E. Bertrand. 1981. Survival of enteroviruses and coliform bacteria in a sludge lagoon. *Appl. Environ. Microbiol.* **41**:459–465.

Hurst, C. J. 1989. Fate of viruses during wastewater sludge treatment processes. *Crit. Rev. Environ. Control* **18**:317–343.

Hurst, C. J., W. H. Benton, and K. A. McClellan. 1989a. Thermal and water source effects upon the stability of enteroviruses in surface freshwaters. *Can. J. Microbiol.* **35**:474–480.

Hurst, C. J., W. H. Benton, and K. A. McClellan. 1989b. Influence of temperature and intrinsic characteristics of surface freshwaters on virus stability, p. 69–79. *In* Proceedings of the Conference on Microbial Aspects of Surface Water Quality, Chicago, Ill., 30 May to 2 June 1989. Water Pollution Control Federation, Alexandria, Va.

Hurst, C. J., and D. A. Brashear. 1987. Use of a vacuum filtration technique to study leaching of indigenous viruses from raw wastewater sludge. *Water Res.* **21**:809–812.

Hurst, C. J., S. R. Farrah, C. P. Gerba, and J. L. Melnick. 1978. Development of quantitative methods for the detection of enteroviruses in sewage sludges during activation and following land disposal. *Appl. Environ. Microbiol.* **36**:81–89.

Hurst, C. J., C. P. Gerba, and I. Cech. 1980. Effects of environmental variables and soil characteristics on virus survival in soil. *Appl. Environ. Microbiol.* **40**:1067–1079.

Chapter 8

Development of Models To Explain the Survival of Viruses and Bacteria in Aerosols

Alan Jeff Mohr

Introduction	160
Physical Properties	162
Particle Size	162
Meteorological Conditions	163
Relative Humidity	163
Temperature	164
Oxygen Concentration	164
Electromagnetic Radiation	164
Open Air Factor	164
Miscellaneous	165
Biological Composition of Microorganisms	165
Microbial Aerosol Stability	166
Bacteria	167
Viruses	170
Bacteriophages	173
Modeling Microbial Aerosols	176
Viability Models	179
Dispersion Models	182
Conclusion	185
References	186

Aerobiology is the study of microbiological particles that have found their way to the airborne state. The significance of biological aerosols to humans depends primarily upon the site of deposition of the aerosol, with the most important being the respiratory tract. Important parameters that influence the infectious nature of airborne microorganisms are their age, origin, and concentration, the meteorological conditions to which they are subjected, and physical properties such as particle size and composition.

Alan Jeff Mohr • Dugway Proving Ground, Dugway, Utah 84022-5000.

Aerobiology is a relatively new discipline which incorporates the fields of microbiology, meteorology, and aerosol physics and engineering. Biological aerosols were not recognized as sources of disease until the early 1930s (Trillat and Beauvillain, 1937), and it was not until the military developed generation and sampling methods during World War II that the nature of the aerosol spread of infection (Rosebury, 1947) began to be understood. Rabies virus was not known to be spread by aerosols until the late 1960s (Constantine, 1967). Only recently has it been determined that oxygen concentration (Cox and Baldwin, 1967) and prehumidification at sampling (Cox, 1966; Hatch and Warren, 1969) can have a significant impact on the overall stability of microbial aerosols. Prehumidification is the vapor-phase rehydration of aerosol particles prior to sampling.

There are relatively few reference sources dealing directly with aerobiology, and often, in the open literature, there are disagreements on the methods of study and on the interpretation of results (Cox, 1987). Application of new technology for systems control of temperature and humidity and the use of computers for modeling promise a better understanding of the parameters that influence inactivation and dispersion in the airborne state.

Biological aerosols are formed naturally from coughing and sneezing and by wind and wave action upon water surfaces (Woodcock, 1955) where microbial matter is present. Aerosols are also produced by anthropogenic means, including the recycling of wastewater for spray irrigation (Teltsch and Katzenelson, 1978; Moore et al., 1979; Bausum et al., 1982), cooling towers (Adams et al., 1979), and wastewater treatment processes including trickle irrigation (Goff et al., 1973) and activated sludge (Fannin et al., 1976). Microbial aerosols are also being generated by humans to control insects (Cook and Baker, 1983), other microorganisms (Thomson et al., 1976), frost damage to plants (Lindow, 1985, 1986), and the weather (bacteria used to increase snow production).

Infectious aerosols have the potential of affecting the unsuspecting population, including health and laboratory workers (Pike, 1978), as well as important agricultural products, plants, and animals. Representatives from different classes of viruses, bacteria, rickettsia, molds, and fungi are all capable of initiating disease and having a significant impact (Pike et al., 1965). Eighty percent of laboratory-acquired infections are caused by unconfirmed sources, but uncontained aerosols are the presumed cause (Richardson and Barkley, 1984). Notable reviews are available that address mechanisms of inactivation for viruses (Spendlove and Fannin, 1982; Akers, 1973) and bacteria (Anderson and Cox, 1967). Many of these review articles deal with organisms that pose no threat upon aerosolization but which were convenient for study purposes.

Many infectious microorganisms are of little interest to aerobiologists because of their poor aerosol stability and subsequent rapid die-off rate. When evaluating an aerosolized microorganism for possible health effects or infectivity, it is most important to determine its survival characteristics under various environmental conditions. The goal of this review is to define and explain the various parameters that influence the aerosol stability of microorganisms, to provide insights for improvement of methodologies which may be applied to dispersion and viability modeling efforts. Generally, factors that influence aerosol stability include physical properties and constraints of aerosols, the environmental conditions to which the aerosol is exposed, and the biological composition of the microorganism.

PHYSICAL PROPERTIES

Aerobiologists are faced with especially intriguing questions as compared with those encountered by toxicologists, industrial hygienists, or physicians, who often challenge test subjects utilizing inanimate aerosol particles. Besides the forces that influence basic physical phenomena like particle size and meteorological effects, the aerobiologist must take into account the viability of microorganisms. Aerobiologists are primarily concerned with particles that are physically capable of entering the respiratory system. In practice, these are particles with aerodynamic diameters between 0.5 and 10.0 μm, although the upper size limit is sometimes extended to 15 to 20 μm (Phalen, 1984). Particles with aerodynamic diameters between 1.0 and 3.0 μm are of primary concern. Aerodynamic diameter is the diameter of a sphere of unit density (1 g/ml) that has the same settling velocity as the particle in question. From this definition, it is easy to see that aerodynamic diameter is dependent on particle size, shape, and density.

Particle Size

Deposition of particles in the human respiratory tract is a function of the aerodynamic particle size. Forces that influence the behavior of an aerosol particle are dependent on the particle size. Deposition of particles can be placed in two important groups (diffusional and inertial) depending on particle dimensions. Diffusional properties include electrical precipitation, charge interactions, interception, and Brownian diffusion. Inertial properties include impaction effects and gravitational settling. Under normal breathing conditions, deposition by inertial impaction is predominant (gravitational settling is secondary) for particles with aerodynamic diameters between 2.5 and 7.0 μm. The major depositional sites are the nasal and oral air passages and the tracheobronchial airways. Deposition and impaction are very dependent on the shape and flow of the individual airway.

Deposition by diffusion is time dependent, and consequently, its efficiency increases with residence time in the airway. For particles with aerodynamic diameters between 0.5 and 2.5 μm, deposition is primarily due to diffusional properties. Brownian diffusion is the primary process, followed by interception and electrical interactions.

The following is a brief discussion of the physical forces which govern the aerodynamic motion of airborne particles.

Brownian diffusion

Deposition and motion increase with temperature and decreasing particle size. Diffusion is important for small particles (around 0.5 μm). The smallest particles strictly follow streamlines (imaginary lines of flow) but may diffuse close enough to an object to be deposited.

Interception

Interception is a diffusional property which acts on particles between 1.0 and 2.5 μm in diameter. If a particle's momentum is sufficient to cross the streamlines (as might occur when an aerosol is required to make a sudden change in direction which could be caused by an obstruction or a bend in a pipe), the particle may contact an object and be lost from the aerosol flow. Laminar flow should be obtained whenever a test system is designed. (A particle in laminar flow follows the precise flow of its predecessors because of the absence of mixing of adjacent layers of air.) Particle loss by inertial impaction can be kept to a minimum by decreasing turbulence and mixing of air streamlines. When a particle is following an air streamline and comes to within a distance of approximately one-half of its diameter to an object, it may be intercepted and lost to the air flow. As the physical size of a particle increases, interception loses importance to inertial properties.

Electrical effects

Substantial electrical gradients are imposed on particles during generation in the laboratory. Particle velocity is increased by net electric charge. Coagulation is increased by induced charging. Physical loss is often increased by charging. Charge neutralization using radioactive sources (such as krypton) can significantly reduce charge effects.

Gravity

Gravity is an inertial property in which deposition depends on aerodynamic diameter. Gravitational deposition is secondary to impaction for particles with diameters between 1.0 and 4.0 μm. Large particles settle much faster than smaller particles. The time it takes a 10.0-μm sphere to fall 1 m is 5.1 min (terminal settling velocity), while a 1.0-μm sphere requires 8.4 h to fall 1 m (density = 1.1 g/ml with no other forces acting).

Impaction

Impaction is also an inertial property which depends on aerodynamic diameter. When aerosols are sampled, inertial losses can be substantial in poorly designed collectors.

METEOROLOGICAL CONDITIONS

It is difficult to separate meteorological conditions from the above physical properties because both are fundamentally physical in nature. Separation is justified here because, when modeling survival and dispersion in the airborne state, meteorological parameters are most often measured (along with particle size) and used as input. Also these parameters can be important in determining aerosol survival. Following are the most frequently used meteorological model inputs (excluding wind data and particle size), listed in decreasing order of importance: relative humidity (RH), temperature, oxygen concentration and free radicals, electromagnetic radiation, and open air factor (OAF).

Relative Humidity

RH is defined as the vapor pressure divided by the saturation vapor pressure with respect to water at the same temperature on a percent basis (McIntosh, 1972). For aerobiologists, RH is the water content or water equilibrium state associated with an aerosol particle. Of all of the measurable meteorological parameters, RH is the most important with respect to aerosol stability (Cox, 1987). Not only is the water content of aerosolized microorganisms a principal determinant of aerosol viability, but the RH of a system will directly affect density, and hence mass and size, which will influence particle size, location of deposition in the respiratory tract, and aerosol sampling results. The effects of RH can also be influenced by the content of the suspension fluid used before aerosolization (Barlow, 1972), the content of the collection fluid (Cabelli, 1962), and prehumidification (Warren et al., 1969). For some microorganisms, shifts in RH after aerosolization have a more profound effect on aerosol stability than does constant RH (Hatch and Dimmick, 1966). These factors will be examined later for the different classification of microorganisms.

Temperature

The vapor pressure, and therefore the RH, of a system is dependent on the temperature, making it somewhat difficult to completely separate individual effects. Results of studies which hold some variable constant while varying the others demonstrate the relative importance of RH. Studies to determine the effect of temperature on aerosol stability have generally shown that increases in temperature tend to decrease the viability of airborne microorganisms (Dimmock, 1967).

Oxygen Concentration

Oxygen has an effect on the aerosol stability and infectivity of some bacteria (Cox et al., 1973). Free radicals of oxygen have been suggested as the cause of this inactivation (Cox, 1987), but negative correlations between oxygen concentration and viability have been found by some investigators (Cox, 1987).

Electromagnetic Radiation

Aerosol inactivation caused by electromagnetic radiation has been shown to be dependent on the wavelength and, hence, the intensity of the radiation. Shorter wavelengths contain more energy and are generally more deleterious to aerosolized microorganisms. RH (Beebe, 1959), water activity (Bridges, 1976; Krinsky, 1976), oxygen concentration (Bridges, 1976), age of the aerosol (Cox, 1987), and the presence of other gases all influence the effect radiation exerts on airborne microorganisms. Short-wave ionizing radiation (X rays, gamma rays) can cause breaks in the DNA of microorganisms (Bridges, 1976). UV radiation acts as an energy source for the production of thymidine dimers. Longer wavelengths in the visible region have been shown to affect cytochromes in the mitochondria of yeasts and bacteria (Krinsky, 1976). Other studies have shown that samplers set out to monitor bacteria from airborne emissions at sewage plants consistently yielded greater numbers at night (Goff et al., 1973).

Open Air Factor

OAF is a term that arose from studies (Druett, 1973b) which showed that inactivation rates for many biological aerosols were significantly increased when the aerosols were challenged with high-efficiency particulate-arresting (HEPA)- filtered air from outdoors, as compared with clean, inert laboratory-supplied air. Cox (1987) suggests that inactivation is primarily caused by reactions between ozone and olefins. OAF inactivation is probably caused by a combination of factors including pollutant concentration, RH, pressure fluctuations, and air ions.

Miscellaneous

Review articles by Cox (1987), Strange and Cox (1976), and Spendlove and Fannin (1982) all address the influences that pollutants, pressure fluctuations, and season may have on aerosol inactivation rates. Atmospheric pollutants which have been studied the most include nitrogen dioxide (NO_2), sulfur dioxide (SO_2), and ozone (O_3). The deleterious effects of these pollutants are greatly influenced by RH. Some microorganisms display greater loss of viability at high RH, while others show the opposite effect at high RH. Some pollutants may react with water to form acids (Ehrlich and Miller, 1971), and the pH of the environment may play a role in acid production and hence inactivation of aerosolized microorganisms. Other pollutants that have been studied include HCOH, CO, HCl, HF, C_2H_2, C_2H_4, and C_2H_8 (Cox et al., 1973; de Mik and de Groot, 1973). These compounds are much less toxic than OAF.

Druett (1973a) showed that when *Escherichia coli* was subjected to rapid pressure fluctuations, there was an enhanced increase in the aerosol decay rate which was related to the denaturation of cell wall components. Certain diseases are more prevalent during particular times of the year; influenza, polio, and foot-and-mouth disease (Smith and Hugh-Jones, 1969) are just a few. This may be due to the presence of the optimum temperatures and RHs which follow natural seasonal variations (Spendlove and Fannin, 1982). Hemmes et al. (1960) hypothesized that low RH in the winter favors survival of influenza virus, while poliovirus survival is enhanced in the summer when RH is higher.

BIOLOGICAL COMPOSITION OF MICROORGANISMS

The relationships between aerosol stability and the biological composition of microorganisms are not well understood. Buckland and Tyrrell (1964) first speculated that virus survival in aerosols might be related to the amounts of lipids in the capsid (envelope or outer shell). There does appear to be a greater aerosol stability associated with higher-lipid-containing viruses at low RH, while viruses with little or no lipid content are more stable at higher RHs. Many viruses exhibit higher survival rates at midrange RH (Harper, 1961), regardless of temperature, while some display better survival at low and high RHs and the lowest survival at midrange RH (Adams et al., 1979; Mohr, 1984). However, survival may depend on the temperature and oxygen content of the test atmosphere at the time of testing. There are some viruses that are stable in the airborne state over broad temperature and RH ranges (Akers et al., 1966).

Bacterial aerosol stability is much more complex than that observed for viruses. Cox (1968) set up an elaborate series of stability experiments to examine the factors that influenced inactivation of *E. coli* B. When *E. coli* B was aerosolized from a suspension of distilled water into highly purified nitrogen, argon, or helium atmospheres, it displayed similar patterns of survival as a function of RH. Under these conditions, survival was virtually complete at low RH, but was critically dependent on RH above values of 80%. Cox (1987) speculated that, under these conditions, the gases slightly modify the water structure through gas hydrate formation and water lattice modification, which affects the stability of biological structures. Cox (1987) then explained the events that may take place as precursors to aerosol inactivation, as follows. When *E. coli* is aerosolized into inert atmospheres at midrange to high RH, the biological membrane constituents become destabilized through loss of water molecules. Additives that supersaturate, such as polyhydroxyl compounds, can stabilize these structures. The polyhydroxyl compounds, by binding to sites on proteins, cause conformational changes in them and thereby stabilize proteins, making them less susceptible to denaturation. This is convincing evidence that the state of proteins on the outer membrane of some microorganisms is critical to the resultant stability profile.

Cox (1987) also explained that Maillard reactions between reducing sugars and amino acids are responsible for some denaturation and unfolding of ovalbumin α-helix which occurs during drying. Additionally, he explained that dried proteins may be stabilized by similar reactions between other sugars and sodium glutamate. Soddu and Vieth (1980) showed that sucrose can bind to collagen membranes with different affinities and produce conformational changes in the protein structure. Cox's (1987) explanation of the sequence of events is as follows. There is little doubt that during the desiccation process, polyhydroxyl compounds and amino acids react together, causing conformational changes that strengthen the overall protein structure. The presence of sugar additives causes conformational changes in the coat proteins, and in the new configuration coat proteins do not react (or react more slowly) with the polyhydroxy coat moieties. In the absence of these sugar additives and free molecules, the coat proteins may react irreversibly, through Maillard reactions, with polyhydroxy coat moieties and cause loss of viability. In addition, the sugar additives could compete with the polyhydroxy coat moieties for the reaction sites of the coat proteins or physically hinder the molecular collisions at those sites. The result in each case would be more aerostable microorganisms. When microorganisms are in their more normal aqueous environment, the Maillard reactions leading to this inactivation may not occur because the reaction sites are separated either by bulk water molecules or by water molecules bound at the reactive sites. Removal of these water molecules by evaporation would then lead to the events proposed above. This possible mechanism for aerosol inactivation has been deduced by inference from mostly unrelated work. There have been many studies on solute concentration effects which may also be applied to help model the molecular events which take place during inactivation. In a practical sense, the process may be viewed as the increase of solute concentration which concomitantly occurs during drying of the aerosol droplet.

MICROBIAL AEROSOL STABILITY

The experimental determination of aerosol stability is dependent on the ability to assay the aerosol for biological activity. Many very sensitive procedures are available which take advantage of the specificity of antigen-antibody reactions. The problem with these assays is that they reveal very little about the viability of the microorganism in question. Spendlove and Fannin (1982) define virus survival or stability in the aerosol particle as a measure of the ability of the virus to infect tissue culture cells of living hosts. For this reason, infectivity and stability should be treated as a single entity.

This section will present data and results from various aerosol stability or infectivity studies on bacteria, animal viruses, bacterial viruses (bacteriophages), and other microorganisms. As stated earlier, the results for many aerosol stability tests performed on the same microorganism often differ because testing techniques have not been standardized. Harper (1963) noted that variations in results can be caused by (i) method of aerosol generation, storage, and sampling procedure; (ii) method of assay; (iii) differentiation of total, physical, and viability decay; (iv) presence or absence of light; (v) methods and extent to which RH and temperature are controlled; and (vi) method of data presentation.

Additionally, variations can occur because of choice of suspending fluid, collection fluid, and content of atmosphere (presence or absence of oxygen and other gases) used for testing.

Bacteria

More variation in aerosol test results has been observed for bacteria than for viruses, phages, and other microorganisms. This is due not only to differences in test procedures and data presentation but also to the greater structural and metabolic complexity of bacteria (cell walls, membranes, and metabolism) as compared with viruses. Much of the published data available cannot be used for comparing aerosol stabilities because of the presence of several stresses which could have acted synergistically in a detrimental manner or possibly enhanced survival in some unexplained way. Some generalities can be made. Spores of many bacteria are extremely resistant to oxygen concentration, RH, and temperature (May et al., 1969) in the aerosol state, including *Bacillus subtilis* subsp. *niger* and *Bacillus anthracis*. Lighthart (1973) found that from low RHs up to 85% RH, aerosol decay of *B. subtilis* subsp. *niger* decreased significantly when measured in a carbon monoxide-containing atmosphere as compared with a non-carbon monoxide-containing atmosphere. The following experimental results show variations in aerosol stability observed for various species of bacteria.

E. coli

More studies have been completed on the aerosol stability of *E. coli* than on any other microorganism. Cox (1968) showed that the stability of *E. coli* B aerosolized in atmospheres containing inert gases was a function of RH. Figures 1, 2, and 3 show the results of those experiments for nitrogen, argon, and helium, respectively. Survival was observed to be high at low RH but was critically dependent on RH above 80%. Cox (1987) postulated that the effect of nitrogen, argon, and helium, as compared with oxygen-containing atmospheres, was caused by formation of a gas hydrate with water, which formed a lattice and subsequently stabilized biological structures. Cox and Baldwin (1967) showed that at high RH, the recovery of *E. coli* was significantly reduced but similar in atmospheres containing oxygen and nitrogen. Cox and Heckley (1973) found that inactivation rates were much lower in an oxygen atmosphere when *E. coli* was frozen first. Cox and Baldwin (1967) also showed that *E. coli* was totally inactivated at RHs above 80% when oxygen was present. Poon (1966) showed that aerosol inactivation of *E. coli* was proportional to the amount of drying that had taken place. With respect to aerosol generation, Cox (1970) found that the stability was similar for both wet and dry dissemination. Webb (1969) and Cox (1966) both observed that the presence of dextrose, glutamine, and inositol in the collection fluid during aerosol collection increased recovery at high RHs. Williamson and Gotaas (1942) demonstrated that there was no temperature effect on the aerosol stability of *E. coli* between 24 and 30°C. Druett (1973a) presented evidence which showed that *E. coli* was very sensitive to air pressure fluctuations; his observations with an electron microscope revealed that

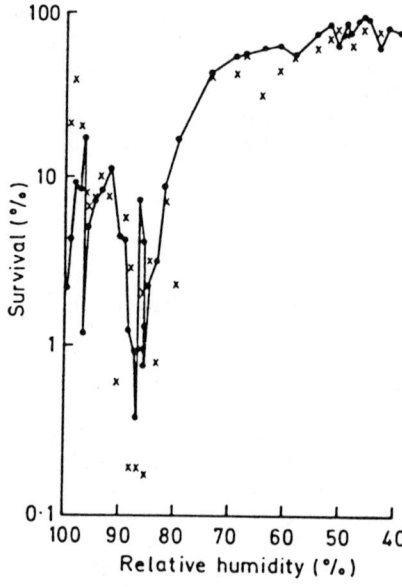

Figure 1. Aerosol survival of *E. coli* B sprayed from distilled water into nitrogen as a function of RH at an aerosol age of 30 min at 26.5°C. ●, Collection by all-glass impinger-30 (AGI-30) into phosphate buffer; ×, collection by AGI-30 into phosphate buffer and 1 M sucrose. Reprinted from Cox (1987) with permission of the publisher.

inactivation was probably due to disruption in the outside surface structures of the organism. Webb (1965) showed that aerosol inactivation by radiation effects increased at RHs of around 70 to 80%. Webb and Booth (1969) illustrated that electromagnetic radiation in the microwave wavelengths altered the growth patterns of *E. coli*.

Klebsiella pneumoniae

Goldberg and Ford (1973) showed that decay rates for *Klebsiella pneumoniae* were the greatest at midrange RHs (around 55%) when the organism was aerosolized in a nitrogen atmosphere from distilled water. When *K. pneumoniae* was sampled from a nitrogen atmosphere, vapor-phase prehumidification exhibited no effect on aerosol stability. When inositol or bovine serum albumin was added to the suspension medium, *K. pneumoniae* was observed to be stable at midrange RH.

Francisella tularensis

Cox (1971) and Cox and Goldberg (1972) presented data which showed that when *Francisella tularensis* was wet-disseminated in air, there was increased inactivation at 80% RH. When *F. tularensis* was dry-disseminated in air, increased inactivation was observed at RHs between 30 and 60%. Beebe and Pirsch (1958) showed that inactivation was proportional to artificial light intensity and decreased with increasing RH, and that a decrease in the decay rate was observed when *F. tularensis* was dry-disseminated. Cox and Goldberg (1972) showed that when *F. tularensis* was produced in culture in the presence of oxygen, a decrease in viability was observed if the organism was aerosolized during the log phase of growth. Beebe (1959) presented data which showed that *F. tularensis* inactivation by UV radiation was dependent on the wavelength (intensity) and RH.

Serratia marcescens

Dimmick et al. (1979) showed evidence for metabolism and formation of new cells for *Serratia marcescens* in aerosols at 95% RH and 30°C. Lighthart (1973) found that carbon monoxide increased aerosol decay rates four- to sevenfold at low RH for *S. marcescens*, but seemed to offer a protecting action at higher RHs. Kethley et al. (1957) showed that increases in decay rates were observed for *S. marcescens* with increasing temperatures from −40 to 32°C. Webb et al. (1963) studied *S. marcescens* decay under artificial light and found that decay rates were higher at low RHs than at high RHs. The effects of oxygen concentration were studied by Cox et al. (1974a), who determined that oxygen inactivation was proportional to oxygen concentration and exposure time. They also found that oxygen was not toxic to *S. marcescens* above 65% RH, but that the toxicity of oxygen increased as RH was decreased. Goodlow and Leonard (1961) found that the recovery of *S. marcescens* was a function of the aerosol age at the temperatures and RHs studied and that recovery decreased with increased time in the aging vessel. Cox et al. (1974a) showed that when *S. marcescens* was grown in culture, there was a significant increase in oxygen inactivation if the organism was aerosolized during the log phase of growth.

Staphylococcus aureus

Williamson and Gotaas (1942) found no influence on aerosol decay rates for *Staphylococcus aureus* between 24 and 30°C. Hatch and Dimmick (1966) studied the effect of sudden RH shifts on the aerosol viability of *S. aureus* and found that inactivation was substantial when RH changed from low to midrange. Druett (1973a) found that *S. aureus* was very sensitive to small air pressure fluctuations. Phillips et al. (1963) studied the effects of air ions on the aerosol stability of *S. aureus* and observed increases of around twofold with positive air ions and about threefold

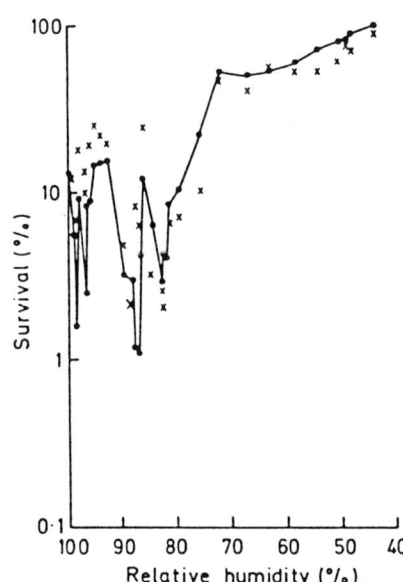

Figure 2. Aerosol survival of *E. coli* sprayed from distilled water into argon as a function of RH at an aerosol age of 30 min at 26.5°C. ●, Collection by AGI-30 into phosphate buffer; ×, collection by AGI-30 into phosphate buffer and 1 M sucrose. Reprinted from Cox (1987) with permission of the publisher.

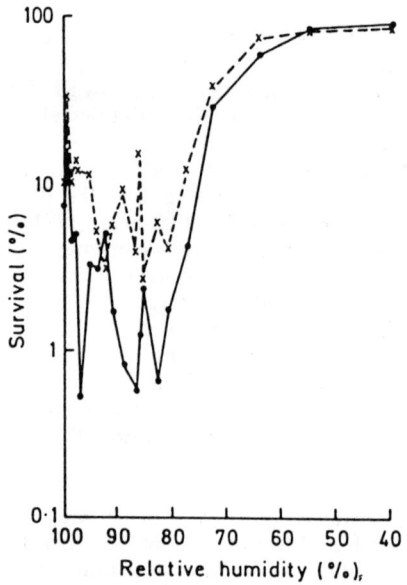

Figure 3. Aerosol survival of *E. coli* B sprayed from distilled water into helium as a function of RH at an aerosol age of 30 min at 26.5°C. ●, Collection by AGI-30 into phosphate buffer; ×, collection by AGI-30 into phosphate buffer and 1 M sucrose. Reprinted from Cox (1987) with permission of the publisher.

for negative air ions. The positive ions were also shown to influence physical decay, while negative ions affected physical and biological decay. Webb (1965) showed that radiation was least effective on aerosol decay rates of *S. aureus* at high RHs.

Viruses

Viruses are very resistant to inactivation by oxygen. This and the relative simplicity of their structure explain why results of aerosol inactivation studies are more consistent for viruses than for bacterial studies. Generalities which can be made about aerosol inactivation rates for viruses include the following: (i) viruses with lipids in their outer coat or capsid are more stable at low RHs than at high RHs; (ii) viruses without lipids are more stable at high RHs than at low RHs; (iii) when viable viruses can no longer be detected after aerosol collection, the nucleic acid can be isolated and is still active (this evidence suggests that aerosol inactivation of viruses is not caused by nucleic acid inactivation but by denaturation of coat proteins); and (iv) prehumidification upon sampling increases recovery of viruses that lack lipids in their outer coat. The following presentation shows the variety of influences that meteorological parameters have on the aerosol stability of various viruses.

Poliovirus

Harper (1963) found that for poliovirus the highest recovery from aerosols was observed at high RH and that different suspending fluids exerted little effect on stability. Dimmock (1967) found that poliovirus was inactivated at low rather than at high temperatures and that at high temperatures RNA activity was greater than might be expected based on the infectivity of the virion. Donaldson (1972) studied radiation inactivation rates and found poliovirus to be very resistant. De Jong and Winkler (1968) found that the best survival for poliovirus was above 60% RH and that where recovery was low at low RH, prehumidification during sampling would help recovery. Benbough (1971) found that salts in the suspending medium would drastically decrease recovery, while the addition of glucose to the suspending medium would increase recovery rates. De Jong (1970) studied aerosol inactivation processes for polioviruses and found that while the virions quickly lost viability at low RH, their RNA could be extracted and remained active, and that aerosolized RNA would retain its activity in the airborne state for a short period.

Rotavirus

Ijaz et al. (1985) studied the aerosol stability of rotavirus and found that maximum instability was observed at midrange RH after 24 h of aging. Moe and Harper (1983) found that rotavirus was stable at high and low RHs and that recovery was higher at 10°C than at 30°C.

Reovirus

Two forms of virus particles are released from reovirus-infected cell cultures, infectious reovirus (IV) and potentially infectious reovirus (PIV). PIV particle forms have a complete outer coat and are not infectious until the outer coat is removed by treatment with proteolytic enzymes. Adams et al. (1982) determined aerosol decay rates for reovirus IV and PIV particles for a range of RHs at 21 and 24°C. At 90 to 100% RH, both particle forms showed less than a 10-fold decrease in infectivity. At lower RHs, the aerosol decay curve showed rapid initial decay, followed by a markedly slower decay rate. Mohr (1984) studied reovirus IV and PIV particles for type 1 Lang, type 2 Jones, and type 3 Dearing at 2, 14, and 27°C and four RHs (20, 40, 60, and 80%). Decay rates were greater for PIV and IV particles at midrange RHs for all three serotypes. Generally, decay rates were greater for IV particles than for PIV particles for the Jones and Dearing serotypes, while the opposite was observed for type 1 Lang. Type 1 Lang exhibited the lowest overall decay rate (0.05%/min) at 2°C and 80% RH. RH exerted a greater effect on decay rates than did temperature. Figures 4, 5, and 6 show the aerosol decay of reovirus type 1 Lang IV particles with respect to RH at 2, 14, and 27°C, respectively. Note the increase in inactivation with aerosol age and that inactivation is greatest at midrange RHs. Figures 7, 8, and 9 show the aerosol decay of reovirus type 1 Lang PIV particles with respect to RH at 2, 14, and 27°C, respectively. Decay profiles were similar, but not identical, for IV and PIV particles.

Mengovirus 37A

Akers and Hatch (1968) found that mengovirus 37A was almost completely inactivated at RHs of less than 70% and that the RNA remained infectious. Prehumidification upon sampling decreased the recovery rates when aerosolized viruses were collected at low RH. Akers and Hatch (1968) also noted that aerosol-induced coat denaturation was irreversible.

Coxsackievirus

Jensen (1964) observed a 99% decrease in recovery rates when coxsackievirus was passed through UV light in a dynamic aerosol test chamber.

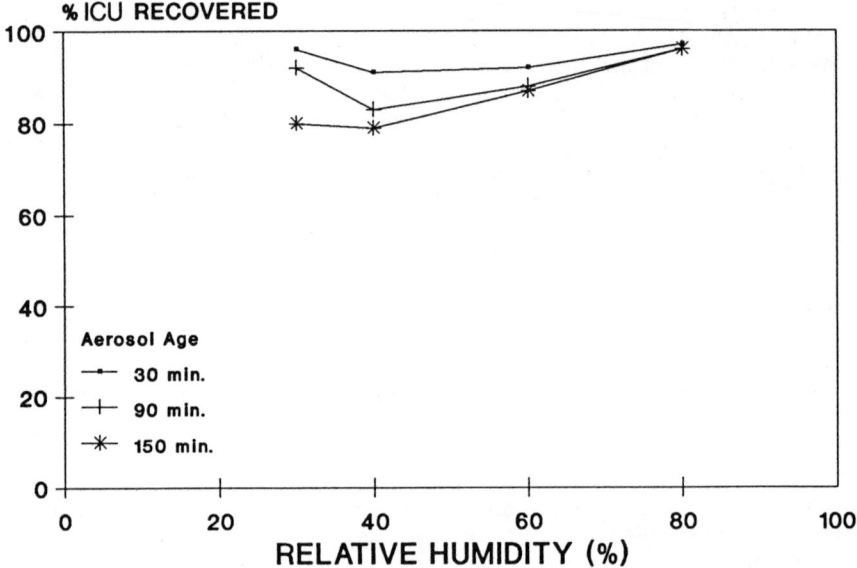

Figure 4. Aerosol survival of reovirus type 1 Lang IV particles at 2°C with respect to various RHs (Mohr, 1984). ICU, Immunofluorescing cell units.

Encephalomyocarditis virus

De Jong (1970) found survival of encephalomyocarditis virus to be the greatest above 60% RH and that prehumidification upon sampling had no influence on the recovery of infectious particles. It was also noted that even after hemagglutinin activity was lost, the viral RNA could be extracted and remained infectious.

Yellow fever virus

Mayhew et al. (1968) studied the aerosol stability of yellow fever virus and found the virus to be unaffected over the temperature range from −1.1 to 26.7°C and RH range from 30 to 80%.

Foot and mouth disease virus

Barlow (1972) and Donaldson (1972) determined that foot and mouth disease virus was relatively stable over a large RH range and was the most stable at high RH. Barlow (1972) found that the addition of polyhydroxyl compounds (inositol, sorbitol, and glucose) to the suspending fluids aided in the recovery of the virus when salts were also present in the suspending fluids. Donaldson and Ferris (1975) found that the virus was not affected by OAF in the dark. Additionally, Donaldson (1972) observed the virus to be very resistant to radiation inactivation.

Semliki Forest virus

Benbough (1971) studied Semliki Forest virus and found the best recovery to be at low RH. Although viability decreased when salts were added to the suspending fluid, recovery could be increased by the addition of bovine serum. De Jong et al.

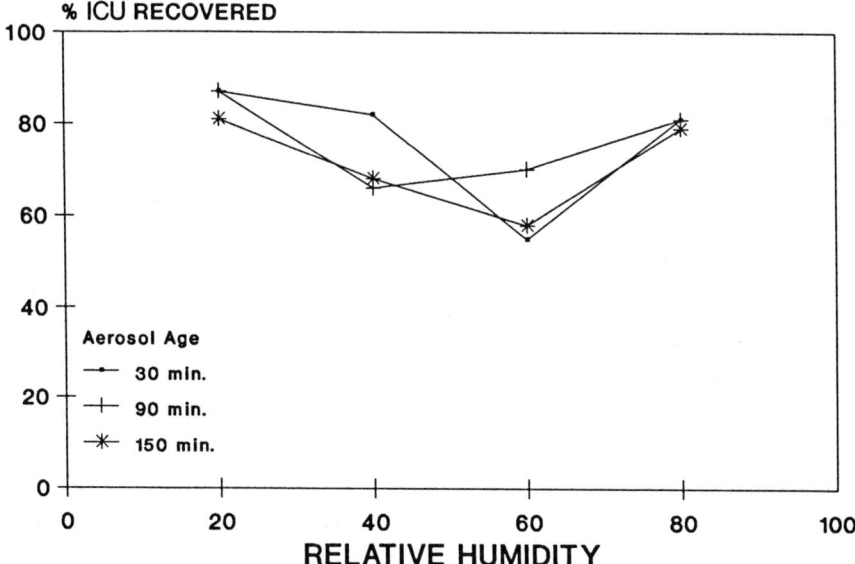

Figure 5. Aerosol survival of reovirus type 1 Lang IV particles at 14°C with respect to various RHs (Mohr, 1984).

(1976) found that the lipoproteins were least stable at midrange RH for Semliki Forest virus. Druett (1973a) found that the virus was very sensitive to small pressure fluctuations, and Benbough and Hood (1971) observed it to be quickly inactivated by OAF.

Venezuelan equine encephalitis virus

Harper (1961) found that Venezuelan equine encephalitis virus was stable in the airborne state from 20 to 80% RH, but that the greatest stability was measured at 20% RH. The study also revealed that when this virus was lyophilized, its stability was decreased at between 20 and 80% RH and that it was very unstable above 29°C. Ehrlich and Miller (1971) aerosolized Venezuelan equine encephalitis virus from a liquid suspension and determined that over the temperature range from −40 to 26°C and RH range from 18 to 90%, the virus was quite stable. Berendt and Dorsey (1971) found the virus to be very sensitive to radiation and that infectivity was quickly lost.

Simian virus 40

The DNA virus simian virus 40 was found to be very stable at all RHs tested and at ambient temperatures (Akers, 1973).

Bacteriophages

Bacteriophages, like other viruses, are not inactivated by the presence of oxygen. Most work has been completed on the T series of coliphages (phages which

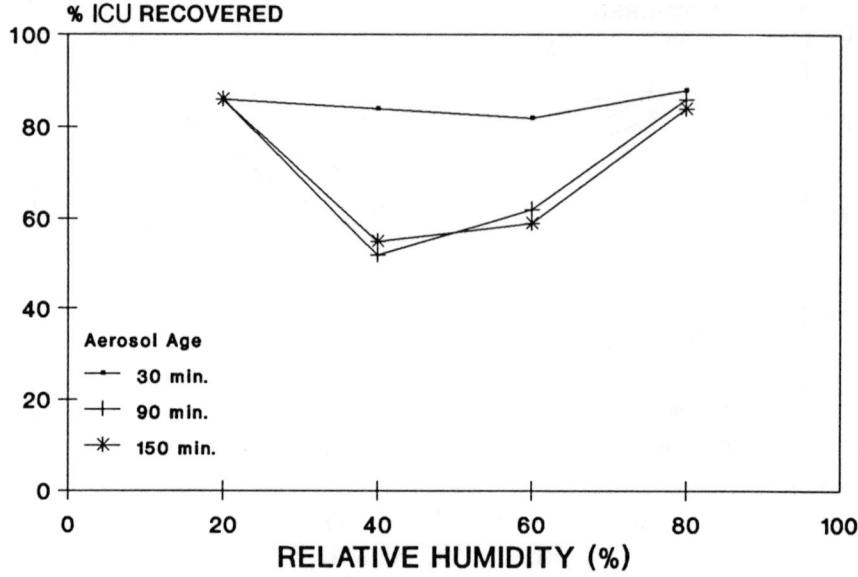

Figure 6. Aerosol survival of reovirus type 1 Lang IV particles at 27°C with respect to various RHs (Mohr, 1984).

infect *E. coli*). Generally, it has been determined that NaCl is toxic to phages, but the effect can be reversed by the addition of peptone to the suspending medium. Aerosol inactivation of complex phages has been shown to be caused by the breakup of the head-tail complex, which could be brought about during the aerosol generation or sampling processes, and prehumidification at sampling often increases recovery rates. Overall, the results for various workers are consistent due to the relative simplicity of the structure of phages.

S-13

Dubovi and Akers (1970) studied the tailless, single-stranded DNA phage designated S-13 and found that it was inactivated in the aerosol state at mid- and low-range RHs. Warren et al. (1969) found that prehumidification was only effective at increasing recovery rates at midrange RH.

ϕX174 phage

Dubovi (1971) studied ϕX174 and determined that inactivation occurred because of denaturation of coat proteins. This conclusion was arrived at because the nucleic acid was still active when sampled and extracted from aerosol-inactivated phages. De Mik and de Groot (1973) also determined that the DNA remained active after the phages lost their infectivity. They also showed that ϕX174 was inactivated by ozone and that the DNA was broken down by ozonized cyclohexene. De Mik and de Groot (1977) later found that ϕX174 was readily inactivated by OAF.

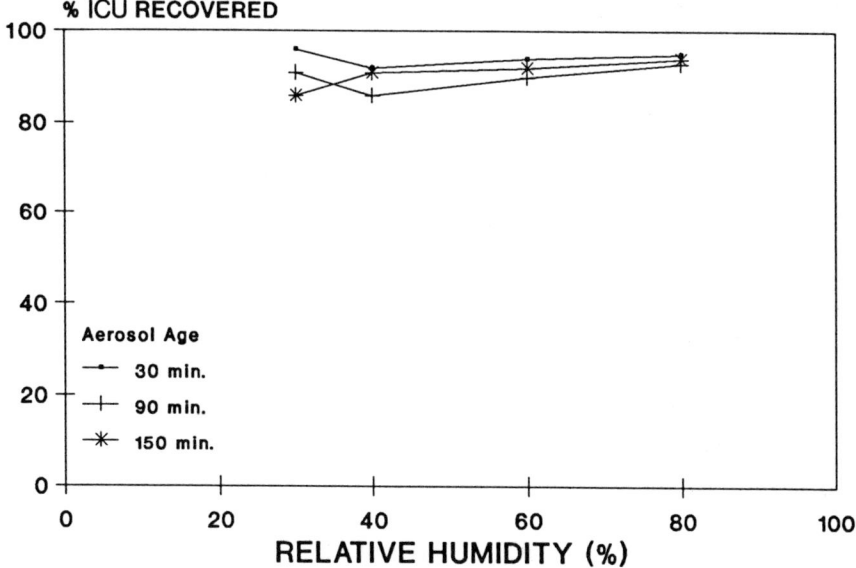

Figure 7. Aerosol survival of reovirus type 1 Lang PIV particles at 2°C with respect to various RHs (Mohr, 1984).

MS-2 phage

MS-2 is a small, tailless RNA coliphage that has been shown to be very stable regardless of RH (Dubovi and Akers, 1970). Dubovi (1971) recovered active RNA from aerosolized phages that were noninfectious, which points to the breakdown of the protein coat as the likely cause of inactivation. It was also shown that salts in the suspending medium caused the phage to be aerosol sensitive and that prehumidification decreased recovery rather than increased recovery rates as usually occurs. Trouwborst et al. (1972) found that peptone and phenylalanine in the suspending fluid protected MS-2 at all RHs tested. MS-2 is used as a simulant for animal viruses in field testing. Figure 10 shows the aerosol decay of MS-2 at 4.5°C and 40 and 75% RH. Figure 11 shows the aerosol decay of MS-2 at 27°C and 35 and 75% RH. Over a 4-h period, MS-2 has been shown to display greater stability at 27°C than at 4.5°C (Mohr, 1984). This stability at warm temperatures makes MS-2 a good viral simulant for outdoor testing in hot climates.

T series of phages

Hatch and Warren (1969) found that all of the T phages (except T1) survived well in pure air and nitrogen at 75% RH and greater; below 75% RH, inactivation was observed to increase. Songer (1966) found that phage T3 was the most stable at 90% RH. Cox et al. (1974b) and Benbough (1971) found that the sampling process would influence recovery rates. Prehumidification upon sampling greatly increased recovery, and subsonic samplers exhibited better recovery rates than the all-glass impinger which collects with a sonic air jet. It was hypothesized that water molecules were replaced during prehumidification. These water molecules were

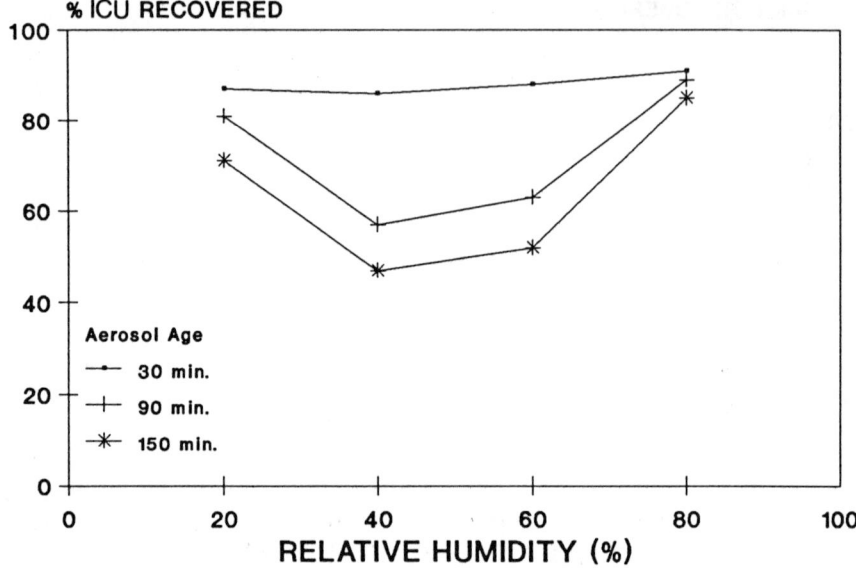

Figure 8. Aerosol survival of reovirus type 1 Lang PIV particles at 14°C with respect to various RHs (Mohr, 1984).

incorporated in the capsid and subsequently strengthened the head-tail complex so that it was not sheared apart during collection. Cox et al. (1974b) and Benbough (1971) further established that the phages T3 and T7 lost viability when they were freeze-dried and that prehumidification did little to increase recovery.

Cox and Baldwin (1966) aerosolized phage T7 and discovered that the phage DNA remained active for up to 2 h after it had been extracted from inactivated phages and injected into an appropriate host. Hatch and Warren (1969) learned that when phage T3 was stored for about 3 h at low RH, the viability slowly decreased. Benbough (1971) studied the T1 phage and found it to be stable at from 20 to 95% RH for 1 h when sprayed from filtered suspending fluid. Trouwborst et al. (1972) observed much lower recovery rates for T phages when salt was added to the suspending medium and hypothesized that inactivation was due to breakup of the head-tail complex. Trouwborst et al. (1972) noted that some of the phage DNA was released at the time of inactivation and concluded that when the T phages were sprayed from suspending fluids with salt present, minimum stability was observed at 70% RH. De Mik and de Groot (1977) found that OAF was very detrimental to phages T1 and T7. Happ et al. (1966) studied the effect of ionized air during sampling and found that recovery was always lower in ionized air, but when the air was charged with positive ions, recovery was better than when it was charged with negative or mixed-polarity ions.

MODELING MICROBIAL AEROSOLS

Accurate predictive models are required to evaluate the diffusion and deposition of airborne microorganisms. Sources of aerosolized microorganisms include

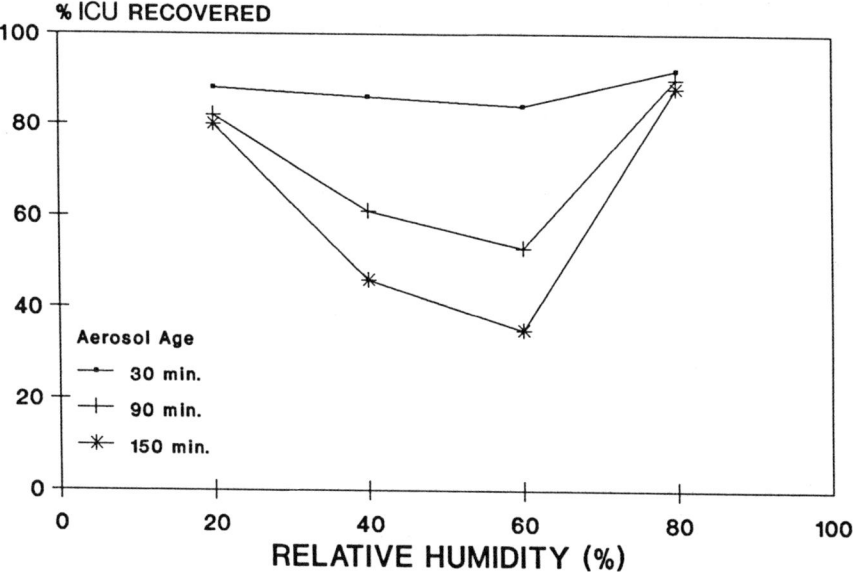

Figure 9. Aerosol survival of reovirus type 1 Lang PIV particles at 27°C with respect to various RHs (Mohr, 1984).

evaporative cooling towers (Adams et al., 1979), spray irrigation practices utilizing wastewater (Bausum et al., 1982; Moore et al., 1979), various wastewater treatment processes (Fannin et al., 1976; Goff et al., 1973), rendering plants (Spendlove, 1974), dairies (Spendlove, 1974), and slaughterhouses (Spendlove, 1974). Recently, interest has been focused on the fate of bacteria which have been intentionally aerosolized to control pests (Cook and Baker, 1983) or to provide protection from freezing (Lindow, 1985). In addition to naturally occurring microorganisms, it has been proposed that genetically engineered microorganisms might offer enhanced competitive abilities and thus be more efficient in controlling pests or providing environmental protection. Whenever a novel entity is introduced into an environment where it will be competing with indigenous flora and fauna, the possible consequences must be evaluated. Concerns about possible environmental effects of genetically engineered microorganisms include uncertainty about their stability, competition with indigenous microorganisms, and the extent of their dispersion (Lindow, 1986).

The primary objectives of modeling aerobiological systems include evaluation of microbial stability after deposition, prediction of ecological impacts, and the determination of the factors which influence these impacts. An accurate understanding of the parameters which influence stability, dispersion, and deposition of microbial aerosols should allow mediation of detrimental results by the modification of a critical condition which would be necessary for viability or by the application of control measures during flight or impact.

Modeling aerobiological systems, with the goal of predicting possible health and environmental consequences, requires integration of diverse information (i.e.,

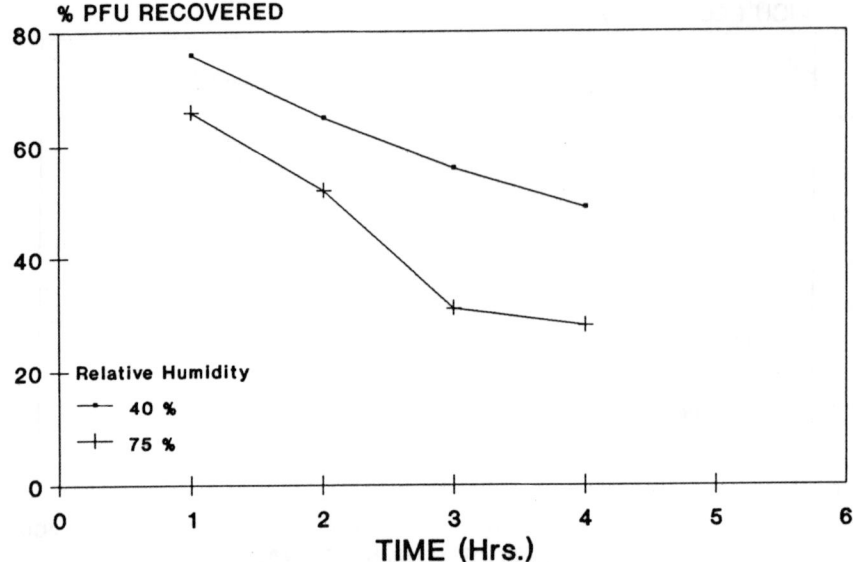

Figure 10. Aerosol survival of MS-2 coliphage at 4.5°C and 40% and 75% RH (Resnick et al., 1987).

the many stages and parameters in a given cycle) into a comprehensive package. Many previous attempts at dispersion modeling have been plagued with insufficient data or inadequate methods of gathering and analyzing data. These problems have made it necessary to develop statistical models which relate the final result to several factors that are responsible for the outcome. Often this technique, termed a regression model, ignores the main functional relationships describing the stages between the cause and effect. Another approach has been the development of simulation models which have predictive results. These models attempt to represent, as completely as possible, the entire aerobiological cycle, including interacting parameters and functional relationships. Problems with this approach include the need for profuse amounts of data and a thorough understanding of the relationships and effects among all the parameters.

The first step in the development of any model should be the conceptualization of all events and parameters which are to be modeled, including the desired final outputs. According to Quentin (1979), a conceptual model should attempt to depict (i) the sequence of events in a complete system cycle, (ii) the significant factors (parameters and variables) that affect the system, and (iii) the interrelationships between the system and other external systems. Based on the time and resources available, the conceptual model aids in the selection of the ultimate type of model to be formulated.

There are several alternatives which may be selected when the boundaries of a conceptual model have been set. This selection depends on numerous factors, including the desired accuracy required for the final results, data requirements (available and those which require collection), and the resources at the investiga-

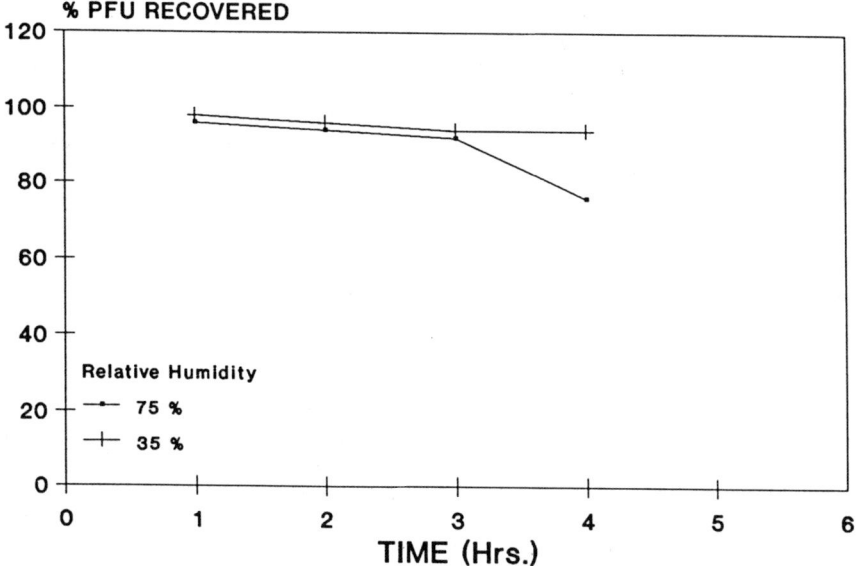

Figure 11. Aerosol survival of MS-2 coliphage at 27°C and 35% and 75% RH (Resnick et al., 1987).

tor's disposal. Development of a deterministic simulation model usually requires a mathematical expression for each event and often must be adjusted for influences caused by other parameters. Alternatively, a statistical model may be developed which relies on the relationship between the final result and important causative factors. Simulation models, even though more complex, offer more in the way of final results and explanation. Additionally, simulation models are often more flexible in that various inputs can be altered and effects can be evaluated. Statistical models can be quite valuable for deterministic models but are often limited to general predictions governed by the origin and circumstances under which the input data were collected.

The value of any model can be judged by comparing its final output with what is observed naturally or under controlled conditions. Model validation is often as time-consuming and complex as model development and if at all possible should be part of the model development.

Various models have been developed which attempt to describe viability and dispersion of airborne microorganisms. Many viability models are empirical, and their predictive power is limited because decay rates are species specific and apply only to a particular type or strain of microorganism.

Viability Models

Exponential decay model

The loss of viability of aerosolized microorganisms is caused by complex physical, meteorological, and cellular interactions. Early attempts to explain aerosol viability relied on the exponential decay model, with mixed results. Even though exponential decay has been shown to be a simplification, application of intricate mathematical expressions often delivered conclusions which were no more accurate than simple expressions. If all of the known parameters were applied to a mathematical expression there would be around 20 inputs, many of which would be dependent on other parameters during specific periods of flight. The expression for exponential decay is

$$V_t = V_0 e^{-K} \qquad (1)$$

where V_t is viability at time t, V_0 is viability at time zero, and K is a decay rate constant.

Kinetic model

Cox, in a series of articles (1987), presented what he called the kinetic model. He was dissatisfied with the lack of explanation of the exponential model, which did not account for the time-dependent decay observed for most microorganisms. Cox's kinetic model supposes that microorganisms contain a molecular species $B(n)H_2O$, the biological activity of which is essential for a microbial cell to replicate or be infectious. $B(n)H_2O$, when exposed to an environment of lowered water activity (or low RH), forms a series of hydrates similar to other biomolecules; some of these hydrates are unstable and spontaneously denature through a first-order process, i.e.,

$$-\partial x/\partial t = kx \qquad (2)$$

where x is the concentration of the species which denatures. The model form is then

$$B(n)H_2O \underset{k_-}{\overset{k_+}{\Longleftrightarrow}} B(n-x)H_2O + xH_2O \Longleftrightarrow B(n-x-y)H_2O + yH_2O$$

$$\Longleftrightarrow B + iH_2O \qquad (3)$$

where $B(n-x)H_2O$ is the denatured form with a rate constant k_x and $B(n-x-y)H_2O$ is the denatured form with a rate constant k_y.

Cox (1987) then applied probability theory to evaluate the likelihood of death, which is related to percent viability, and set up appropriate boundary conditions and integrated the results. When denaturation follows a first-order pattern the final form of the equation is written

$$\ln V + K_1[B(n-x)H_2O]_0(e^{-kt} - 1) + \ln 100 \qquad (4)$$

where k is a first-order denaturation constant, t is time, K_1 is the probability constant, and V is viability. Cox (1987) has analyzed several hundred viability-time curves and has found very good agreement with experimental results for dehydration inactivation. Figure 12 shows the comparison between experimental results

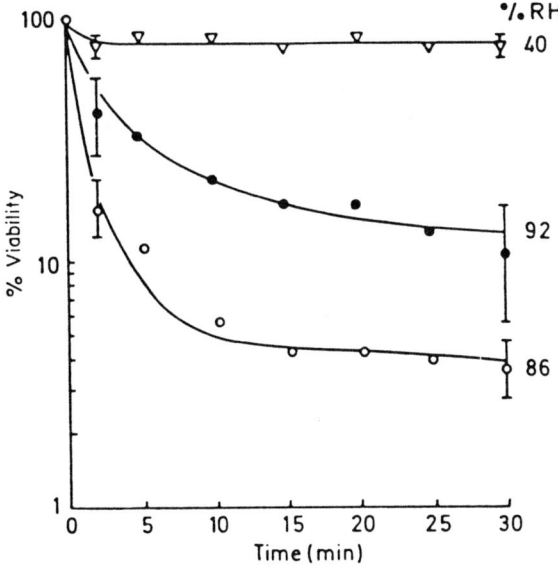

Figure 12. Aerosol survival of *E. coli* Jepp wet-disseminated into nitrogen. Points are experimental data; lines are calculated from equation 4. Reprinted from Cox (1987) with permission of the publisher.

and results generated by equation 4 for *E. coli* B. Figure 13 shows the comparison between experimental results and results generated by equation 4 for Semliki Forest virus.

Additionally, Cox (1987) applied kinetic and probability theories to model inactivation caused by oxygen concentration, free radicals, and OAF. These viability models are more complex than the first-order kinetics observed for dehydration inactivation, but closely follow experimental survival profiles. Prediction capabilities of aerosol survival for additional microorganisms have yet to be established for these kinetic and probability models.

Catastrophe model

Catastrophe theory was formulated to explain the nature of discontinuous events such as the breaking of waves on a beach, the crash of the stock market, or the sudden aggression displayed by an animal. Overall reactions involving large numbers of molecules appear to behave in a continuous fashion because discontinuities are smoothed, but on an individual level, the reactions are discontinuous. If the number of reactions is relatively low, as is the case when aerosol viability is being considered, discontinuity is appropriately represented and aerosol inactivation rates may be predicted. On an individual level, the loss of infectivity of an airborne microorganism is a sudden discontinuous event and this change in state is termed the catastrophe. The basis of catastrophe theory is related to the potential energy and therefore equilibrium of the system. The potential energy is a function

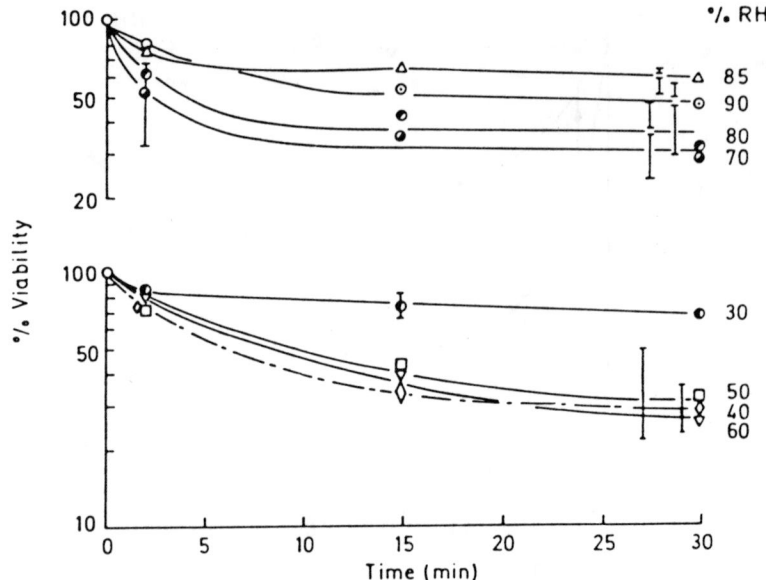

Figure 13. Aerosol survival of Semliki Forest virus suspended in medium 199 + 10% (vol/vol) calf serum into air. Points are experimental data; lines are calculated from equation 4. Reprinted from Cox (1987) with permission of the publisher.

of what are termed control parameters which govern the equilibrium of an event. Within a certain range, equilibrium will not be influenced by variations in the control parameters, but some small critical change can cause a shift in the potential energy and result in inactivation. Cox (1987) combined catastrophe theory and kinetics to explain loss of viability caused by desiccation, temperature, oxygen, and OAF. The calculated viability curves agreed very well for inactivation caused by oxygen concentration and OAF and were more accurate than predictions based on probability theory for denaturation induced by desiccation. Figure 14 shows the comparison between an experimentally derived aerosol stability profile and one generated utilizing catastrophe theory. The results show the effect of temperature (second-order denaturation) on the aerosol survival of *Flavobacterium* species.

Dispersion Models

Models have been developed to predict dispersion and deposition of microbial aerosols with respect to penetration of structures (Spendlove, 1975), to predict infectious microbial concentrations downwind from known sources (Sorber et al., 1976; Lembke and Kniseley, 1980; Zeterberg, 1973; Adams and Spendlove, 1970; Teltsch et al., 1980; Lighthart, 1984), and to evaluate the spread of plant pathogens outdoors (Cook and Baker, 1983). Many of the early models were based on treatments of atmospheric diffusion which were created to predict the fate of air pollutants. The inert particle dispersion model by Pasquill (1961) is empirical and

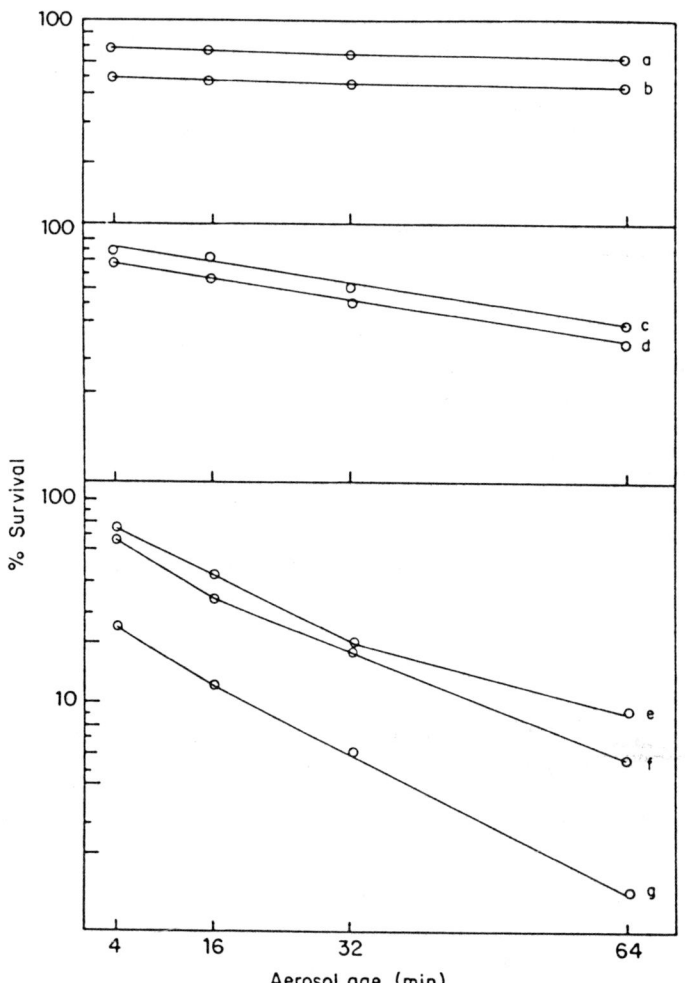

Figure 14. Effect of temperature (a, 18°C; b, 40°C; c, 24°C; d, 2°C; e, 29°C; f, 38°C; g, 49°C) on the aerosol survival of *Flavobacterium* sp. Points are experimental data; lines are calculated from catastrophe theory for a second-order denaturation process. Reprinted from Cox (1987) with permission of the publisher.

based on observations of the deposition of inanimate particles. Results from its application yield average distributions of airborne microbes, have a somewhat limited downwind range, and should be applied to flat terrain with steady wind conditions. Assumptions which are made for this model include the following: (i) Gaussian distribution of the plume in the horizontal and vertical planes; (ii) total reflection of particles from the ground; (iii) particle emission from the source at a constant rate; (iv) constant wind velocity and direction; (v) flat terrain; (vi) particles smaller than 20 μm, making gravitational effects negligible; and (vii) negligible diffusion downward as well as negligible difference in wind velocity between the source and real wind. The classical form of the inert particle dispersion model is:

$$X(x,y,z:H) = Q/[2\pi(\sigma_y\sigma_z)u \times \exp[-0.5(Y/\sigma_y)^2$$
$$\times \{\exp[-0.5(\{Z - H\}/\sigma_y)^2] + [-0.5(\{Z + H\}/\sigma_z)^2]\} \quad (5)$$

where X is the number of particles per cubic meter at downwind location x,y,z; Q is the number of particles emitted from the source per second; u is the mean air speed in meters per second; σ_y is the standard deviation of the horizontal concentration (at the downwind distance); σ_z is the standard deviation of the vertical concentration (at the downwind distance); and H is the height of the source plus the plume rise.

The equation can be simplified for ground level ($z = 0$) and becomes

$$X(x,y,0:H) = Q/\pi(\sigma_y\sigma_z)u \exp[-(Y^2/2\sigma_y^2 + H^2/2\sigma_z^2)] \quad (6)$$

For a ground-level source ($H = 0$) and analysis along the center line ($y = 0$), the equation becomes

$$X(x,y,z:H) = Q/\pi(\sigma_y\sigma_z)u \quad (7)$$

As explained earlier, these equations are for inanimate particles and do not take into account microbial viability decay. Lighthart and Frisch (1976) expanded on the inert particle diffusion model, added a biological death constant (α), and created a graphic method which would estimate ground-level concentrations when microbial death rate, mean wind speed, atmospheric stability class, source height, and downwind sampled distance were known. Microbial death rate was determined under laboratory conditions and applied to the model. The equation for biological death (BD), after modification, becomes

$$X(x,y,z:H)_{BD} = X(x,y,z:H) \exp(-\alpha t) \quad (8)$$

where t was approximated by x/u.

Biological decay and the rate constants associated with inactivation must be determined under dynamic conditions in the laboratory. Teltsch et al. (1980) presented methodology to estimate the death constant under field conditions to determine inactivation and final concentrations downwind from sprinklers using wastewater. Death constants must be evaluated over a wide range of environmental conditions for accurate results, and the determined decay constants cannot be applied for predictive uses because of the specific meteorological conditions under which the study took place.

Lighthart and Mohr (1987) applied best-fit laboratory-generated decay constants to the Gaussian plume model using microbial source strength and local hourly mean weather data to drive the model through a summer and winter day cycle. For near-source locations, higher wind speeds or short travel times exerted a major modulating effect during the day because time was inadequate for inactivation. Additionally, as travel time increased because of low wind speed or long distances, modulation was due more to solar irradiance, RH, and temperature than to wind speed.

Spendlove (1975) applied various models (box, test tube) to determine the penetration of structures by microbial aerosols. Models showed that for a single-story dwelling, without air conditioning, the aerosol dose received by people was the same inside as outside and that air conditioning units would act to moderately decrease inside concentrations (depending on ventilation rates).

Recently, circumstances have combined which may change the methods by which airborne microorganisms are modeled. These changes include the application of computers to sort out numerous data inputs; new interest in the fate and deposition of genetically altered microorganisms intended to moderate destruction caused by climate, other microorganisms, and insects; and a better understanding of the parameters that influence aerosol inactivation and dispersion. Lighthart and Kim (1989) have recently presented a simulation model which describes the dispersion of individual droplets of water containing viable microbes. By repeating the modeling process many times for individual droplets, an aerosol cloud can be simulated. The model accounts for the physical, chemical, biological, and measured meteorological parameters of each droplet for many time increments. The droplet model is separated into five submodels which include aerosol generation, evaporation, dispersion, deposition, and microbial death, all of which are calculated at chosen time intervals in the trajectory for each drop. The results show that evaporation is an important factor in determining deposition sites because of particle size dependence, chemical reactions, and protection offered by large droplets. Wind gust data are required because average wind velocity data tend to "smooth" and oversimplify what is occurring on a micrometeorological scale.

CONCLUSION

The mechanisms of inactivation for airborne microorganisms are complicated and often act codependently and synergistically. Complete explanations of the basis of microbial injury and death have been elusive because of the difficulties encountered when trying to separate the numerous events responsible for aerosol inactivation. The major stress encountered is dehydration, but several other complicating factors are always present, confusing evaluation of aerosol stability. Aerosol generation (content of suspending fluids), aging (RH, temperature, oxygen content, electromagnetic radiation), and collection (prehumidification, content of collection fluids) can all play pivotal roles in aerosol inactivation. The ability of some microbes to exhibit levels of recovery during and after aerosolization offers new areas of study concerning reversible injury. The selection of the most sensitive and unobstrusive assay procedure can be critical when evaluating aerosol decay rates.

The number of airborne pathogens necessary to initiate infection is known for very few microorganisms. The effects on the environment of genetically engineered microorganisms used for crop protection and as insecticides must be known before large-scale applications can proceed. The fate of aerosolized microorganisms originating from the use of wastewater for spray irrigation and other industrial processes must be assessed. For these reasons it is important to continue efforts to improve the accuracy and predictability of viability and dispersion models. It has been shown that some viability models agree very well with experimental results, but it must be reiterated that there are no standardized generation, aging, collection, or assay methodologies for comparison. It is imperative that standardized methods be adopted so that results can be more easily compared. Application of new detection methodologies will no doubt increase assay sensitivities. Inert particle dispersion models have probably run their course of usefulness because it is now possible, with the aid of computers, to mathematically confront critical physical, biological, and chemical parameters and observe, with adequate spatial resolution, the interactions of these inputs. There is still much to be learned about the mechanisms and interactions acting on airborne microorganisms, but careful application of new methods of control and monitoring will no doubt make the task an easier one.

REFERENCES

Adams, A. P., and J. C. Spendlove. 1970. Coliform aerosols emitted by sewage treatment plants. *Science* **169**:1218–1220.

Adams, A. P., J. C. Spendlove, and M. Garbett. 1979. Emission of microbial aerosols from vents of cooling towers. *Dev. Ind. Microbiol.* **20**:13–33.

Adams, D. J., J. C. Spendlove, R. S. Spendlove, and B. B. Barnett. 1982. Aerosol stability of infectious and potentially infectious reovirus particles. *Appl. Environ. Microbiol.* **44**:903–908.

Akers, T. G. 1973. Survival, damage and inactivation in aerosols, p. 73–86. *In* J. F. Hers and K. C. Winkler (ed.), *Airborne Transmission and Infection*. John Wiley & Sons, Inc., New York.

Akers, T. G., S. Bond, and L. J. Goldberg. 1966. Effect of temperature and relative humidity on survival of airborne Colombia SK group viruses. *Appl. Microbiol.* **14**:361–364.

Akers, T. G., and M. T. Hatch. 1968. Survival of a picornavirus and its infectious ribonucleic acid after aerosolization. *Appl. Microbiol.* **16**:1811–1813.

Anderson, J. D., and C. S. Cox. 1967. Microbial survival, p. 203–226. *In* T. R. Gray and J. R. Postgate (ed.), *The Survival of Vegetative Microbes*. Cambridge University Press, Cambridge.

Barlow, D. F. 1972. The effects of various protecting agents on the inactivation of foot-and-mouth disease virus in aerosols and during freeze-drying. *J. Gen. Virol.* **17**:281–288.

Bausum, H. T., S. A. Schaub, K. F. Kenyon, and M. J. Small. 1982. Comparison of coliphage and bacterial aerosols at a wastewater spray irrigation site. *Appl. Environ. Microbiol.* **43**:28–38.

Beebe, J. M. 1959. Stability of disseminated aerosols of *Pasteurella tularensis* subjected to simulated solar radiations and various humidities. *J. Bacteriol.* **78**:18–24.

Beebe, J. M., and G. W. Pirsch. 1958. Response of airborne species of *Pasteurella* to artificial radiation simulating sunlight under different conditions of relative humidity. *Appl. Microbiol.* **6**:127–138.

Benbough, J. E. 1971. Some factors affecting the survival of airborne viruses. *J. Gen. Virol.* **10**:209–220.

Benbough, J. E., and A. M. Hood. 1971. Viricidal activity of open air. *J. Hyg.* **69**:619–626.

Berendt, R. F., and E. L. Dorsey. 1971. Effect of simulated solar radiation and sodium fluorescein on the recovery of Venezuelan equine encephalomyelitis virus from aerosols. *Appl. Microbiol.* **21**:447–450.

Bridges, B. A. 1976. Survival of bacteria following exposure to ultraviolet and ionizing radiations. *Symp. Soc. Gen. Microbiol.* **26**:183–208.

Buckland, F. E., and D. A. J. Tyrrell. 1964. Loss of infectivity on drying various viruses. *Nature* (London) **195**:1063–1064.

Cabelli, V. J. 1962. The rehydration of aerosolized bacteria: compounds which enhance the survival of rehydrated *Pasteurella tularensis*. Technical report no. 314. U.S. Army, Dugway Proving Ground, Dugway, Utah.

Constantine, D. G. 1967. Rabies transmission by air in bat caves. CDC Monograph, Public Health Service publication no. 1617. Center for Disease Control, Atlanta, Ga.

Cook, R. J., and K. F. Baker. 1983. *The Nature and Practice of Biological Control of Plant Pathogens*. American Phytopathological Society Press, St. Paul, Minn.
Cox, C. S. 1966. The survival of *Escherichia coli* atomized into air and into nitrogen from distilled water and from solution of protecting agents as a function of relative humidity. *J. Gen. Microbiol.* **43**:383–399.
Cox, C. S. 1968. The aerosol survival of *Escherichia coli* B in nitrogen, argon, and helium atmospheres and the influence of relative humidity. *J. Gen. Microbiol.* **50**:139–147.
Cox, C. S. 1970. Aerosol survival of *Escherichia coli* B disseminated from the dry state. *Appl. Microbiol.* **19**:604–607.
Cox, C. S. 1971. Aerosol survival of *Pasteurella tularensis* disseminated from the wet and dry states. *Appl. Microbiol.* **21**:482–486.
Cox, C. S. 1987. *The Aerobiological Pathway of Microorganisms*. John Wiley & Sons, Inc., Chichester, United Kingdom.
Cox, C. S., and F. Baldwin. 1966. The use of phage to study causes of loss of viability of *Escherichia coli* in aerosols. *J. Gen. Microbiol.* **44**:15–22.
Cox, C. S., and F. Baldwin. 1967. The toxic effect of oxygen upon the aerosol survival of *Escherichia coli* B. *J. Gen. Microbiol.* **49**:115–117.
Cox, C. S., J. Baxter, and B. J. Maidment. 1973. A mathematical expression for oxygen-induced death in dehydrated bacteria. *J. Gen. Microbiol.* **75**:179–185.
Cox, C. S., S. J. Gagen, and J. Baxter. 1974a. Aerosol survival of *Serratia marcescens* as a function of oxygen concentration, relative humidity and time. *Can. J. Microbiol.* **20**:1529–1534.
Cox, C. S., and L. J. Goldberg. 1972. Aerosol survival of *Pasteurella tularensis* and the influence of relative humidity. *Appl. Microbiol.* **23**:1–3.
Cox, C. S., W. J. Harris, and J. Lee. 1974b. Viability and electron microscope studies of phages T3 and T7 subjected to freeze-drying, freeze-thawing and aerosolization. *J. Gen. Microbiol.* **81**:207–215.
Cox, C. S., and R. J. Heckley. 1973. Effects of oxygen upon freeze-dried and freeze-thawed bacteria: viability and free radical studies. *Can. J. Microbiol.* **19**:189–194.
de Jong, J. C. 1970. Decay mechanism of polio and EMC viruses in aerosols, p. 210–245. *In* I. H. Silver (ed.), *Third International Symposium on Aerobiology*. Academic Press, Inc., New York.
de Jong, J. C., M. Harnsen, A. D. Plantinga, and T. Trouwborst. 1976. Aerosol stability functions of Semliki Forest virus. *Appl. Environ. Microbiol.* **32**:315–319.
de Jong, J. C., and K. C. Winkler. 1968. The inactivation of poliovirus in aerosols. *J. Hyg.* **66**:557–565.
de Mik, G., and I. de Groot. 1973. Effect of gases on the aerosol stability of various microorganisms, p. 155–158. *In* J. F. P. Hers and K. C. Winkler (ed.), *Airborne Transmission and Infection*. John Wiley & Sons, Inc., New York.
de Mik, G., and I. de Groot. 1977. The germicidal effect of the open air in different parts of the Netherlands. *J. Hyg.* **78**:175–180.
Dimmick, R. L., H. Wolochow, and M. A. Chatigny. 1979. Evidence that bacteria can form new cells in airborne particles. *Appl. Environ. Microbiol.* **37**:924–927.
Dimmock, N. L. 1967. Differences between the thermal inactivation of picornaviruses at "high" and "low"temperatures. *Virology* **31**:338–353.
Donaldson, A. I. 1972. Effect of radiation on selected virus. *Vet. Bull.* **48**:83–94.
Donaldson, A. I., and N. P. Ferris. 1975. The survival of foot-and-mouth disease virus in open air conditions. *J. Hyg.* **74**:409–415.
Druett, H. A. 1973a. Effect on the viability of microorganisms in aerosols of the rapid rarification of the surrounding air, p. 90–94. *In* J. F. P. Hers and K. C. Winkler (ed.), *Airborne Transmission and Infection*. John Wiley & Sons, Inc., New York.
Druett, H. A. 1973b. The open air factor, p. 141–151. *In* J. F. P. Hers and K. C. Winkler (ed.), *Airborne Transmission and Infection*. John Wiley & Sons, Inc., New York.
Dubovi, E. J. 1971. Biological activity of the nucleic acid extracted from bacterial viruses. *J. Appl. Microbiol.* **21**:624–630.

Dubovi, E. J., and T. G. Akers. 1970. Airborne stability of tailless bacterial viruses S-13 and MS-2. *J. Appl. Microbiol.* **19**:624–630.

Ehrlich, R., and S. Miller. 1971. Effect of relative humidity and temperature on airborne Venezuelan equine encephalitis virus. *J. Appl. Microbiol.* **22**:194–200.

Fannin, K. F., J. C. Spendlove, K. W. Cochran, and J. J. Gannon. 1976. Airborne coliphages from wastewater treatment facilities. *Appl. Environ. Microbiol.* **31**:705–710.

Goff, G. D., J. C. Spendlove, A. P. Adams, and P. S. Nicholes. 1973. Emission of microbial aerosols from sewage treatment plants that use trickling filters. *Health Serv. Rep.* **88**:640–652.

Goldberg, L. J., and I. Ford. 1973. The function of chemical additives in enhancing microbial survival in aerosols, p. 86–89. In J. F. P. Hers and K. C. Winkler (ed.), *Airborne Transmission and Infection*. John Wiley & Sons, Inc., New York.

Goodlow, R. G., and F. A. Leonard. 1961. Viability and infectivity of microorganisms in experimental airborne infection. *Bacteriol. Rev.* **25**:182–187.

Happ, J. W., J. B. Harstad, and L. M. Buchanan. 1966. Effect of air ions on submicron T1 bacteriophage aerosols. *Appl. Microbiol.* **14**:888–891.

Harper, G. J. 1961. Airborne microorganisms: survival tests with four viruses. *J. Hyg.* **59**:479–486.

Harper, G. J. 1963. The influence of environment on the survival of airborne virus particles in the laboratory. *Arch. Gesamte Virusforsch.* **13**:64–71.

Hatch, M. T., and R. L. Dimmick. 1966. Physiological responses of airborne bacteria to shifts in relative humidity. *Bacteriol. Rev.* **30**:597–603.

Hatch, M. T., and J. C. Warren. 1969. Enhanced recovery of airborne T_3 coliphage and *Pasteurella pestis* bacteriophage by means of a presampling humidification technique. *Appl. Microbiol.* **17**:685–689.

Hemmes, J. H., K. C. Winkler, and S. M. Kool. 1960. Virus survival as a seasonal factor in influenza and poliomyelitis. *Nature* (London) **188**:430–438.

Ijaz, M. K., S. A. Sattar, C. M. Johnson-Lussenburg, and V. S. Springthorpe. 1985. Comparison of the airborne survival of calf rotavirus and poliovirus type 1 (Sabin) aerosolized as a mixture. *Appl. Environ. Microbiol.* **49**:289–293.

Jensen, M. M. 1964. Inactivation of airborne viruses by ultraviolet irradiation. *Appl. Microbiol.* **12**:418–420.

Kethley, T. W., W. B. Crown, and E. L. Fincher. 1957. The nature and composition of experimental bacterial aerosols. *Appl. Microbiol.* **5**:1–17.

Krinsky, N. I. 1976. Cellular damage initiated by visible light. *Symp. Soc. Gen. Microbiol.* **26**:209–239.

Lembke, L. L., and R. N. Kniseley. 1980. Coliforms in aerosols generated by a municipal solid waste recovery system. *Appl. Environ. Microbiol.* **40**:888–891.

Lighthart, B. 1973. Survival of airborne bacteria in a high urban concentration of carbon monoxide. *Appl. Microbiol.* **25**:86–91.

Lighthart, B. 1984. Microbial aerosols: estimated contribution of combine harvesting to an airshed. *Appl. Environ. Microbiol.* **47**:430–432.

Lighthart, B., and A. S. Frisch. 1976. Estimation of viable airborne microbes downwind from a point source. *Appl. Environ. Microbiol.* **31**:700–704.

Lighthart, B., and J. Kim. 1989. Simulation of airborne microbial transport. *Appl. Environ. Microbiol.* **55**:2349–2355.

Lighthart, B., and A. J. Mohr. 1987. Estimating downwind concentrations of viable airborne microorganisms in dynamic atmospheric conditions. *Appl. Environ. Microbiol.* **53**:1580–1583.

Lindow, S. E. 1985. Ecology of *Pseudomonas syringae* relevant to the field use of Ice^- deletion mutants constructed in vitro for plant frost control, p. 23–35. In H. O. Halvorson, D. Pramer, and M. Rogul (ed.), *Engineered Organisms in the Environment: Scientific Issues*. American Society for Microbiology, Washington, D.C.

Lindow, S. E. 1986. Strategies and practice of biological control of ice nucleation active bacteria on plants, p. 293–311. In N. Fokkema (ed.), *Microbiology of the Phyllosphere*. Cambridge University Press, Cambridge.

May, K. R., H. A. Druett, and L. P. Packman. 1969. Toxicity of open air to a variety of microorganisms. *Nature* (London) **221**:1146–1147.

Mayhew, C. J., W. D. Zimmerman, and N. Hahon. 1968. Assessment of aerosol stability of yellow fever virus by fluorescent-cell counting. *Appl. Microbiol.* **16**:263–266.
McIntosh, D. H. 1972. *Meteorological Glossary.* Chemical Publishing, New York.
Moe, K., and G. J. Harper. 1983. The effect of relative humidity and temperature on the survival of calf rotavirus in aerosol. *Arch. Virol.* **76**:211–216.
Mohr, A. J. 1984. Aerosol stability of reovirus. Ph.D. Dissertation. Utah State University, Logan.
Moore, B. E., B. P. Sagik, and C. A. Sorber. 1979. Procedure for the recovery of airborne human enteric viruses during spray irrigation of treated wastewater. *Appl. Environ. Microbiol.* **38**:688–693.
Pasquill, F. 1961. The estimation of the dispersion of windborne material. *Meteor. Mag.* **90**:33–49.
Phalen, R. F. 1984. *Inhalation Studies: Foundations and Techniques.* CRC Press, Inc. Boca Raton, Fla.
Phillips, G., G. J. Harris, and M. W. Jones. 1963. The effect of ions on microorganisms. *Int. J. Biometeorol.* **8**:27–37.
Pike, R. M. 1978. Past and present hazards of working with infectious agents. *Arch. Pathol. Lab. Med.* **102**:333–336.
Pike, R. M., S. E. Sulkin, and M. L. Schulze. 1965. Continuing importance of laboratory-acquired infections. *Am. J. Public Health* **55**:190–199.
Poon, C. P. C. 1966. Studies on the instantaneous death of airborne *Escherichia coli. Am. J. Epidemiol.* **84**:1–9.
Quentin, G. H. 1979. Modeling of aerobiological systems, p. 311–329. *In* R. L. Edmonds (ed.), *Aerobiology: the Ecological Systems Approach.* Dowden, Hutchinson, and Ross, Inc., Stroudsburg, Pa.
Resnick, I. G., A. J. Mohr, and B. G. Harper. 1987. Production, purification and use of MS2 bacteriophage as a viral aerosol simulant. Abstr. Annu. Meet. Am. Soc. Microbiol. 1987, Q23, p. 285.
Richardson, J. H., and W. E. Barkley (ed.). 1984. *Biosafety in Microbiological Laboratories,* p. 1–4. Publication no. (CDC) 84-839. Centers for Disease Control, Atlanta, Ga.
Rosebury, T. 1947. *Experimental Air-Borne Infection.* The Williams & Wilkins Co., Baltimore.
Smith, L. P., and M. E. Hugh-Jones. 1969. The weather factor in foot-and-mouth disease epidemics. *Nature* (London) **223**:712–715.
Soddu, A., and W. R. Vieth. 1980. The affect of sugars on membranes. *J. Mol. Catal.* **7**:491–500.
Songer, J. R. 1966. Influence of relative humidity on the survival of some airborne viruses. *Appl. Microbiol.* **15**:1–16.
Sorber, C. A., H. T. Bausum, S. A. Schaub, and M. J. Small. 1976. A study of bacterial aerosols at a wastewater irrigation site. *J. Water Pollut. Control Fed.* **48**:2367–2379.
Spendlove, J. C. 1974. Industrial, agricultural, and municipal microbial aerosol problems. *Dev. Ind. Microbiol.* **15**:20–27.
Spendlove, J. C. 1975. Penetration of structures by microbial aerosols. *Dev. Ind. Microbiol.* **16**:20–27.
Spendlove, J. C., and K. F. Fannin. 1982. Methods of characterization of virus aerosols, p. 261–329. *In* C. P. Gerba and S. M. Goyal (ed.), *Methods in Environmental Virology.* Marcel Dekker, Inc., New York.
Strange, R. E., and C. S. Cox. 1976. Survival of dried and airborne bacteria. *Symp. Soc. Gen. Microbiol.* **26**:111–154.
Teltsch, B., and E. Katzenelson. 1978. Airborne enteric bacteria and viruses from spray irrigation with wastewater. *Appl. Environ. Microbiol.* **35**:290–296.
Teltsch, B., H. I. Shuval, and J. Tadmor. 1980. Die-away kinetics of aerosolized bacteria from sprinkler application of wastewater. *Appl. Environ. Microbiol.* **39**:1191–1197.
Thomson, S. V., M. N. Schroth, W. J. Moller, and W. O. Reil. 1976. Efficacy of bactericides and saprophytic bacteria in reducing colonization and infection of pear flowers by *Erwinia amylovora. Phytopathology* **66**:1457–1459.

Trillat, A., and A. Beauvillain. 1937. Studies of airborne transmission of influenza in ferrets by the pulmonary and ocular routes. *C. R. Acad. Sci. Paris* **205**:1186–1188. (In French.)

Trouwborst, T., J. C. de Jong, and K. C. Winkler. 1972. Mechanism of inactivation in aerosols of phage T_1. *J. Gen. Virol.* **15**:235.

Warren, J. C., T. G. Akers, and E. J. Dubovi. 1969. Effect of prehumidification on sampling of selected airborne viruses. *Appl. Microbiol.* **18**:893–896.

Webb, S. J. 1965. *The Role of Bound Water in the Maintenance of the Integrity of a Cell or Virus.* Charles C Thomas, Publisher, Springfield, Ill.

Webb, S. J. 1969. The effects of oxygen on the possible repair of dehydration damage by *Escherichia coli*. *J. Gen. Microbiol.* **58**:317–326.

Webb, S. J., R. Bather, and R. W. Hodges. 1963. The effect of relative humidity and inositol on air-borne viruses. *Can. J. Microbiol.* **9**:87–94.

Webb, S. J., and A. D. Booth. 1969. The effect of radiation on *Escherichia coli*. *Nature* (London) **222**:1199–1200.

Williamson, A. E., and H. B. Gotaas. 1942. Aerosol sterilization of air-borne bacteria. *Ind. Med.* **11**:40–45.

Woodcock, A. H. 1955. Bursting bubbles and air pollution. *Sewage Ind. Wastes* **27**:1189–1192.

Zeterberg, J. M. 1973. A review of respiratory virology and the spread of virulent and possibly antigenic viruses via air conditioning systems. *Ann. Allergy* **31**:228–234.

Chapter 9

Models for the Survival of Bacteria Applied to the Foliage of Crop Plants†

Guy R. Knudsen

Why Model Bacterial Survival on Foliage? 191
Epiphytic Bacteria and the Phylloplane: the Descriptive Approach. 193
Analytical Models 197
Simulation Models 199
Vishnu/Siva: a Different Bacterial Population Model 201
Borrowings from Population and Community Ecology 209
Conclusions. 211
References. 212

WHY MODEL BACTERIAL SURVIVAL ON FOLIAGE?

Until recently, most interest in the topic of bacterial survival on plant foliage derived from the role of some bacterial species as foliar pathogens of plants. That concern is as important as ever, but more and more we are looking at populations of foliar bacteria that fulfill, at least from an anthropocentric viewpoint, a more benign role towards plants. Strains of bacteria in several genera, whether naturally occurring or modified by recombinant DNA techniques, offer promise for foliar application to crops for control of insect pests, plant pathogens, or weeds or to protect plants from frost injury (Blakeman and Brodie, 1976; Falcon, 1971; Freeman and Charudattan, 1980; Lindemann, 1985; Lindow, 1986; Spurr, 1981).

The first genetically engineered microorganisms to be released legally in the United States were strains of *Pseudomonas syringae* and *Pseudomonas fluorescens* from which the ice nucleation gene was deleted (Lindow, 1985, 1986; Supkoff et al., 1988). These Ice⁻ strains were intended to competitively exclude indigenous ice nucleation-active bacteria involved in frost damage on plant surfaces. To date, small-scale field releases of aerosolized Ice⁻ bacteria have been undertaken by

†Published as Idaho Agricultural Experiment Station Paper no. 90750.

Guy R. Knudsen • Plant Pathology Division, Department of Plant, Soil and Entomological Sciences, University of Idaho, Moscow, Idaho 83843.

Advanced Genetic Sciences, Inc., and the University of California, Berkeley. There has been a large investment in time, money, and effort in attempts to determine dispersal patterns and survival of the applied bacteria (Knudsen, 1989; Lindow et al., 1988; Supkoff et al., 1988). There have been increasing numbers of applications for experimental-use permits to release genetically engineered bacteria in small-scale field trials. These include requests to apply microbial pest control agents to foliage of agricultural crops, since research with these organisms has progressed to the point where field tests are needed to evaluate their performance. As future tests are performed, the availability of predictive models for bacterial dispersal and survival could reduce the cost and improve the utility of sampling efforts.

In the field of entomology, insect pathologists have found several species of bacteria, both sporeformers and nonsporeformers, to be pathogenic to insects (Falcon, 1971). Insect pest control with the most widely used bacterial agent, *Bacillus thuringiensis*, relies largely on a crystalline toxin acting as a microbial insecticide. However, with most other potential bacterial biocontrol agents, and perhaps with *B. thuringiensis*, the ability of the organism to survive (as spores or vegetative cells) and multiply is important for control. Most pathogens in the genus *Bacillus* are ingested as spores, which germinate in the gut and subsequently invade the hemocoel, causing a bacterial septicemia (Falcon, 1971). Most nonsporeforming bacterial entomopathogens do not multiply readily in the gut, but multiply and cause a septicemia in the hemocoel once they have invaded it (Falcon, 1971). Ignoffo (1985) discussed the need for quantitative models of the pathogen-insect complex and development of epizootiological models in model ecosystems. He pointed out that the most important attributes of insect pathogens, i.e., their ability to establish, disperse, multiply, and persist, need to be researched and quantified in order to evaluate their potential in specific agricultural ecosystems.

To plant pathologists, biological control of foliar plant pathogens is an attractive prospect. It offers the potential to reduce chemical pesticide use and, perhaps, to establish self-perpetuating disease control agents on plant surfaces. There have been numerous reports of epiphytic bacteria that are antagonistic to phytopathogens in vitro, on excised leaves, or on whole plants in controlled environments (Blakeman and Brodie, 1976; Blakeman and Fraser, 1971; Knudsen and Hudler, 1981; Leben et al., 1965; Sleesman and Leben, 1976; Spurr, 1981). Unfortunately, few of these preliminary observations have been followed by reports of consistently successful disease control in field trials (Blakeman and Brodie, 1976; Knudsen and Spurr, 1987; Leben et al., 1965). Lack of disease control in the field has been attributed to both the increased complexity of the field system and the inability of bacterial antagonists to survive in sufficient numbers in fluctuating environments (Knudsen and Hudler, 1987; Leben et al., 1965; Newhook, 1951; Sleesman and Leben, 1976). However, there has been relatively little quantitative description of the environmental factors affecting bacteria applied to plant surfaces.

Knudsen and Spurr (1988) pointed out that biological control is a function of complexity, but paradoxically, complexity hampers efforts to understand and apply biological control. Cause-and-effect relationships can be obscured by interactions that make them inappropriate for simple statistical analysis. Field evaluation of biocontrol agents thus becomes a black box: we have some idea of what goes in (e.g., antagonist and pathogen) and what comes out (the level of disease control), but the internal chain of events remains at least partially a mystery.

Evaluation of an ecological system requires more than determining the

Bacterial Survival on Crop Plant Foliage 193

interactions in the system and quantifying them. To understand the system, and to make predictions about it, the form in which information is organized is as important as the information itself (Shrum, 1982). As Albert Einstein remarked, referring to Kepler's discovery of the geometric form of planetary orbits, "It seems that the human mind has first to construct forms, independently, before we can find them in things." Although there may be a good understanding of selected elements of a system (e.g., pathogenesis and mechanisms of inhibition), a comprehensive framework (model) is needed to order knowledge into testable hypotheses and theories. With natural systems it is often difficult to define the boundaries of the system to be modeled. The flows in and out of the boundaries of the system are often diffuse and change over time (Dommergues et al., 1978) and with scale, but this exercise of defining the system is one of the most useful attributes of the modeling approach.

Some important development criteria and philosophical aspects of modeling in microbiology were summarized by Dommergues et al. (1978). First, a model should be held in perspective as only one of several tools used to analyze a system. As much as possible, the model's validity should be testable. Finally, although a model abstracts from reality, it should be able to capture essential features of the system and express them as a system of mathematical relationships that are able to imitate some subset of the original system's behavior (Reichle et al., 1973).

There are few models that are specifically identified for epiphytic bacterial populations. However, there is a fairly large body of literature in which knowledge about bacteria on plant surfaces has been assembled and used in a descriptive or predictive manner; this is a broad interpretation of the term "model." In this review, I will consider various ways in which quantitative information about bacteria on plant surfaces can be organized. These include so-called descriptive or statistical models, applications of traditional analytical population models, some simulation efforts, and, finally, possibilities of borrowing from other areas of ecological theory.

EPIPHYTIC BACTERIA AND THE PHYLLOPLANE: THE DESCRIPTIVE APPROACH

The existence of epiphytic populations of phytopathogenic *Pseudomonas* spp., *Xanthomonas* spp., and *Erwinia* spp. has been extensively documented (Hirano and Upper, 1983; Leben, 1974). Reported epiphytic bacteria that were not specifically identified as pathogens also include species of *Pseudomonas*, *Xanthomonas*, *Chromobacterium*, and *Klebsiella* (Campbell, 1989).

Seasonal variation in temperature and rainfall causes cyclical patterns of multiplication and survival of epiphytic bacteria. A regular, annual succession of microorganisms apparently occurs on foliage of many plants (Jensen, 1971). Such seasonal effects have been described for ice nucleation-active (INA) bacterial colonists of fruit trees (Gross et al., 1983). In the spring, developing and mature flowers are rapidly colonized, after which resident bacterial populations decline to low or undetectable levels during the warm, dry summer. Ephemeral sources of nutrients, such as aphid honeydew or pollen, may also contribute to variability in the composition of the phylloplane microbiota. In autumn, numbers of bacteria increase gradually until leaf abcission. During winter dormancy of the fruit tree, bacterial numbers decline slowly until new vegetative growth begins in the spring. Thus, an overall cyclical pattern can be observed (Gross et al., 1983).

Cameron (1970) also reported a cyclical pattern of colonization of sweet cherry trees by *P. syringae*, and Lindow et al. (1978) found that populations of INA bacteria in *Prunus* species were higher during the spring than autumn. Gross et al. (1983) suggested that patterns of growth and colonization of INA bacteria correlate with those times when frosts are likely to occur and also when highly frost-sensitive plant tissues, such as developing buds, flowers, and fruit, are present.

Where do epiphytic bacteria overwinter? Much of the current information on this subject comes from the plant pathology literature. For instance, epidemics of bacterial brown spot disease of bean, caused by *P. syringae* pv. *syringae*, have been attributed to infected seed (Hoitink et al., 1968), inoculum overwintering in debris from infected plants (Hagedorn and Patel, 1965), or epiphytic bacteria present on nonhost leguminous plants (Ercolani et al., 1974). Leben (1974) offered three "constructs" as a framework to think about the survival of plant pathogenic bacteria: (i) long-term survival of bacterial pathogens takes place only in association with living or dead plant tissues; (ii) long-term survival usually requires cells to be in aggregates or otherwise protected in association with living plant tissues; (iii) pathogens in a state of reduced metabolism, or hypobiosis, are more likely to survive than are active cells.

The term "hypobiosis" is taken from the microbiological literature (Linton, 1971); a related term, "bacteriostasis," has also been used (Brown, 1973). Hypobiotic cells may survive for long periods without added nutrients and are more able to withstand physical and chemical stresses that would be fatal to cells with high metabolic activity (Leben, 1974). For example, cells of *Pseudomonas aeruginosa* were more sensitive to desiccation in the exponential growth phase than when cells were 7 days old (Skaliy and Eagon, 1972). Hypobiotic cells of some bacterial plant pathogens can survive for many years in dry plant tissues (Leben, 1974).

Populations of epiphytic bacteria differ on upper and lower leaf surfaces, as well as within plant canopies depending on exposure to wind, rain, and solar radiation (Campbell, 1989). On a larger scale, climate and cropping patterns affect distribution patterns of epiphytic bacteria. Gross et al. (1983) compared populations of INA bacteria in Yakima Valley and Wenatchee (Washington State) area orchards and in the Hood River valley. They concluded that INA bacterial populations were not uniformly distributed and that their presence was strongly influenced by local climatic conditions. Lindemann et al. (1984a) studied the influence of cropping patterns on epiphytic populations of *P. syringae* pv. *syringae* on bean in bean-growing areas as compared with areas where there was no commercial bean production. Differences in *P. syringae* pv. *syringae* on both host and nonhost plants, along with differences in brown spot disease incidence, were attributed to the intensive cropping of snap bean in some areas.

Mew and Kennedy (1971) sprayed suspensions of different pathogenic strains of *P. syringae* pv. *glycinea* onto soybean leaves and monitored bacterial populations over 10 days. They observed population increases on leaves of susceptible soybean cultivars, but not on resistant cultivars, and concluded that cultivar specificity of *P. syringae* pv. *glycinea* correlates with the resident phase of the bacteria on leaf surfaces of soybeans. Unfortunately, the environmental conditions prevailing over the 10-day incubation period and the variability in observed populations were not reported. However, subsequent work by Surico et al. (1981) confirmed that *P. syringae* pv. *glycinea* in aerosols can initiate extensive epiphytic growth under some conditions and perhaps can contribute to field epidemics. Populations on leaflets showed an initial rapid decrease with a subsequent increase. Survival of about 50

CFU of *P. syringae* pv. *glycinea* on soybean leaves after exposure to the aerosol was sufficient to generate populations of 10^3 to 10^4 CFU cm^{-2} after 9 to 14 days of incubation.

Survival and colonization by introduced bacterial inocula to establish enzootic or epizootic insect diseases have not been extensively studied. In soil, spores of *Bacillus popilliae*, a pathogen of Japanese beetle grubs, became established after introduction (Falcon, 1971). Brand and Pinnock (1981) and Pinnock et al. (1978) monitored the persistence of *B. thuringiensis* spores on foliage and related these population levels to the mortality of insects feeding on the foliage. Armstrong et al. (1987) followed populations of a recombinant strain of *Pseudomonas cepacia* sprayed onto radish leaves which were then eaten by cutworm larvae. The bacteria accumulated in the insects' foreguts, but passage completely through the larval digestive system was not observed. However, these observations suggested that insects might serve as reservoirs for recombinant bacteria sprayed onto plant surfaces.

Environmental factors severely restrict the growth of microorganisms on leaf surfaces. Physical factors, of which solar radiation, moisture, and temperature are most important, vary on a large scale with climate, on a smaller scale within the plant canopy, and on a microscopic scale over the leaf surface. Solar radiation in the range of 300 to 320 nm is destructive to exposed bacteria (Ignoffo, 1985). Unprotected spores of *Bacillus* spp. exposed to direct sunlight have been reported to have a half-life of less than 1 day (Ignoffo, 1985). Water is frequently limiting on plant surfaces under temperate and arid conditions, and growth of epiphytic microbes may only occur following rain, periods of dew, or at least high humidity (Campbell, 1989). Water affects survival, multiplication, and dispersal of plant surface bacteria, and free water is probably necessary for diffusion of antimicrobial compounds from epiphytic bacteria used as biocontrol agents (Knudsen and Hudler, 1987). There is an extensive body of literature in the area of agricultural and forest meteorology, appropriate for monitoring and modeling components of the physical environment at various levels of scale (Monteith, 1973; Pennypacker, 1978; Wallin, 1963; Zadoks and Schein, 1979).

Both temperature and solar radiation have direct effects (e.g., mortality) and indirect effects (metabolic rates, mutations, etc.) on numbers and types of bacteria on above-ground plant parts. There are also numerous interactions among physical factors. For example, cells with higher metabolic activity due to higher temperatures are likely to be more susceptible to UV radiation. On the other hand, high relative humidity may reduce adverse effects of UV radiation. Drying kills metabolically active cells of most bacteria, but as previously mentioned, metabolically inactive cells usually are much more able to survive desiccation (e.g., Skaliy and Eagon, 1972).

Wilkinson (1966) determined survival curves for *Serratia marcescens*, *Sarcina lutea*, *Francisella tularensis*, and *Yersinia pestis* cells that were deposited from aerosols onto metallic surfaces (which may be partially analogous to leaf surfaces) and then incubated at different humidity and temperature levels. With the exception of *S. marcescens*, which survived best in a saturated atmosphere, cells remained alive longest in a dry (below 20% relative humidity) atmosphere. Survival of *S. marcescens*, *S. lutea*, and *Y. pestis* was lowest at intermediate humidity (20 to 80% relative humidity), whereas survival of *F. tularensis* was lowest in a saturated atmosphere. At the least, these results suggest the level of variability in survival among different species of bacteria. Wilkinson (1966) also emphasized the importance of variable physical factors (aerosolization, cell clumping, trapped moisture, residual medium, etc.) that influence cellular mechanisms and thereby influence the efficiency of recovery of living cells from test surfaces. Although general survival curves for each species at particular temperature and humidity levels were repeatable, Wilkinson (1966) observed a large variability, sometimes as much as 100-fold, in individual observations about means for each sample time. He emphasized that, despite constant environmental stress, significant variability arises from differential effects of environmental stress on individual cells within a population. In subsequent discussion of population models, I will return to the important point that bacterial populations in nature are not homogeneous, but are made up of individuals in differing physiological states and in different microhabitats.

Numerous experiments have pointed out differences between laboratory and field results with epiphytic bacterial populations. Knudsen and Hudler (1987) applied a rifampin-resistant strain of *P. fluorescens* to red pine seedlings in a nursery, then monitored populations at 1- to 12-h intervals over a 36-h period by washing shoots and then assaying the washings on a selective medium. Populations of *P. fluorescens* declined slowly during the night when leaf surfaces remained wet with dew, but declined rapidly during the following day, which was sunny and warm. At each sample time, recovery was greater from seedlings that had been sprayed with bacteria in dilute nutrient solution than from those treated with bacteria in water. The decline of bacterial populations in these field trials was in contrast to previous growth chamber experiments (Knudsen and Hudler, 1987), in which bacteria persisted and even increased over several days if leaf wetness was maintained. Knudsen and Spurr (1987) also observed an approximately exponential decline in viability of two biocontrol agents, *P. cepacia* and *B. thuringiensis*, formulated as lyophilized cells or spores in a wettable powder and sprayed onto peanut foliage in the field. However, over a series of 2-week periods, the numbers of *P. cepacia* recovered were generally more variable than those of *B. thuringiensis* and were sometimes observed to increase during longer periods of leaf wetness or high humidity.

There have been several prototype forecast models for bacterial plant pathogens on aerial plant parts. Thomson et al. (1975) monitored populations of *Erwinia amylovora* on individual pear flowers and determined that low or undetectable populations of the pathogen correlated with low levels of disease, suggesting that monitoring of epiphytic populations in flowers may be useful for timing bactericide applications.

Lindemann et al. (1984b) made the observation that although theoretically a single pathogenic bacterium can produce a lesion on a susceptible host, the probability is quite small, so that commonly a minimum effective dose of inoculum, or "infection threshold" (Layne, 1968), is needed to obtain bacterial infection in experiments. For example, Weller and Saettler (1980) reported an infection threshold of 5×10^6 epiphytic *Xanthomonas campestris* pv. *phaseoli* per bean leaf for development of bean common blight symptoms in the field, based on the lowest population of *X. campestris* pv. *phaseoli* recovered from washings of individual leaves with visible lesions.

Starting with the assumption that epiphytic bacterial population size frequencies are usually log-normally distributed (Hirano et al., 1982), Lindemann et al. (1984b) calculated $\log_{10}$ values of *P. syringae* populations on individual bean leaflets and plotted them on a probability scale at appropriate cumulative probability values expected for samples taken from a normal distribution. Expected frequencies of leaflets with populations greater than or equal to a particular value were estimated graphically and compared with observed levels of brown spot disease incidence in the field. Brown spot was not detected at sites where $\log_{10}$ of the epiphytic *P. syringae* population size was <4.0 on every bean leaflet sampled, and thus 10^4 CFU/g of leaflet tissue was estimated to be an apparent infection threshold population of *P. syringae* on bean. Lindemann et al. (1984b) suggested that predictive models based on infection thresholds are preferable to models based on mean pathogen populations, because "infections occur on individual plant parts, rather than on some theoretical mean plant part."

Rouse et al. (1985) furthered this concept by developing a theoretical model relating foliar disease incidence to bacterial population size, specifically incorporating the distribution of population size frequencies. Their model, derived for bacterial brown spot of bean, again assumed a log-normal distribution of the bacterial pathogen (*P. syringae* pv. *syringae*) on individual leaflets. They coupled this population model with a probit dose-response function (i.e., the probability of disease occurrence increases with the logarithm of bacterial numbers on a leaflet) (Ercolani, 1973). In their model, the probability that an individual leaflet was diseased was dependent on the mean and variance of the distribution of $\log_{10}$ of bacterial numbers on leaflets and the mean and variance of the dose-response function. The model was then used to relate pathogen population frequency to disease incidence 4 to 11 days later.

Spatial analysis of bacterial populations has received relatively little attention. Yuen et al. (1987) used monoclonal antibodies to detect and differentiate strains of *X. campestris* pv. *campestris* that had been sprayed onto cabbage seedlings, which were then transplanted into field plots. In this way, they were able to follow the progress of black rot disease caused by two serologically distinct strains originating from different inoculum sources and to quantify infection rates and spatial patterns for each strain. Spatial spread of one strain corresponded to the direction of surface water flow after heavy rainfall.

Although spatial patterns have been analyzed for several plant diseases, including some caused by bacteria, there have seldom been attempts to characterize changes in spatial pattern over the course of an epidemic (Madden, 1989). Madden (1989) discussed an assortment of relatively new statistical techniques, including spatiotemporal autocorrelation models, available for this type of analysis. Another approach is to combine simulation methods and spatial statistical analysis by simulating processes that generate spatiotemporal patterns. The resultant patterns at different points in time can be analyzed using geostatistical techniques (e.g., Isaaks and Srivastava, 1989; Yates et al., 1986). Although to my knowledge this approach hasn't been applied to microbial populations, recent papers by Knudsen and Schotzko (submitted) and Schotzko et al. (submitted) describe a simulation model for aphid population growth and dispersal within a host plant population, along with geostatistical analysis of resultant patterns of insects and plant damage. This approach, in which process and pattern are modeled concurrently, should be applicable to the ecology of epiphytic bacteria.

ANALYTICAL MODELS

The analytical approach to quantitative microbial ecology involves estimation of constants or parameters in equations chosen to represent processes such as substrate depletion, growth, or surface colonization. Several analytical models are available to describe bacterial growth and kinetics in simple systems such as laboratory culture. Probably the simplest model states that change in growth of the microbial population ($\partial N/\partial t$) is directly proportional to the size of the population. Thus, $\partial N/\partial t = rN$, where r is the growth rate constant; this is exponential growth and is based upon first-order chemical reaction kinetics and assumes that no increasing growth retardation is operative (Lambrecht et al., 1988). If the value of r is negative, the size of the population declines exponentially. Such a model (exponential decay) may reasonably fit a variety of microbial death processes. Brand and Pinnock (1981) and Pinnock et al. (1978) used exponential decay models to predict the persistence of *B. thuringiensis* spores on foliage and to relate these population levels to the mortality of insects feeding on the foliage.

Bacterial population size in a nonreplenished medium usually approaches a maximum limit or carrying capacity (K), at which time the specific growth rate approaches zero. The logistic model is frequently used to describe this aspect of growth. The model, $\partial N/\partial t = rN[(K-N)/K]$, assumes that growth increase is retarded linearly with increasing population size (Lambrecht et al., 1988). The logistic model has drawbacks as a general model of population growth, since its predicted asymptotic stable density is probably almost never achieved by natural populations (Krebs, 1985). Knudsen and Hudler (1987) incorporated logistic growth into their simulation model for *P. fluorescens* on pine foliage and remarked on the unrealistic limitations imposed by the invariant carrying capacity (asymptote) of the logistic model.

Knudsen and Spurr (1988) combined simple analytical models to predict survival of *P. cepacia* and *B. thuringiensis* on peanut foliage, where bacteria were applied as aqueous suspensions of freeze-dried vegetative cells (*P. cepacia*) or spores (*B. thuringiensis*). The model for *P. cepacia* assumed that mortality was very high (95%) on the day that cells were applied, but that thereafter cells died at a slower exponential rate during periods of relative humidity less than 95% and that the population increased slowly (logistic growth) under conditions of high humidity. The model for *B. thuringiensis* assumed a simple exponential decline over time and did not distinguish between endospores and vegetative cells. Model predictions, especially for *P. cepacia*, were a reasonably good fit to field observations, potentially providing a basis for improving field performance of these biocontrol agents by modifying application rates or frequency.

Bacteria inoculated into liquid medium usually exhibit a lag phase, which has been defined as a transition period during which r increases to the maximum value characteristic of the culture environment (Pirt, 1975). In general, a lag in growth may result from changes in nutrition or the physical environment, presence of inhibitory factors, spore germination, or age of the culture (Delaquis et al., 1989). The logistic model can be readily modified to incorporate time lags (Wangersky and Cunningham, 1956). For example, a lag between a change in the environment and the corresponding change in the rate of population growth can be incorporated into the term $(K-N)/K$ as follows: $\partial N/\partial t = rN[(K-N_{t-w})/K]$, where w is the reaction time lag (Krebs, 1985).

Attachment to surfaces may cause variations in the specific growth rate r even under constant environmental conditions, which may be relevant to colonization of leaves and other aerial plant parts. Delaquis et al. (1989) studied attachment, detachment, and rates of growth of *P. fluorescens* on glass coverslips in petri dishes and observed that the specific growth rate of unattached cells in biofilms was almost twice that determined for the total population.

Another commonly used microbial growth model is the Monod equation, which expresses growth rate as a function of substrate concentration. Other models, such as the Blackman bilinear (Blackman, 1905) and Powell diffusional (Powell, 1967) models, may better describe growth for some organisms than the Monod equation (Robinson, 1985). Robinson (1985) described the use of nonlinear parameter estimation techniques (also referred to as nonlinear regression analysis or nonlinear optimization) to estimate microbial kinetic parameters. Nonlinear parameter estimation methods involve recursive techniques in which parameter estimates are sequentially improved until the estimates that minimize differences between observed and predicted values are determined (Robinson, 1985). To my knowledge, the above models and techniques have not yet been used to describe bacterial population changes on plant surfaces.

SIMULATION MODELS

The marble index of a mind for ever voyaging
Through strange seas of thought alone . . .
—W. Wordsworth

Wordsworth did not write the above couplet in reference to the art of simulation, but about Isaac Newton, who preceded the development of simulation modeling by several centuries. The quote seems appropriate, though, since for many biologists simulation modeling embodies the seductive concept of "voyaging through strange seas of thought alone." However, even imaginary ships sometimes sail through rocky waters, and within complex simulation models the rocks can be hard to discern. Some recent simulation attempts are described below, along with a suggested different approach to modeling bacterial populations.

Various methods have been used for simulating microbial processes in nature. One basic systems approach to modeling is initially to divide a system into a number of compartments, which are usually represented as pools of biomass and pools of various inorganic and organic entities. Next, it is necessary to define the processes that allow interaction of one compartment with another (Dommergues et al., 1978; Zadoks, 1971). To look specifically at limiting factors, a somewhat different approach from that used for ecosystem analysis is needed. First, the model is designed to study a specific, measurable microbial population. Second, the population studied should have a definable measurable activity or behavior (Dommergues et al., 1978). Examples of such a measurable attribute would be population size or metabolic activity.

Most models that fit the above descriptions have been developed for mixed populations in chemostat cultures or, in some cases, for bacteria in aquatic or soil systems. For example, Beek and Frissel (1973) developed a complex soil system model that incorporated organic matter decomposition processes and nitrogen cycle transformations; microbial growth rates in soil could be limited by environmental factors including temperature, pH, and moisture. Generally, however, few models have tried to deal specifically with the way in which such factors may affect growth rates (Dommergues et al., 1978). I know of no true ecosystem models for foliar bacteria.

In recent years, plant pathologists have made significant progress in organizing epidemiological knowledge into models of varying complexity (Berger, 1977;

Zadoks, 1971). Berger (1977) listed epidemiological strategies for plant disease control and noted that the same strategies and techniques will be useful in efforts to induce epidemics rather than control them. The same principles should largely apply to manipulating populations of bacteria on foliage. Berger (1977) also advocated the use of computer simulation and modeling techniques to assess epidemics. Computer simulation of epidemics usually imitates the temporal progression of an epidemic in small steps. A dynamic system is thus specified as a set of state variables, which are measurable entities (bacterial cells per unit leaf area, lesions, etc.), and rate variables, which define state changes over time. This approach ideally generates testable predictions about populations. Disease simulators have been developed for several plant diseases, but there are few examples thus far of simulations for bacteria on plant surfaces.

Knudsen and Hudler (1987) developed a computer simulation model (INHIB-SIM) to predict population changes of *P. fluorescens* applied to red pine seedlings and the influence of the bacteria on germination of conidia of the plant pathogen *Gremmeniella abietina*, under various environmental conditions. Effects of temperature, leaf wetness, and added nutrients were incorporated into the simulation. Bacterial population growth was modeled as a temperature- and nutrient-dependent logistic increase in the presence of free water (leaf wetness) and an exponential decline in the absence of free water. Germination of conidia was a dose-response function of the number of surviving bacteria at any time. The model was tested under controlled environment and field conditions. In simulation runs as well as validation experiments, recoverable populations of bacteria were extremely sensitive to leaf wetness. Even short dry periods (3 to 4 h) caused a large, rapid decrease in numbers of recoverable bacteria. Bacteria increased on the leaf surface following restoration of leaf wetness. However, once populations were depleted, the rate of increase was insufficient to adequately restore them to the large numbers needed for significant inhibition of fungal spore germination. Bacterial populations were relatively insensitive to temperature fluctuations. Thus, experimental results and model sensitivity analysis suggested that even under optimum conditions for survival of *P. fluorescens*, disease control would be inadequate, and pointed out the need for increased efficacy of microbial antagonists for successful biological control in that system.

Knudsen et al. (1988) described a simulation model to predict dynamics of survival and conjugative plasmid transfer of genetically engineered donor and nonrecombinant recipient strains of *P. cepacia* in simple rhizosphere and phylloplane microcosms. Rate parameters were derived from previous laboratory and field studies (Walter et al., 1987; Knudsen and Spurr, 1985, 1987), and the model was tested (i) in a microcosm planted with radish seeds and inoculated with donor and recipient strains and (ii) on leaf surfaces of radish and bean plants also growing in microcosms. The computer simulation approach allowed predicted survival rates to vary in response to a changing environment.

MICROBE-SCREEN, a simulation model developed for the U.S. Environmental Protection Agency (Reichenbach et al., 1987), was designed to assess the potential fate of passively dispersed bioengineered microorganisms released to the environment. Although MICROBE-SCREEN was initially developed without a vegetation component, it contains modules to predict aerial survival, dispersal, and deposition kinetics, as well as survival and dispersal in soil and surface water. MICROBE-SCREEN would be a likely candidate for inclusion of a submodel to describe deposition, survival, and dispersal on plant surfaces.

Population models, including simulation models, can serve as tools for computer-aided design of methods to genetically manipulate and apply bacteria to foliage of crop plants (Spurr and Knudsen, 1985). The modeling approach allows data to be assembled from in vitro and controlled-environment experiments into a logical framework in which a potential control agent can be subjected to an infinite variety of simulated weather conditions. Knudsen and Spurr (1985) suggested the use of an "expert systems" approach which couples a population model with a data base containing information on attributes of bacterial strains, formulations, and weather data. To date, however, this approach has not been fully implemented.

VISHNU/SIVA: A DIFFERENT BACTERIAL POPULATION MODEL

Bacterial population models based on exponential, Monod, logistic growth, or similar descriptive (i.e., statistically derived) models may not be especially applicable in a spatially and temporally heterogeneous environment. The problem is compounded when such models are used as a basis for prediction of microbial effects, since these will also vary with environmental and physiological influences. Individual bacterial cells in nature may be in a condition ranging from physiological dormancy to active metabolism and cell division. Environmental effects on survival, growth, and potential for ecological effects will vary with the physiological state of the cells (Caldwell et al., 1989; Olsen and Bakken, 1987; Reeve et al., 1984; Stotzky, 1980; Tempest et al., 1983; Walmsley, 1976). Furthermore, populations of cells tend to become asynchronous over time with respect to physiological state, due to changes in microenvironment as well as genetic differences among individuals. Ideally, a useful and flexible model would need to account for variability in bacterial populations as described above as well as for the interaction between environment and physiological status of bacterial cells.

How might such a model be constructed? One possible implementation is suggested below. To help envision this model, I would first like to present the general concept metaphorically. In the Hindu pantheon, at the highest level alongside Brahma, the creator, stand two other deities: Vishnu the preserver and Siva the destroyer. They are manifest at all times and are inseparable, since preservation and destruction are like two sides of the same coin. "Morning dies to give birth to noon. Noon dies when night is born. In this chain of birth and death the day is maintained. In this perpetual process of creation and destruction, the universe is maintained" (Parthasarathy, 1983). What does this have to do with bacterial populations? The equilibrium between Vishnu and Siva, which is a balance between a tendency toward existence and a tendency toward annihilation, is suggested here as a metaphor for bacterial life and as the basis for an approach to modeling vegetative bacterial populations.

To visualize this model, consider an endless path representing the physiological state of a bacterial cell (Fig. 1). To move leftward on this path is to drift into an ever-decreasing stage of metabolic activity; this is the realm of Siva, whose name derives from the root "sin," meaning "to sleep." Siva's power is represented by the eternal night in which all goes to sleep (Danielou, 1963). In contrast, movement to the right on the path is towards increasing metabolism and reproduction: the realm of Vishnu, the preserver of life.

What moves a cell along this imaginary pathway? In general, it is the cell's endogenous reserves and ability to use exogenous nutrient sources, both mediated by constitutive factors and environmental conditions. At some point, a cell that

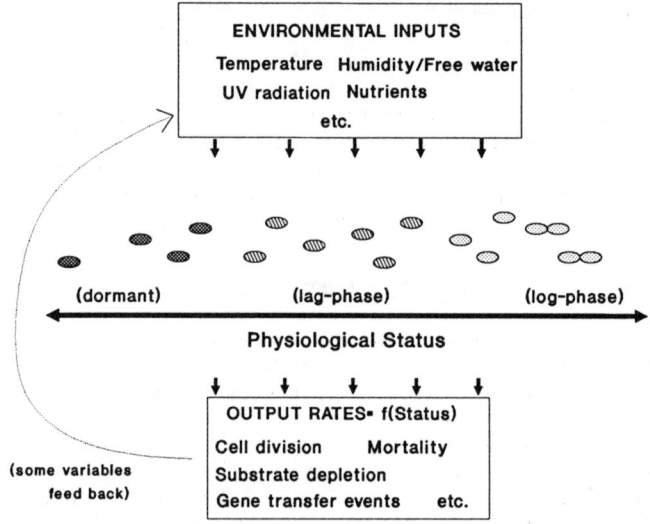

Figure 1. Diagrammatic representation of the Vishnu/Siva model for bacterial population dynamics.

moves too far into dormancy may never return and is functionally dead. Cell division by binary fission produces one parent and one daughter cell, one or both of which may require a time lag before a high level of activity is resumed; this may be envisioned as a move to the left on the path for one or both of these cells. Paraphrasing from the Bhagavadgita (9.21), "Having enjoyed these immense heavens, once their accumulated merits have been spent, they come back to the world . . . and, following the path of merit, those seekers of enjoyment keep on coming and going endlessly" (Danielou, 1963).

There are three attributes of the Vishnu/Siva metaphorical model that may make it useful for modeling bacterial populations, as follows: (i) environmental factors can influence the metabolic activity of a bacterial cell, and the model has flexibility for different factors to work additively, synergistically, or in opposition; (ii) environmental factors can affect different cells in a population differently, especially when the cells are at different points on the path; and (iii) not only does the environment move a cell along the path, but environmental effects can be made dependent on the metabolic status of the cell. For example, metabolically active cells may be more receptive to an influx of nutrients, but may also be more susceptible to killing by UV radiation. I would like first to discuss why these attributes may be important for modeling bacterial populations and then to describe a simple implementation of the Vishnu/Siva model.

Microbial ecology can be defined as the study of the interrelations between microbes and their biotic and abiotic environment. Change in average specific growth rate in response to environmental change, observed at the population level, has its origin at the metabolic level (Konings and Veldkamp, 1980). The bacterial cell cycle is at present best understood at the molecular level in *Escherichia coli* (Dow and Whittenbury, 1980). The cell cycle of *E. coli* entails a period of variable length that precedes chromosome replication (I period), followed by more constant periods during which the chromosome is replicated and the cell physically divides (C and D periods) (Donachie et al., 1976; Begg and Donachie, 1977). For bacteria in nutrient-limiting natural environments, doubling time may greatly exceed the sum of the C plus D periods for a particular species. Thus, a large proportion of the cells at any time will be in the I phase (Dow and Whittenbury, 1980) or, in other words, in a lag phase of growth. For doubling times greater than about 1 h (probably the case on leaf surfaces for most bacteria), *E. coli* cells resulting from cell division are physiologically distinct (Donachie and Begg, 1970; Begg and Donachie, 1977). The parent cell may initiate a new round of replication immediately, while the daughter cell may first need to undergo a lag period (Dow and Whittenbury, 1980).

Below some minimum growth rate level, bacteria stop energy-consuming processes such as biosynthesis, ATP synthesis, or motility and become dormant or hypobiotic. Continued degradation of cellular reserves may result in loss of viability. The relationship between threshold values of nutrients and minimum growth rates is not clear. Konings and Veldkamp (1980) described growth response in terms of environmental effects on the proton motive force across the cytoplasmic membrane and its coupled energy-transducing processes.

Because bacterial cells have a genetic constitution that is not immutable, most "pure" cultures are in reality genetically heterogeneous. When the environment changes, it may select for variants within that population. A model should ideally allow for that possibility rather than treating the population as a homogenous group of cells. As an example, Knudsen and Spurr (1988) described transitions between life stages (spores, vegetative cells, cells with endospores) for populations of *B. thuringiensis* applied as spore suspensions to peanut leaf disks. Within a few days of constant environmental conditions, all life stages were present in the population. Heterogeneous constitutive and environmental influences ensure that transitions between life stages generally do not remain synchronous within a bacterial population. There would be obvious advantages to a model that allowed variation among cells in a population and among microsites in an environment. One approach is therefore to model environmental effects on metabolism of many individual cells and then to extrapolate to the larger microbial population. Another advantage of such an approach is that it allows for an environment that is spatially heterogeneous with respect to chemical, physical, and biotic conditions.

In our laboratory, a simulation model is under development to meet these needs. The Vishnu/Siva model treats change in physiological status as a Brownian motion process with drift. Drift refers to the movement of a cell along the physiological "path" under the influence of environmental factors. Brownian motion describes the random variability inherent in this process. Such an approach, not previously used in this context, is biologically appealing and readily amenable to computer implementation and a modular program structure. Direction and rate of drift, i.e., movement towards either dormancy or increased metabolic activity, are influenced by environmental factors (temperature, moisture, etc.). Also, environmental factors in the model affect both mortality and ecological effects (e.g., antibiotic production, nitrogen transformations) according to the status of the cell. Each of these processes is stochastic, and the model tracks a cohort of bacterial cells through time. The structure of the model is also designed to accommodate different environmental influences on spatially dispersed segments of the population.

A simplified implementation of this model is presented for a hypothetical

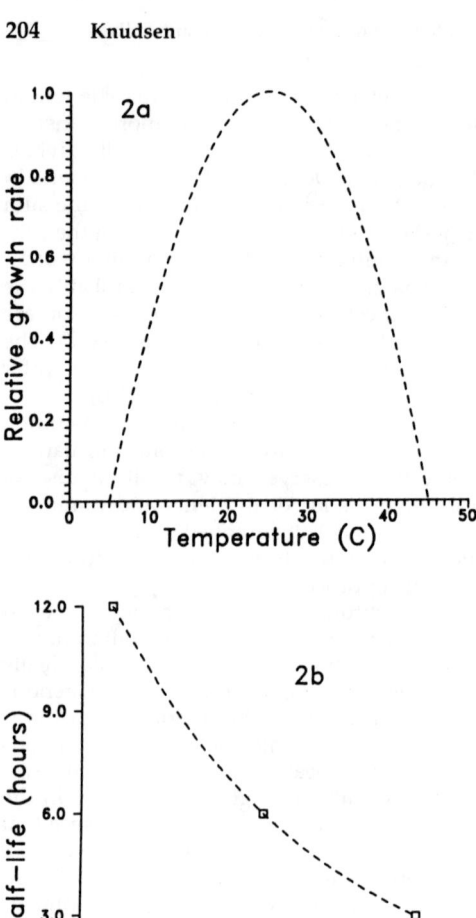

Figure 2. Temperature (a) and solar radiation (b) response curves for a hypothetical epiphytic bacterial strain, as used in simulations.

bacterial population. First, we will consider effects of only three environmental factors: temperature, nutrients, and solar radiation. The form of these effects is somewhat arbitrary, although with support in the literature (e.g., Ignoffo, 1985; Leben, 1974; Skaliy and Eagon, 1972). The temperature optimum for our hypothetical organism is 25°C, with minimum and maximum growth temperatures of 5 and 45°C, respectively. The temperature response curve is quadratic in shape (Fig. 2a). Response to temperature (when nutrients aren't limiting) is realized as drift, with the greatest movement towards increased physiological activity occurring at the temperature optimum of 25°C. Temperature response is determined in the model for each member of a cohort of cells on an hourly basis. There is random variability about each calculated drift value, determined by a Monte Carlo process (Swartzman and Kaluzny, 1987). Parameters were chosen to generate a mean cell doubling time of 4 h at the optimum growth temperature with nutrients not limiting. No other effect of temperature was included at this point.

Nutrients are modeled as a finite pool capable of supporting active growth of a particular size population of cells for a certain number of hours. Thus, larger populations use up the nutrient pool faster than do smaller populations. If the nutrient pool is used up, bacterial cells drift towards increasing dormancy. In these simulations, the nutrient pool is not replenished.

Solar radiation has the effect of killing cells; in the model this is a random process with a relatively high probability for log-phase cells, lower probability for lag-phase cells, and much lower probability for hypobiotic cells (Fig. 2b). Although a cell's physiological status is a continuous variable, the cell is placed in its appropriate class (dormant, lag phase, log phase) for determination of radiation effects.

Finally, cells reaching the highest level of physiological activity in this model divide by binary fission. The resultant parent cell remains in log phase, and the daughter cell starts in lag phase (Dow and Whittenbury, 1980).

In a series of simulations, a constant temperature of 25°C was stipulated along with a nutrient pool sufficient to support growth for 10^7 "cell-hours." In the first simulation (Fig. 3a), a cohort of 300 lag-phase cells was introduced into a spatially homogeneous environment with no solar radiation. The model was run for a simulated 5-day period. The result is a characteristic S-shaped population growth curve, which goes through a lag phase as cells get physiologically "up to speed," a short phase of exponential growth, and then a period where the relative rate of growth per cell slows due to depletion of the nutrient pool. Figure 3b shows the same growth process, with population curves for cells in each of the three physiological classes shown separately. It can be seen that shortly after 48 h in the simulation, all cells are in a dormant state resulting from unavailability of nutrients.

Figure 4 presents results of a simulation run that was identical to the previous one, except that a 12-h photoperiod was stipulated, with full sunlight during hours 1 to 12, 25 to 36, etc. The killing effect of solar radiation is apparent. Other noteworthy results are the decrease in daytime death rate after about 48 h and the coincident inability of populations to recover overnight, both due to most or all of the population being in a dormant state (Fig. 4b).

In a final simulation run, a heterogeneous environment was simulated in the form of a plant canopy that provides different levels of shading from solar radiation. Three canopy layers were stipulated, providing respectively 0%, 50%, and 75% protection (shade) from solar radiation. One hundred cells were randomly assigned to each canopy layer, and it was assumed that progeny remained on the same canopy layer as their parent cells. The same diurnal conditions were simulated. Comparing the results in Fig. 5a with results from the full-sunlight simulation (Fig. 4a), the strong buffering effect of the canopy is apparent. Again, daytime death rates decreased over time, but populations were unable to recover at night due to lack of nutrients. These population trends are similar to those reported from field trials with *P. fluorescens* on pine needles (Knudsen and Hudler, 1987) and with *P. cepacia* on peanut leaves (Knudsen and Spurr, 1987). Figure 5b shows results of the last simulation, with population curves at each canopy level shown separately. Due to shading, populations in the different canopy levels became greatly divergent over time.

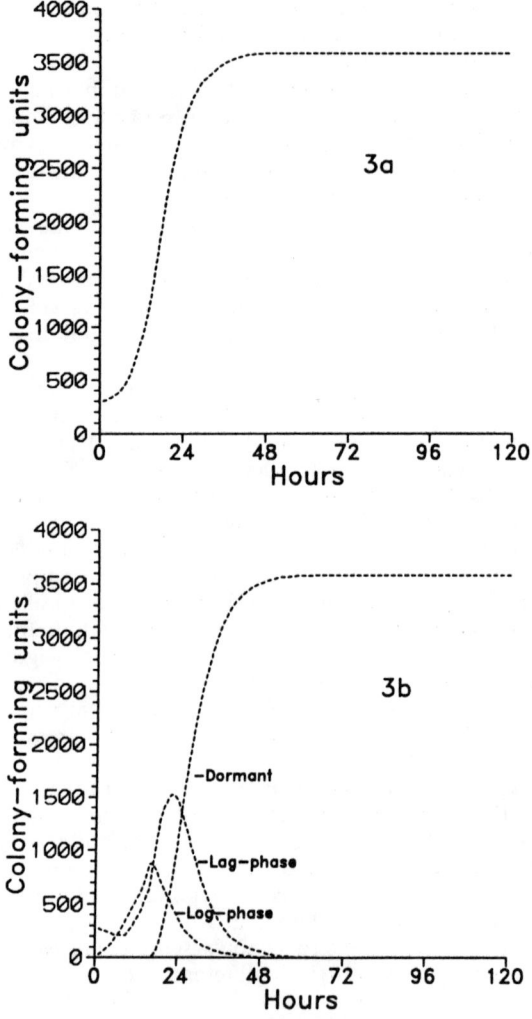

Figure 3. Simulated growth curves for bacteria on leaf surfaces with constant temperature (25°C), limited nutrients, and no solar radiation. (a) Total population; (b) numbers of bacteria in dormant, lag, and log phase.

The Vishnu/Siva model of bacterial populations in the environment is still at a beginning stage. Parameters need to be determined and tested for specific strains of bacteria. However, this model may be useful for conceptualizing environmental effects on bacterial populations and as a framework to quantify these effects and, eventually, to make valid predictions about bacteria in nature.

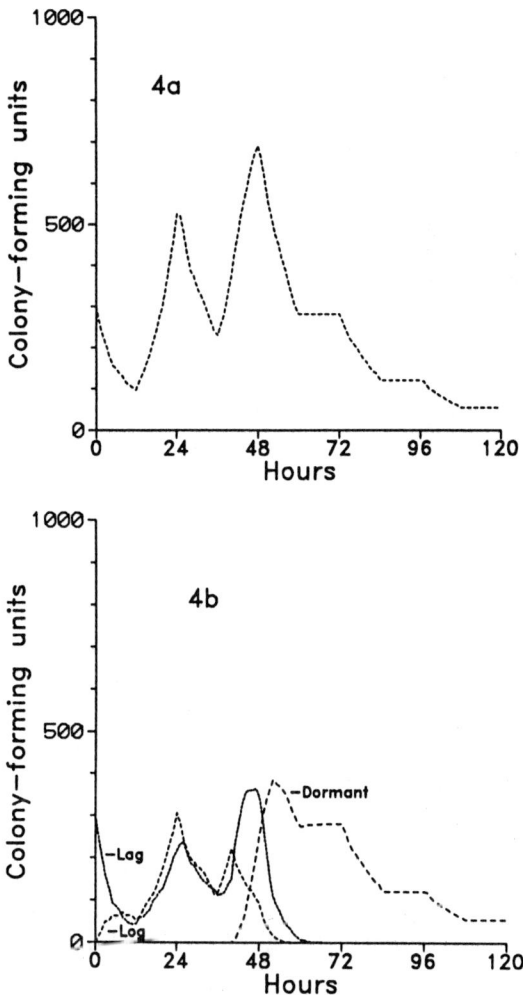

Figure 4. Simulated growth curves for bacteria on leaf surfaces with constant temperature (25°C), limited nutrients, and a 12:12 (light:dark) photoperiod, without shading. (a) Total population; (b) numbers of bacteria in dormant, lag, and log phase.

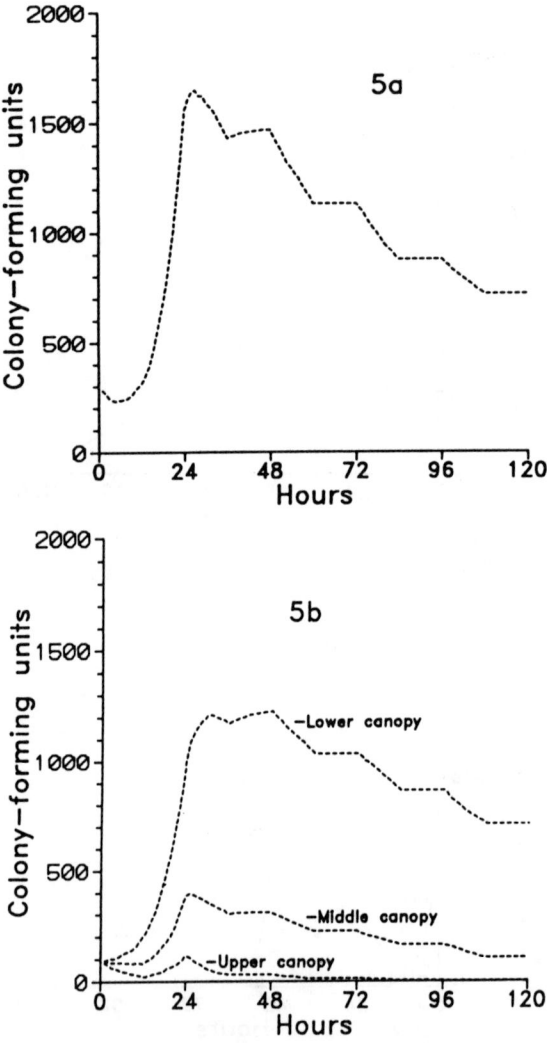

Figure 5. Simulated growth curves for bacteria on leaf surfaces with constant temperature (25°C), limited nutrients, and a 12:12 (light:dark) photoperiod. Equal numbers of bacteria were initially assigned to the upper canopy (0% shade), middle canopy (50% shade), or lower canopy (75% shade). (a) Total population; (b) numbers of bacteria in lower, middle, and upper canopy.

BORROWINGS FROM POPULATION AND COMMUNITY ECOLOGY

> Nature may be compared to a surface covered with ten thousand sharp wedges . . . representing different species, all packed closely together and driven in by incessant blows, . . . sometimes a wedge of one form, sometimes another being struck; the one driven deeply in forcing out others; with the jar and shock often transmitted very far to other wedges in many lines of direction.
> —Charles Darwin (1859)

Bacteria applied to plant foliage are like the driven wedges described above. Crick (1981) suggested that life on earth itself may have arisen from single-celled invaders (bacteria or blue-green algae) that traveled here, 4 billion years ago, in a spaceship. The theory of "Panspermia" describes how these invaders arrived in an unmanned spaceship sent to earth by a higher civilization which had developed elsewhere, billions of years ago (Crick, 1981). If true, such an event probably represents one of the few instances where bacterial invaders have encountered virgin territory. Like most groups of human colonists, bacterial colonists probably encounter considerable antagonistic activity from indigenous populations. As Kinkel et al. (1989) noted, "immigration is necessary, but not sufficient, for the establishment of a microorganism within the phylloplane community."

Immigration is a population-regulating attribute of all communities, including the microbial community on the phylloplane. Application of microbes to leaf surfaces is therefore equivalent to altering immigration patterns on leaves (Kinkel et al., 1989). Working with epiphytic fungi, Kinkel et al. (1989) investigated influences of introduced populations on leaf surface communities. They artificially generated "uncolonized" leaf surfaces by surface-disinfecting apple leaves with hydrogen peroxide and were able to significantly enhance total fungal community size by introducing populations to leaf surfaces. However, long-term influences on the leaf surface fungal community were not significant, suggesting the existence of a "carrying capacity" or limit to the number of individuals a leaf can support.

Solar radiation, intermittent water, and low availability of nutrients create a highly stressed and difficult environment into which to introduce antagonists and have them survive and multiply (Campbell, 1989). In most cases introduced organisms are not maintained in numbers sufficient to be effective, so that where persistence has been demonstrated in the field it has usually involved raising the carrying capacity of the leaf surface by adding nutrients and/or maintaining high humidity and possibly free water on the leaf (Campbell, 1989).

Morris and Rouse (1985) considered the availability of nutrients on the phylloplane and utilization of these nutrients by epiphytic bacteria. By applying different combinations of nutrients to foliage in field plots, they modified the composition of the bacterial community, altered the size of fluorescent pseudomonad populations, and in some cases reduced disease caused by *P. syringae*. Knudsen and Spurr (1988) suggested the potential of using novel synthetic chemical compounds as nutrient sources to selectively enhance biocontrol bacteria engineered to have novel catabolic genes.

The ephemeral nature of phylloplane resources probably favors the most "*r*-selected" strains of bacteria, i.e., those able to reproduce most rapidly and use up the resources before other, competing strains can exploit the habitat (Kurihara et al., 1990; Wilson and Bossert, 1971). However, most bacterial species are able to grow under a range of environmental conditions, of which only a limited number allow growth faster than that of competing species. When a variety of organisms with similar ecological niches coexist on the phylloplane, it may be largely due to the heterogeneous and fluctuating chemical and physical microenvironments of the bacterial cells. Thus, each strain or species is able to grow at certain times faster than its competitors. In this context, Boyce (1984) criticized over-reliance on the r- and K-selection model and pointed out the relative importance of frequency-dependent natural selection rather than density-dependent K selection in the evolution of competitive ability. Stephanopoulos and Frederickson (1979) provided a mathematical description of the effects of spatial heterogeneity on the coexistence of competing microbial populations.

Epiphytic populations of bacterial species may be highly prone to periodic extinction in parts of their ranges through competitive exclusion and perturbations in the physical environment. Pielou (1969) discussed mathematical models to predict the likelihood of extinction for a population undergoing stochastic changes in size. Random detrimental effects become more important as population size becomes smaller (Krebs, 1985). Persistence of species under these conditions may depend on the existence of a discontinuous "metapopulation" from which extinct or declining populations may be replenished (Harrison and Quinn, 1989).

Population sampling for analysis of discontinuous metapopulations presents statistical difficulties. The importance of scale in sampling and analyzing phylloplane ecosystems has been alluded to above. Hirano et al. (1982) and Hudelson et al. (1989) demonstrated the importance of scale in sampling epiphytic bacterial populations. Although a log-normal distribution of $P. syringae$ on individual leaves has been demonstrated (Hirano et al., 1982), the universality of this phenomenon should not be assumed uncritically, especially for bacteria that are sprayed onto plant foliage. It is likely that, initially, aerosolized bacteria would be distributed more uniformly than log-normally on individual leaves. Whether the distribution evolves to log-normality or not, or to some other distribution, will depend on both macro- and microenvironmental effects. If the distribution becomes highly aggregated or "patchy" (a likely event), the sampling scale will profoundly affect the degree of spatial autocorrelation observed within and between patches. Several references are available on the subject of statistics for ecological modeling, including the use of geostatistical analysis (Isaacs and Srivastava, 1989; Ludwig and Reynolds, 1988; Madden, 1989; Parberry et al., 1981).

A related field of study, that of landscape ecology, potentially offers insights and tools for analysis of microbial ecosystems. Landscape ecology focuses on spatial and temporal patterns of landscapes, including the development of spatial heterogeneity and its influence on biotic and abiotic processes (Risser, 1987). Although landscapes are conventionally considered at the spatial scale of tens to hundreds of kilometers (Risser, 1987), they share fundamental characteristics with systems at a smaller scale. Forman and Godron (1986) defined these characteristics as (i) structure: the spatial relationships that determine distribution of energy, materials, and species in relation to the sizes, shapes, numbers, kinds, and configurations of ecosystems; (ii) function: how the spatial elements interact via the flow of energy, materials, and species; and (iii) change: how the structure and function of the ecological mosaic change over time. Landscape ecology focuses largely on the fundamental role of disturbance in the evolution of natural systems (Pickett, 1980; Bormann and Likens, 1981) and, conversely, on the role of landscape heterogeneity in the spread and inhibition of disturbance (Risser, 1987). Thus, landscape ecology may provide valuable theory and tools for prediction about agricultural systems, which depend on repeated disturbance for their maintenance.

Problems of resolution and scale present difficulties in the development of predictive models. Adequate validation of a predictive model of bacterial growth and effects may require extensive sampling resources. Some techniques that can provide high resolution, such as scanning electron microscopy, serological methods, gene probes, or polymerase chain reaction amplification of gene sequences, may be too time-consuming or expensive to use on a large enough scale for statistical interpretation. But beyond technical limitations, are there fundamental limits on predictability which are inherent in systems? The reductionist philosophy that dominates most current scientific thought suggests that, with enough resolution and replication of experiments, the fundamental workings of any system can be understood and predicted. Contrary evidence may be emerging from the mathematical field of topological dynamics, which seeks to describe how systems change over time (Peterson, 1988). This approach suggests that direct cause-and-effect models, or physical systems that are reflected by such models, can undergo transitions to randomlike behavior, termed chaos.

Although chaotic behavior is governed by strict mathematical determinism, small differences in starting conditions are amplified, so that even though the behavior is predictable in the short term, it may fluctuate wildly in the long run (Peterson, 1988). The validity of chaos theory in ecology is still tenuous (Pool, 1989). A frequently cited example of a chaotic system is a population whose growth follows the difference form of the logistic equation. Although chaotic behavior of this model is readily demonstrated, the question remains whether populations in nature ever truly follow that model. However, chaos theory offers insight into possible origins of what has otherwise been called random "noise" in the behavior of systems. Chaos theory also suggests how systems which behave stably for a wide range of initial conditions may suddenly shift into chaotic behavior when one parameter moves through a critical value (Peterson, 1988).

CONCLUSIONS

A variety of approaches to the modeling of epiphytic bacterial populations have been discussed in this chapter. As Jeger (1986) noted, the rich body of theory that has been developed in population dynamics as a whole makes very few references to microbial pathogens (or nonpathogens) of plants. Certainly this includes epiphytic bacteria. The list of published models for bacterial populations is remarkably short. However, there is a significant and increasing body of quantitative description of epiphytic populations and their changes over time. The attention paid to statistical considerations in much of this research is exemplary (Hirano et al., 1982; Hudelson et al., 1989; Parberry et al., 1981) and provides powerful tools for analyzing population patterns.

The use of simulation techniques in conjunction with statistical interpretation is promising, since it allows process and pattern to be modeled concurrently. A preliminary version of a simulation model has been presented as one possibility for exploring both temporal and spatial dynamics of bacterial populations in nature.

Finally, there is a rich body of ecological theory that can lend insight to studies of epiphytic bacteria. With only a few exceptions (e.g., Andrews et al., 1987; Kinkel et al., 1989), this literature has received scant attention. A very small sample of potentially related topics has been presented, but the possibilities are endless.

REFERENCES

Andrews, J. H., L. L. Kinkel, F. M. Berbee, and E. V. Nordheim. 1987. Fungi, leaves, and the theory of island biogeography. *Microb. Ecol.* **14**:277–290.

Armstrong, J. L., G. R. Knudsen, and R. J. Seidler. 1987. Survival of recombinant bacteria associated with plants and herbivorous insects in a microcosm. *Curr. Microbiol.* **15**:229–232.

Beek, J., and M. J. Frissel. 1973. *Simulation of Nitrogen Behavior in Soils.* Simulation Monographs, Center for Agricultural Publishing and Documentation, Wageningen, The Netherlands.

Begg, K. J., and W. D. Donachie. 1977. The growth of the *Escherichia coli* surface. *J. Bacteriol.* **129**:1524–1535.

Berger, R. D. 1977. Application of epidemiological principles to achieve plant disease control. *Annu. Rev. Phytopathol.* **15**:165–183.

Blackman, F. F. 1905. Optima and limiting factors. *Ann. Bot.* (London) **19**:281–295.

Blakeman, J. P., and I. J. S. Brodie. 1976. Inhibition of pathogens by epiphytic bacteria on aerial plant surfaces, p. 529–557. *In* C. H. Dickinson and T. F. Preece (ed.), *Microbiology of Aerial Plant Surfaces.* Academic Press, Inc. (London), Ltd., London.

Blakeman, J. P., and A. K. Fraser. 1971. Inhibition of *Botrytis cinerea* spores by bacteria on the surface of chrysanthemum leaves. *Physiol. Plant Pathol.* **1**:45–54.

Bormann, F. H., and G. E. Likens. 1981. *Pattern and Process in a Forested Ecosystem.* Springer-Verlag, New York.

Boyce, M. S. 1984. Restitution of r- and K-selection as a model of density-dependent natural selection. *Annu. Rev. Ecol. Syst.* **15**:427–447.

Brand, R. J., and D. E. Pinnock. 1981. Application of biostatistical modelling to forecasting the results of microbial control trials, p. 655–693. *In* H. D. Burges (ed.), *Microbial Control of Pests and Plant Diseases 1970–1980.* Academic Press, Inc. (London), Ltd., London.

Brown, M. E. 1973. Soil bacteriostasis limitation in growth of soil and rhizosphere bacteria. *Can. J. Microbiol.* **19**:165–199.

Caldwell, B. A., C. Ye, R. P. Griffiths, C. L. Moyer, and R. Y. Morita. 1989. Plasmid expression and maintenance during long-term starvation-survival of bacteria in well water. *Appl. Environ. Microbiol.* **55**:1860–1864.

Cameron, H. R. 1970. *Pseudomonas* content of cherry trees. *Phytopathology* **60**:1343–1346.

Campbell, R. 1989. *Biological Control of Microbial Plant Pathogens.* Cambridge University Press, Cambridge.

Crick, F. H. C. 1981. *Life Itself: Its Origin and Nature.* Simon and Schuster, New York.

Danielou, A. 1963. *Hindu Polytheism.* Routledge and Kegan Paul, London.

Darwin, C. 1859. *On the Origin of Species.* John Murray, London.

Delaquis, P. J., D. E. Caldwell, J. R. Lawrence, and A. R. McCurdy. 1989. Detachment of *Pseudomonas fluorescens* from biofilms on glass surfaces in response to nutrient stress. *Microb. Ecol.* **18**:199–210.

Dommergues, Y. R., L. W. Belser, and E. L. Schmidt. 1978. Limiting factors for microbial growth and activity in soil. *Adv. Microb. Ecol.* **2**:49–104.

Donachie, W. D., and K. J. Begg. 1970. Growth of the bacterial cell. *Nature* (London) **227**:1220–1224.

Donachie, W. D., K. J. Begg, and M. Vincente. 1976. Cell length, cell growth and cell division. *Nature* (London) **264**:328–333.

Dow, C. S., and R. Whittenbury. 1980. Prokaryotic form and function, p. 391–417. *In* D. C. Ellwood, J. N. Hedger, M. J. Latham, J. M. Lynch, and J. H. Slater (ed.), *Contemporary Microbial Ecology.* Academic Press, Inc. (London), Ltd., London.

Ercolani, G. L. 1973. Two hypotheses on the aetiology of response of plants to phytopathogenic bacteria. *J. Gen. Microbiol.* **75**:83–95.

Ercolani, G. L., D. J. Hagedorn, A. Kelman, and R. E. Rand. 1974. Epiphytic survival of *Pseudomonas syringae* on hairy vetch in relation to epidemiology of bacterial brown spot of bean in Wisconsin. *Phytopathology* **64**:1330–1339.

Falcon, L. A. 1971. Use of bacteria for microbial control, p. 67–95. *In* H. D. Burges and N. W.

Hussey (ed.), *Microbial Control of Insects and Mites*. Academic Press, Inc. (London), Ltd., London.
Forman, R. T. T., and M. Godron. 1986. *Landscape Ecology*. John Wiley & Sons, Inc., New York.
Freeman, T. E., and R. Charudattan. 1980. Biological control of weeds with plant pathogens. Prospectus—1980, p. 293–299. In E. S. Del Fosse (ed.), *Proceedings of the V International Symposium on Biological Control of Weeds*. CSIRO, Melbourne.
Gross, D. C., Y. S. Cody, E. L. Proebsting, Jr., G. K. Radamaker, and R. A. Spotts. 1983. Distribution, population dynamics, and characteristics of ice nucleation-active bacteria in deciduous fruit tree orchards. *Appl. Environ. Microbiol.* **46**:1370–1379.
Hagedorn, D. J., and P. N. Patel. 1965. Halo blight and bacterial brown spot of bean in Wisconsin in 1964. *Plant Dis. Rep.* **49**:591–595.
Harrison, S., and J. F. Quinn. 1989. Correlated environments and the persistence of metapopulations. *Oikos* **56**:293–298.
Hirano, S. S., E. V. Nordheim, D. C. Arny, and C. D. Upper. 1982. Lognormal distribution of epiphytic bacterial populations on leaf surfaces. *Appl. Environ. Microbiol.* **44**:695–700.
Hirano, S. S., and C. D. Upper. 1983. Ecology and epidemiology of foliar bacterial plant pathogens. *Annu. Rev. Phytopathol.* **21**:243–269.
Hoitink, H. A. J., D. J. Hagedorn, and E. McCoy. 1968. Survival, transmission and taxonomy of *Pseudomonas syringae* von Hall, the causal agent of bacterial brown spot of bean (*Phaseolus vulgaris* L.). *Can. J. Microbiol.* **14**:437–441.
Hudelson, B. D., M. K. Clayton, K. P. Smith, D. I. Rouse, and C. D. Upper. 1989. Nonrandom patterns of bacterial brown spot in snap bean row segments. *Phytopathology* **79**:674–681.
Ignoffo, C. M. 1985. Manipulating enzootic-epizootic diseases of arthropods, p. 243–262. In M. A. Hoy and D. C. Herzog (ed.), *Biological Control in Agricultural IPM Systems*. Academic Press, Inc., Orlando, Fla.
Isaaks, E. H., and R. M. Srivastava. 1989. *An Introduction to Applied Geostatistics*. Oxford University Press, New York.
Jeger, M. J. 1986. The potential of analytic compared with simulation approaches to modeling in plant disease epidemiology, p. 255–281. In K. J. Leonard and W. E. Fry (ed.), *Plant Disease Epidemiology*, vol. 1. Macmillan, New York.
Jensen, V. 1971. The bacterial flora of beech leaves, p. 463–469. In C. H. Dickinson and T. F. Preece (ed.), *Ecology of Leaf Surface Micro-organisms*. Academic Press, Inc. (London), Ltd., London.
Kinkel, L. L., J. H. Andrews, and E. V. Nordheim. 1989. Microbial introductions to apple leaves: influences of altered immigration on fungal community dynamics. *Microb. Ecol.* **18**:161–173.
Knudsen, G. R. 1989. Model to predict aerial dispersal of bacteria during environmental release. *Appl. Environ. Microbiol.* **55**:2641–2647.
Knudsen, G. R., and G. W. Hudler. 1981. Germination of *Gremmeniella abietina* inhibited by epiphytic pseudomonads. *Phytopathology* **71**:562.
Knudsen, G. R., and G. W. Hudler. 1987. Use of a computer simulation model to evaluate a plant disease biocontrol agent. *Ecol. Modell.* **35**:45–62.
Knudsen, G. R., and D. J. Schotzko. Submitted for publication.
Knudsen, G. R., and H. W. Spurr, Jr. 1985. Computer simulation as a tool to evaluate foliar biocontrol strategies. *Phytopathology* **75**:1343.
Knudsen, G. R., and H. W. Spurr, Jr. 1987. Field persistence and efficacy of five bacterial preparations for control of peanut leaf spot. *Plant Dis.* **71**:442–445.
Knudsen, G. R., and H. W. Spurr, Jr. 1988. Management of bacterial populations for foliar disease biocontrol, p. 83–92. In K. G. Mukerji (ed.), *Biocontrol of Plant Diseases*. CRC Press, Boca Raton, Fla.
Knudsen, G. R., M. V. Walter, L. A. Porteous, V. J. Prince, J. L. Armstrong, and R. J. Seidler. 1988. A predictive model of conjugative plasmid transfer in the rhizosphere and phyllosphere. *Appl. Environ. Microbiol.* **54**:343–347.
Konings, W. N., and H. Veldkamp. 1980. Phenotypic responses to environmental change, p.

161–191. *In* D. C. Ellwood, J. N. Hedger, M. J. Latham, J. M. Lynch, and J. H. Slater (ed.), *Contemporary Microbial Ecology*. Academic Press, Inc. (London), Ltd., London.

Krebs, C. J. 1985. *Ecology: the Experimental Analysis of Distribution and Abundance*, 3rd ed. Harper and Row, Inc., New York.

Kurihara, Y., S. Shikano, and M. Toda. 1990. Trade-off between interspecific competitive ability and growth rate in bacteria. *Ecology* **71**:645–650.

Lambrecht, R. S., J. F. Carriere, and M. T. Collins. 1988. A model for analyzing growth kinetics of a slowly growing *Mycobacterium* sp. *Appl. Environ. Microbiol.* **54**:910–916.

Layne, R. E. C. 1968. A quantitative local lesion bioassay for *Corynebacterium michiganense*. *Phytopathology* **58**:534–535.

Leben, C. 1974. *Survival of Plant Pathogenic Bacteria*. Special Circular 100. Ohio Agricultural Research and Development Center, Wooster.

Leben, C., G. C. Daft, J. D. Wilson, and H. F. Winter. 1965. Field tests for disease control by an epiphytic bacterium. *Phytopathology* **55**:1375–1376.

Lindemann, J. 1985. Genetic manipulation of microorganisms for biological control, p. 116–170. *In* C. E. Windels and S. L. Lindow (ed.), *Biological Control on the Phylloplane*. American Phytopathological Society, St. Paul, Minn.

Lindemann, J., D. C. Arny, and C. D. Upper. 1984a. Epiphytic populations of *Pseudomonas syringae* pv. *syringae* on snap bean and nonhost plants and the incidence of bacterial brown spot disease in relation to cropping patterns. *Phytopathology* **74**:1329–1333.

Lindemann, J., D. C. Arny, and C. D. Upper. 1984b. Use of an apparent infection threshold population of *Pseudomonas syringae* to predict incidence and severity of brown spot of bean. *Phytopathology* **74**:1334–1339.

Lindow, S. E. 1985. Ecology of *Pseudomonas syringae* relevant to the field use of Ice⁻ deletion mutants constructed in vitro for plant frost control, p. 23–35. *In* H. O. Halvorson, D. Pramer, and M. Rogul (ed.), *Engineered Organisms in the Environment: Scientific Issues*. American Society for Microbiology, Washington, D.C.

Lindow, S. E. 1986. Strategies and practice of biological control of ice nucleation active bacteria on plants, p. 293–311. *In* N. Fokkema (ed.), *Microbiology of the Phyllosphere*. Cambridge University Press, Cambridge.

Lindow, S. E., D. C. Arny, and C. D. Upper. 1978. Distribution of ice nucleation-active bacteria on plants in nature. *Appl. Environ. Microbiol.* **36**:831–838.

Lindow, S. E., G. R. Knudsen, R. J. Seidler, M. V. Walter, V. W. Lambou, P. S. Amy, D. Schmedding, V. Prince, and S. Hern. 1988. Aerial dispersal and epiphytic survival of *Pseudomonas syringae* during a pretest for the release of genetically engineered strains into the environment. *Appl. Environ. Microbiol.* **54**:1557–1563.

Linton, A. H. 1971. Influence of external factors on viability of micro-organisms, p. 193–217. *In* L. E. Hawker and A. H. Linton (ed.), *Micro-Organisms, Function, Form and Environment*. American Elsevier Publishing Co., Inc., New York.

Ludwig, J. A., and J. F. Reynolds. 1988. *Statistical Ecology*. John Wiley & Sons, Inc., New York.

Madden, L. V. 1989. Dynamic nature of within-field disease and pathogen distributions, p. 96–126. *In* M. J. Jeger (ed.), *Spatial Components of Plant Disease Epidemics*. Prentice Hall, Englewood Cliffs, N.J.

Mew, T. W., and B. W. Kennedy. 1971. Growth of *Pseudomonas glycinea* on the surface of soybean leaves. *Phytopathology* **61**:715–716.

Monteith, J. L. 1973. *Principles of Environmental Physics*. Arnold, London.

Morris, C. E., and D. I. Rouse. 1985. Role of nutrients in regulating epiphytic bacterial populations, p. 63–82. *In* C. E. Windels and S. E. Lindow (ed.), *Biological Control on the Phylloplane*. American Phytopathological Society, St. Paul, Minn.

Newhook, F. J. 1951. Microbiological control of *Botrytis cinerea* Pers. I. The role of pH changes and bacterial antagonism. *Ann. Appl. Biol.* **38**:169–184.

Olsen, R. A., and L. R. Bakken. 1987. Viability of soil bacteria: optimization of plate-counting technique and comparison between total counts and plate counts within different size groups. *Microb. Ecol.* **13**:59–74.

Parberry, I. H., J. F. Brown, and V. J. Bofinger. 1981. Statistical methods in the analysis of

phylloplane populations, p. 47–65. *In* J. P. Blakeman (ed.), *Microbial Ecology of the Phylloplane.* Academic Press, Inc. (London), Ltd., London.
Parthasarathy, A. 1983. *The Symbolism of Hindu Gods and Rituals.* Shailesh Printers, Bombay.
Pennypacker, S. P. 1978. Instrumentation for epidemiology, p. 97–118. *In* J. G. Horsfall and E. B. Cowling (ed.), *Plant Disease: an Advanced Treatise*, vol. 2. *How Disease Develops in Populations.* Academic Press, Inc., New York.
Peterson, I. 1988. *The Mathematical Tourist: Snapshots of Modern Mathematics.* W. H. Freeman and Co., New York.
Pickett, S. E. T. 1980. Non-equilibrium coexistence in plants. *Bull. Torrey Bot. Club* **1007**:238–248.
Pielou, E. C. 1969. *An Introduction to Mathematical Ecology.* Wiley Interscience, New York.
Pinnock, D. E., R. J. Brand, J. E. Milstead, M. E. Kirby, and N. F. Coe. 1978. Development of a model for prediction of target insect mortality following field application of a *Bacillus thuringiensis* formulation. *J. Invertebr. Pathol.* **31**:31–36.
Pirt, S. J. 1975. *Principles of Microbe and Cell Cultivation.* John Wiley & Sons, Inc., New York.
Pool, R. 1989. Chaos theory: how big an advance? *Science* **245**:26–28.
Powell, E. O. 1967. The growth rate of microorganisms as a function of substrate concentration, p. 34–56. *In* E. O. Powell, C. G. T. Evans, R. E. Strange, and D. W. Tempest (ed.), *Microbial Physiology and Continuous Culture.* Her Majesty's Stationery Office, London.
Reeve, C. A., A. T. Bockman, and A. Martin. 1984. Role of protein degradation in the survival of carbon-starved *Escherichia coli* and *Salmonella typhimurium*. *J. Bacteriol.* **157**:758–763.
Reichenbach, N., G. B. Wickramanayake, B. A. Lordo, and D. M. Hetrick. 1987. Biotechnology model: MICROBE-SCREEN. Final report. Prepared for Contract 68-02-4246, Task 2-1, U.S. Environmental Protection Agency, Washington, D.C.
Reichle, D. E., R. V. O'Neill, S. V. Kay, P. Sollins, and R. S. Booth. 1973. Systems analysis as applied to modeling ecological processes. *Oikos* **24**:337–345.
Risser, P. G. 1987. Landscape ecology: state of the art, p. 3–14. *In* M. G. Turner (ed.), *Landscape Heterogeneity and Disturbance.* Springer-Verlag, New York.
Robinson, J. A. 1985. Determining microbial kinetic parameters using nonlinear regression analysis. *Adv. Microb. Ecol.* **8**:61–114.
Rouse, D. I., E. V. Nordheim, S. S. Hirano, and C. D. Upper. 1985. A model relating the probability of foliar disease incidence to the population frequencies of bacterial plant pathogens. *Phytopathology* **75**:505–509.
Schotzko, D. J., G. R. Knudsen, and C. M. Smith. Submitted for publication.
Shrum, R. D. 1982. Creating epiphytotics, p. 113–136. *In* R. Charudattan and H. L. Walker (ed.), *Biological Control of Weeds with Plant Pathogens.* John Wiley & Sons, Inc., New York.
Skaliy, P., and R. G. Eagon. 1972. Effect of physiological age and state on survival of desiccated *Pseudomonas aeruginosa*. *Appl. Microbiol.* **24**:763–767.
Sleesman, J. P., and C. Leben. 1976. Microbial antagonists of *Bipolaris maydis*. *Phytopathology* **66**:1214–1218.
Spurr, H. W., Jr. 1981. Experiments on foliar disease control using bacterial antagonists, p. 369–381. *In* J. P. Blakeman (ed.), *Microbial Ecology of the Phylloplane.* Academic Press, Inc. (London), Ltd., London.
Spurr, H. W., Jr., and G. R. Knudsen. 1985. Biological control of leaf diseases with bacteria, p. 45–62. *In* C. E. Windels and S. E. Lindow (ed.), *Biological Control on the Phylloplane.* American Phytopathological Society, St. Paul, Minn.
Stephanopoulos, G., and A. G. Frederickson. 1979. Effect of spatial inhomogeneities on the existence of competing microbial populations. *Biotechnol. Bioeng.* **21**:1491–1498.
Stotzky, G. 1980. Activity, ecology, and population dynamics of microorganisms in soil, p. 57–135. *In* A. I. Laskin and H. Lechevalier (ed.), *Microbial Ecology.* Chemical Rubber Co., Cleveland.
Supkoff, D. M., L. G. Bezark, and D. Opgenorth. 1988. Monitoring of the winter 1987 field release of genetically engineered bacteria in Contra Costa County. Report BC 88-1. California Department of Food and Agriculture, Sacramento.
Surico, G., B. W. Kennedy, and G. L. Ercolani. 1981. Multiplication of *Pseudomonas syringae*

pv. *glycinea* on soybean primary leaves exposed to aerosolized inoculum. *Phytopathology* **71**:532–536.

Swartzman, G. L., and S. P. Kaluzny. 1987. *Ecological Simulation Primer*. Macmillan, New York.

Tempest, D. W., O. M. Neijssel, and W. Zevenboom. 1983. Properties and performance of microorganisms in laboratory culture: their relevance to growth in natural ecosystems. *Symp. Soc. Gen. Microbiol.* **34**:119–152.

Thomson, S. V., M. N. Schroth, W. J. Moller, and W. O. Reil. 1975. Occurrence of fire blight of pears in relation to weather and epiphytic populations of *Erwinia amylovora*. *Phytopathology* **65**:353–358.

Wallin, J. R. 1963. Dew: its significance and measurement in phytopathology. *Phytopathology* **53**:1210–1216.

Walmsley, R. H. 1976. Temperature dependence of mating-pair formation in *Escherichia coli*. *J. Bacteriol.* **126**:222–224.

Walter, M. V., K. Barbour, M. McDowell, and R. J. Seidler. 1987. A method to evaluate survival of genetically engineered microorganisms in soil extracts. *Curr. Microbiol.* **15**:193–197.

Wangersky, P. J., and W. J. Cunningham. 1956. On time lags in equations of growth. *Proc. Natl. Acad. Sci. USA* **42**:699–702.

Weller, D. M., and A. W. Saettler. 1980. Colonization and distribution of *Xanthomonas phaseoli* and *Xanthomonas phaseoli* var. *fuscans* in field-grown navy beans. *Phytopathology* **70**:500–506.

Wilkinson, T. R. 1966. Survival of bacteria on metal surfaces. *Appl. Microbiol.* **14**:303–307.

Wilson, E. O., and W. H. Bossert. 1971. *A Primer of Population Biology*. Sinauer, Sunderland, Mass.

Yates, M. V., S. R. Yates, A. W. Warrick, and C. P. Gerba. 1986. Use of geostatistics to predict virus decay rates for determination of septic tank setback distances. *Appl. Environ. Microbiol.* **52**:479–483.

Yuen, G. Y., A. M. Alvarez, A. A. Benedict, and K. J. Trotter. 1987. Use of monoclonal antibodies to monitor the dissemination of *Xanthomonas campestris* pv. *campestris*. *Phytopathology* **77**:366–370.

Zadoks, J. C. 1971. Systems analysis and the dynamics of epidemics. *Phytopathology* **61**:600–610.

Zadoks, J. C., and R. D. Schein. 1979. *Epidemiology and Plant Disease Management*. Oxford University Press, Oxford.

IV. DISINFECTION

Chapter 10

Virus Inactivation by Disinfectants

James M. Vaughn and James F. Novotny

Introduction 217
General Properties of Viruses............................... 218
Viruses as Environmentally Transmitted Agents 219
    Enterovirus.. 220
    Rotavirus.. 220
    Calicivirus 220
    Norwalk-Like Viruses................................... 220
    Astrovirus 221
    Adenovirus 221
    Other Viruses.. 221
Halogen-Based Disinfectants 221
    Chlorine... 221
    Chlorine Dioxide 226
Nonhalogen Disinfectants 228
    Ozone 228
    UV Light 234
References... 236

Virtually since the advent of studies which elucidated viral structure and function, disease potential, and the importance of aquatic systems in the passive transmission to susceptible human hosts of many important viral species, much attention has been focused on the development of efficient methods for the elimination of viruses through the use of chemical disinfectants. Through the years, these efforts have been compromised by the unique biology of these organisms, which differ radically from previously studied procaryotic and eucaryotic microbial pathogens (i.e., bacteria, fungi, protozoans). These differences are thought to contribute to the decreased sensitivities, often observed with viruses, to many of the disinfectants commonly used to control the incidence and transmission of other microbes through the environment. Rudimentary experiments specifically addressing the question of viral stability to disinfectant challenge were begun during the late

---

James M. Vaughn and James F. Novotny • Department of Microbiology, University of New England College of Medicine, Biddeford, Maine 04005.

1940s. Although rather crudely designed, owing to the paucity of routine laboratory procedures for the conduct of such studies, early experiments yielded considerable preliminary information. From these modest beginnings, studies were expanded and improved during the next four decades, engendering the substantial data base presently available.

The present review will focus on pertinent aspects, gleaned from numerous experimental studies, of the virucidal properties of several commonly used halogen- (i.e., chlorine) and non-halogen-based water disinfectants in terms of their inactivation dynamics and probable mechanisms of action.

GENERAL PROPERTIES OF VIRUSES

Viruses are infectious agents possessing properties which radically distinguish them from other microorganisms. It was noted as early as the late 1800s that suspect agents were not retained on bacterial filters and that the resulting filtrates were capable of inducing disease in specific animal and plant hosts. This first distinguishing characteristic of the group, i.e., small size (20 to 300 nm), became the initial basis upon which viruses were classified (Iwanowski, 1892). These agents could not be observed by normal light microscopy techniques and could not be grown on available microbiological media. These observations led to the development of theories which concluded that viral infections were actually diseases produced by toxins, perhaps of bacterial origin. The word "virus" is derived from the Greek *ios*, meaning poison. Other early studies demonstrated that the infectious agents could be passed from host to host with great dilution prior to each passage, indicating that the amount of infectious agent was actually increasing and could not simply be a microbial toxin (Loeffler and Frosch, 1898). It was further noted that the infection or disease caused by viruses was only observed when infectious agents were directly associated with the cells and tissues of living organisms. Through this property, the parasitic nature of viruses was determined. The initial operational definition of this group was presented as follows: viruses were infectious agents which could pass through bacteria-retaining filters; they were too small to be seen by existing microscopy; they could not be artificially cultivated; and they were strict intracellular parasites.

With the development of electron microscopy and improved analytical biochemical techniques, small, obligately intracellular parasitic bacteria were identified which had previously been grouped as viruses. These observations demonstrated that viruses could not be classified merely on the basis of size and parasitic properties. Analytical studies conducted during the 1930s to 1950s on a variety of viruses led to the demonstration that they consisted largely of protein and nucleic acid. Since the nature and activity of genetic material was, as yet, not fully understood, the significance of these findings was uncertain. Hershey and Chase (1952) first demonstrated the independent functions of viral protein and nucleic acid. In another study, Fraenkel-Conrat and Singer (1957) confirmed the hereditary role of viral nucleic acid by demonstrating that hybrid virions, formed when two different strains of virus were associated with each other, retained the morphology and infectivity characteristic of the nucleic acid contained within the original particles. The proof that viral nucleic acid was indeed the controlling factor was reported by Gierer and Schramm (1956). In these studies, purified viral RNA was shown to attack host cells, with the resulting infection yielding progeny virions containing all of the characteristics of the virus from which the RNA had been extracted. Viruses were also found to contain only a single type of nucleic acid, as opposed to all other known organisms, which possess both DNA and RNA (Fraenkel-Conrat, 1970).

The foregoing provides a basis upon which viruses can be defined and compared with other microorganisms. Lwoff (1957) portrayed viruses as strict intracellular pathogenic entities with an infectious phase; as possessing only one type of nucleic acid, which multiplied in the form of their genetic material; as being unable to grow and undergo binary fission; and as being devoid of a system of enzymes for energy production. Viruses are considered to be elements of genetic material that can encode, in the appropriate host cells, the biosynthesis of materials for the construction of new virus progeny and their ultimate transfer to new host cells (Luria, 1958). This definition stresses the role of the host cell (i.e., the ability of viruses to use host synthetic machinery to direct the synthesis of new infectious particles) in the virus replication cycle. Viruses intrinsically behave as blocks of genetic material ("genetic parasites") which may be transmitted from host cell to host cell, producing infection and disease within infected tissues. Viruses cannot be replicated without first taking over the biosynthetic apparatus of the host cell and are essentially inanimate particles when outside of their specific host cell.

Chemical agents that are frequently used as disinfectants against traditional cellular microorganisms act by dissolving lipids from cell membranes, by damaging nucleic acids or structural proteins (Davis et al., 1990), or by inactivating essential enzymes and coenzymes (e.g., Joret et al., 1986). Although viruses do not possess cellular membranes, they do contain both nucleic acids and proteins. These moieties, however, are not metabolically active in non-cell-associated virions (i.e., those occurring in environmental waters), and their resistance to acids and alkalis, salts, heavy metals, halogens, alkylating agents, surfactants, phenols, and alcohols differs considerably from that of higher microorganisms which are physiologically and metabolically active within the environment. The protein shell (capsid) surrounding the genome of most viruses may contain several polypeptide sequences, often arranged in multiple layers (Alberts et al., 1989). Many viruses possess a lipid envelope which encloses the capsid. The configuration of capsid components and lipid envelope is a consequence of the diverse infection cycles of different viruses and imparts unique resistance properties to each virus type. The efficiency with which a particular virion withstands challenge by a specific disinfectant is directly related to these chemical and structural properties.

VIRUSES AS ENVIRONMENTALLY TRANSMITTED AGENTS

While the host range of viral agents includes virtually every cellular species known, the discussion here will focus on agents of human disease commonly transmitted through the environment, principally via water. Myriad authors have contributed volumes of information regarding the occurrence of human viruses in aquatic systems and the subsequent transmission of human disease through these systems. This information has been the subject of numerous reviews (Slade and Ford, 1983; Block, 1983; Farrah and Schaub, 1983; Vaughn and Landry, 1983; Richards, 1985; Craun, 1986; Gerba, 1988) and need not be reiterated here.

The human viruses most likely to occur in natural waters and wastewater, i.e., those of greatest concern for the use of water disinfectants, include a variety of species commonly released from the human alimentary and urinary tracts. Most of these appear to be involved in, but not necessarily limited to, human gastrointestinal diseases. The listing presented below, derived from several recent sources (Tyrrell, 1982; Blacklow and Cukor, 1982; Holmes, 1982; Kapikian et al., 1982; Caul and Egglestone, 1982; Centers for Disease Control, 1990), includes those virus types known to be associated with the human gastrointestinal tract which may gain entry to environmental waters (we hasten to note that not all have been proven to be water borne).

Enterovirus

Enteroviruses are small (25- to 30-nm) viruses containing a single-stranded RNA genome surrounded by a protein coat. The group contains the much-studied poliovirus, echovirus, and coxsackieviruses A and B, etiologic agents for diseases including poliomyelitis, meningitis, fever, respiratory disease, herpangina, and myocarditis, as well as the new enterovirus types 68 to 71, which cause meningitis, encephalitis, conjunctivitis, and respiratory diseases. Also included is the agent of hepatitis type A (enterovirus 72), one of the most important waterborne viral infections.

Rotavirus

Rotaviruses are typically 80-nm, double-coated icosahedral particles which contain a double-stranded RNA that is usually divided into 11 segments. These agents are the most common cause of acute nonbacterial gastroenteritis in children. Acute rotavirus infections are characterized by the excretion of large numbers of infectious particles ($>10^9$/ml of stool). Although the virus is most commonly spread by person-to-person transmission, water- and food-related outbreaks have been observed.

Calicivirus

Caliciviruses are small spherical viruses (30 to 40 nm) composed of a single-layered capsid and single-stranded RNA. Members of this group, which when viewed by electron microscopy following negative staining procedures resemble a "Star of David," are agents of acute diarrhea, usually in children. Transmission modes include person to person and wastewater-contaminated shellfish, drinking water, and foods.

Norwalk-Like Viruses

The Norwalk-like viruses are a heterogeneous group, also called the small round-structured viruses, whose many members have been designated according to the geographic locales where outbreaks have occurred (e.g., Taunton, Hawaii, Snow Mountain, Sapporo). This group has been associated with epidemic viral gastroenteritis, and several food- and waterborne outbreaks have been documented.

Astrovirus

The 28-nm particles of astrovirus contain surface structures which resemble a star when viewed in the electron microscope. They are thought to contain a single-stranded RNA genome. Members of this group are agents of nonspecific gastroenteritis. Transmission through contaminated food and water has been documented.

Adenovirus

Adenoviruses are 70 to 90 nm in diameter and contain double-stranded DNA. Two members of this group, types 40 and 41, have been associated with childhood diarrheal disease. To date, food and potable water have not been implicated as vehicles.

Other Viruses

Several other viral agents of human gastroenteritis are readily shed in the feces and offer potential for waterborne transmission. These include members of the coranavirus, pestivirus, picobirnavirus, torovirus, and parvovirus groups.

HALOGEN-BASED DISINFECTANTS

Chlorine

Chlorine is the predominant halogen used for the disinfection of water in the United States. It was used as a deodorizing substance long before the discovery of its antimicrobial activity, when chlorinated lime was used to deodorize London sewage during the 1800s. The first use of chlorine as a disinfectant in a wastewater treatment plant occurred in Hamburg, Germany, in 1893. The town of Middelkerke, Belgium, is reported to have continuously chlorinated drinking water since 1902. Approximately 3% of the chlorine currently produced in the United States is used for sanitation purposes.

Chlorine is usually transported to application sites in the liquid state and is released into solution as a gas under controlled conditions. Sodium or calcium hypochlorite [$NaOCl$, $Ca(OCl)_2$] is often used instead of chlorine gas; these sater forms of the halogen can be applied as a powder or in solution (Chambers, 1971). As chlorine is dissolved in water, hypochlorous acid ($HOCl$) is produced. Sodium hypochlorite and calcium hypochlorite salts ionize into hypochlorite ion (OCl^-), which combines with hydrogen ions in water to produce hypochlorous acid. Hypochlorous acid is a potent oxidizing agent at pH 6.5 to 7.5. Many environmental factors profoundly influence the chlorine disinfection process, and it becomes apparent after reviewing the literature that not all studies can be directly correlated with one another. The following includes a brief review of some of the more common variables affecting the germicidal activity of chlorine.

The concentration of hydrogen ions within chlorine solutions has a significant effect on the disinfecting power of chlorine. At pH levels below 6.0, the available chlorine is present in the form of hypochlorous acid; at pH 6.0, 95% of the available chlorine assumes this form. When pH values are above 8.0, nearly all of the chlorine will be ionized to form hypochlorite ion (Clark et al., 1956; Weidenkopf, 1958; Engelbrecht, 1978). Hypochlorous acid has been thought to be the more potent germicide. For example, Clark et al. (1956) reported that adenovirus type 3 was approximately seven times more sensitive to this chlorine species than to hypochlorite ion. Weidenkopf (1958), one of the first to document pH-dependent chlorine inactivation processes, reported a 10-fold improvement in poliovirus inactivation with the hypochlorous acid form.

The presence of inorganic ions such as iron, manganese, nitrite, and ammonia has also been shown to adversely affect the activity of oxidizing disinfectants, such as chlorine (Sawyer and McCarty, 1978). Organic matter, particularly containing sulfur or unsaturated bonds, reduces germicidal potency by competing with the target microbes for available chlorine (Sung, 1974). Such competition for disinfectant molecules by soluble or suspended organic molecules has significantly impeded the development and use of water disinfectants. In wastewater, the desired concentration of hypochlorous acid is short-lived because of reaction with ammonia and ammonialike compounds to form chloramines. This reaction is of utmost importance, as the disinfecting power of chloramines is approximately 25 to 100 times less than that of the hypochlorous acid species of chlorine (Chambers, 1971).

Variations in temperature have also been shown to radically alter the germicidal potential of chlorine. Scarpino et al. (1979), for example, demonstrated that poliovirus inactivation rates were increased by more than a factor of 2 when the temperature was increased from 5 to 25°C.

Chlorine-induced inactivation can be influenced by several other environmental properties. When microbes associate with suspended solids, they become adsorbed to or encapsulated within these solids, which afford them protection against attack by disinfecting chemicals (Moore et al., 1975). Another phenomenon, closely associated with this property, involves the fact that viruses frequently exist as aggregates which interfere with the penetration of disinfecting chemicals (Wallis and Melnick, 1967).

Inactivation process

During the past 20 to 25 years, many studies have addressed the virucidal activity of chlorine, due in part to the nearly exclusive use of this chemical for disinfection of potable and wastewater in the United States. Because a wide range of susceptibility to chlorine disinfection is displayed by different viruses, it has become recognized that no single virus type can be relied upon as a general indicator for evaluating disinfection processes. As a result, varying susceptibilities to inactivation have led to investigations of many families and serotypes of human viruses and bacteriophages.

Weidenkopf (1958) first illustrated pH-dependent inactivation of enteroviruses. He reported a 99% inactivation of poliovirus 1 by a free chlorine residual of 0.1 mg/liter following a 10-min exposure at pH 6.0 (0°C). At pH 7.0, the same degree of inactivation required a 15-min exposure time. The author correlated this increased exposure requirement with a pH-dependent decrease in the hypochlorous-hypochlorite ion ratio. As the pH was raised to 9.0, a sixfold increase in the time necessary to produce the 99% inactivation was observed, demonstrating that hypochlorite ion was a less effective virucidal agent. The survival of T2 bacteriophage and poliovirus type 1 in chlorine-treated raw wastewater, secondary effluent, and stormwater overflow was reported by Lothrop and Sproul (1969). They concluded that chlorine input levels of 28 and 40 mg/liter were needed to effect a 99.99% inactivation of T2 bacteriophage and poliovirus type 1, respectively, during a 30-min contact time. It was also concluded that existing recommended chlorine residual levels (1 mg/liter) were inadequate for high-level virus inactivation.

Liu et al. (1971) examined 20 different strains of human enteric virus for their response to free chlorine (0.5 mg/liter; pH 7.0) in partially treated Potomac River water. A wide range of virus resistance to chlorine was noted, with reovirus type 1 requiring a 2.7-min exposure to effect a 99.99% reduction, whereas poliovirus 2 required 40 min of exposure to achieve the same level of inactivation. The results additionally showed that echovirus type 12 required a contact time of more than 60 min for a 4-log reduction. The authors concluded that as specific groups, polioviruses and coxsackieviruses were the most resistant, followed by adenoviruses and echoviruses. Reoviruses appeared to be the least resistant to chlorine treatment.

Inactivation experiments conducted prior to 1970 often appeared to be equivocal because uniform testing conditions were not used. Scarpino et al. (1972) conducted a well-controlled study using highly purified virus pools, chlorine-demand-free reagents and equipment, standardized methods for titrating free chlorine residuals, and standardized microbial assays. Poliovirus type 1 and *Escherichia coli* were selected as test microorganisms as they represented microbial organisms of sanitary significance. Using a pH 6.0 phosphate buffer system and a pH 10.0 borate buffer system, the authors designed experiments which utilized chlorine as hypochlorous acid (pH 6, >95% HOCl) or hypochlorite ion (pH 10, 99.7% OCl^-). The 99% inactivation times for each organism were used to demonstrate concentration-time relationships of first-order inactivation mechanisms. Poliovirus was approximately 130 times more resistant than *E. coli* to hypochlorous acid, while the bacterium was more than three times more resistant to hypochlorite ions than the enterovirus. The data appeared to contradict other work which had indicated the hypochlorous acid form of chlorine to be the better virucide (Liu et al., 1971; Engelbrecht et al., 1980). The apparent conflict was resolved by Sharp and Leong (1980), who noted that under the high-salt buffer conditions employed by Scarpino et al. (0.1 M KCl), hypochlorite ion became an effective virucide. They further reported that at pH 10, where the only active disinfecting species was hypochlorite ion, the rate of virus inactivation in dilute phosphate-carbonate buffer was one-fifth as fast as that observed at pH 6.0, where the predominant form was hypochlorous acid. The effects of the addition of 0.1 M NaCl to the buffer were also evaluated. The experiment verified that the time required to inactivate 99% of the virus in the absence of salt was 31 times greater than when the salt was present. At pH 10.0, disinfection was more rapid with salt added than it was at pH 6.0 whether or not salt was present. It was noted that overall inactivation could be augmented 30- to 150-fold when NaCl was present. The authors suggested that high pH, in combination with high ionic strength, had sufficiently weakened the virus protein capsid to render it more susceptible to the penetration of hypochlorite ions.

The effects of NaCl on the amplification of virus inactivation by hypochlorite ion were also observed by Jensen et al. (1980), who noted that the comparatively weak disinfecting action of hypochlorite ion at pH 10.0 was enhanced 20- to 30-fold by the addition of 0.1 M NaCl. However, when the inactivation rates of two strains of coxsackievirus by hypochlorous acid at pH 6.0, in the presence of NaCl, were compared with inactivation rates of similarly treated poliovirus type 1 (Sharp and Leong, 1980), it was found that the coxsackievirus rates were not influenced by NaCl. These results served to reiterate the premise that broad generalizations related to specific aspects of virus inactivation by chlorine cannot be made with impunity.

The kinetics of chlorine inactivation were found to be linear with relation to time (Scarpino et al., 1972). Sharp and Leong (1980) corroborated this observation in reporting that the rate of poliovirus type 1 inactivation occurred as a linear dose-response function, with increased inactivation resulting as the chlorine concentration was raised from 10 to 40 μM at pH 6.0. Jensen et al. (1980) reported that inactivation of coxsackievirus B5 followed first-order kinetics over a 10- to 30-μM range at pH 6.0. In this study, plots of virus survival versus time were linear, despite the fact that the virus strain had a tendency to aggregate at all pH ranges employed in the study. The effect of viral aggregation on inactivation was also noted by Sharp et al. (1975) and Sharp (1976). Aggregation-dependent effects appeared to be influenced by both the virus type and the ionic strength of the suspending medium.

Variations in the sensitivities of viruses to disinfectants and the recognition that the behavior of one type does not necessarily represent others were noted by Harakeh and Butler (1984). Poliovirus type 1, coxsackievirus B5, echovirus 1, simian rotavirus SA-11, and a human rotavirus extracted from feces were evaluated for chlorine resistance in dilute activated sludge effluent. The authors reported that a threshold disinfectant concentration could be produced above which total inactivation could be achieved within 5 min. The threshold concentration for human rotavirus inactivation was 7.0 mg/liter. At lower concentrations (e.g., 4.1 mg/liter), a biphasic reaction was observed in which most of the inactivation occurred within 5 min, with little further loss of infectivity even though residual disinfectant was present. This phenomenon was also noted for poliovirus type 1, coxsackievirus B5, echovirus 1, and SA-11. Human rotavirus appeared to be the most resistant of the viruses tested. The results, however, may have been more a function of the presence of chlorine-binding fecal material than any innate strain resistance. Vaughn et al. (1986) investigated chlorine-induced inactivation in purified, single-particle preparations of simian rotavirus SA-11 and culturable, human rotavirus type 2 (Wa) over a range of disinfectant concentrations and pH levels. Inactivation rates of SA-11 were quite rapid when chlorine concentrations of $\geq$0.1 mg/liter were applied at acid or neutral pH. pH-related effects were noted, with considerable reduction of virucidal activity with increasing alkalinity. Human rotavirus inactivation proceeded in a similar manner, except that no pH-related effects were observed at chlorine concentrations of $\geq$0.1 mg/liter. The overall resistance of human rotavirus to chlorine treatment was somewhat greater than that of SA-11. Both viruses were rapidly inactivated at chlorine concentrations as low as 0.3 mg/liter. The data suggested that human rotavirus strains were no more resistant to chlorine action than previously tested enteric viruses, earlier findings to the contrary having likely resulted from the presence of fecal material.

The inactivation of a persistent strain of hepatitis A virus was later studied by the same authors (Vaughn et al., 1987b). Here, single-particle preparations of virus, suspended in an ionically defined buffer, were challenged with various chlorine concentrations over a range of pH levels. Considerable chlorine tolerance was noted, with a 3-mg/liter chlorine residual requiring a 10-min contact period to effect a 99.9% reduction in virus titer at pH 7.0. The study demonstrated that hepatitis A virus was highly tolerant to free chlorine at all pH levels tested and that its resistance was significantly greater than that of any previously tested human viruses.

Inactivation mechanisms

Few studies have specifically addressed the mechanisms of viral inactivation by chlorine. Most of the studies to date favor the involvement of sites on the viral nucleic acid, with the subsequent alteration of the viral genome. Olivieri et al. (1975) observed that chlorine was able to inactivate purified bacteriophage f2 RNA at a rate which paralleled the inactivation of intact virions. RNA and intact virus particles were rapidly inactivated at lower pH levels and more slowly at alkaline levels. This latter finding led the authors to speculate that at alkaline pH, the protein coat of the virion functioned as a barrier between the viral RNA and hypochlorite ion. The most compelling evidence for nucleic acid involvement came from a series of studies by Dennis et al. (1979a, 1979b), who reported that the rate of f2 phage inactivation by free chlorine precisely paralleled the rate of ^{36}Cl incorporation into the bacteriophage genome (Dennis et al., 1979a). Both ^{36}Cl incorporation and overall particle inactivation were higher at pH 5.6 than at pH 9.9. At the lower pH, 87% of the labeled chlorine was found to be associated with the viral RNA, whereas at the higher pH, only 60% was bound. Chlorine-protein interactions were not considered to be significant, since no correlation was noted between the oxidation-reduction reactions between chlorine and protein and the overall rate of f2 inactivation. Redox reactions were noted at pH levels where virus inactivation was diminished, suggesting that chlorine reaction with protein might actually inhibit virus inactivation. The hypothesis that chlorine reacted with viral RNA was later verified by studies which demonstrated that chlorine could specifically react with the nucleotides of RNA (Dennis et al., 1979b). The authors found that chlorine-mediated destruction of AMP and CMP nucleotides was inversely proportional to pH and corresponded directly to the rate of virus inactivation. By analyzing the UV absorption spectrum of chlorine-reacted nucleotides, they observed that the greatest spectrum changes occurred at low pH levels, where hypochlorous acid was the predominant species of chlorine. As the pH was increased, AMP-chlorine and CMP-chlorine interactions decreased, whereas hypochlorite ion reactions with GMP increased. UMP appeared to be unreactive at all pH levels tested.

Results obtained from studies using human viruses correlated well with the above observations. O'Brien and Newman (1979) noted that the inactivation of poliovirus type 1 involved a conversion of the intact 156S virus particle to 80S empty capsids. The capsids maintained the same isoelectric and cell-adsorption properties of intact viruses, indicating that no major conformational changes had occurred in the protein. They also found that poliovirus RNA was rapidly cleaved and released from the viral capsid. Nucleic acid cleavage, considered to be the primary mode of viral inactivation, appeared to occur prior to its release. Taylor (1982), studying the virucidal properties of chlorine and other disinfectants in demineralized water, reported that significant morphological changes occurred when poliovirus type 1 was exposed for 1 min to various concentrations of chlorine. When a chlorine concentration of 45 μM was used, 90% of the virus was inactivated and complete virions were observed in the preparation. An 85-μM concentration produced a 99% inactivation, and the viral suspension was shown to contain a mixture of full and empty capsids. A 150-μM concentration inactivated more than 99.9% of the virus with no intact virions detected in the suspension.

Changes in virus structure and composition caused by chlorine treatment were also reported by Alvarez and O'Brien (1982a). Through the use of radioactive labeling and rate zonal centrifugation procedures, specific changes in the virus structure were correlated with inactivation. At chlorine concentrations of up to 0.4 mg/liter, 38% of the virus was inactivated, with no major changes detected in its physical properties. At chlorine concentrations of 0.8 mg/liter, there was a significant shift in the sedimentation coefficient of the virus, indicating that viral capsids were without their RNA components. However, the percentage of radioisotope counts associated with empty capsids did not correlate with inactivation, as only 56% inactivation was noted. At a 2-mg/liter residual concentration, a 97% inactivation was correlated with empty capsid counts of 47%. The authors concluded that chlorine concentrations of greater than 0.8 mg/liter produced reactions within the virion which caused separation of viral components. At lower chlorine concentrations, however, the integrity of the virion was not altered. A series of experiments conducted by Vaughn and Chen (unpublished data) demonstrated that chlorine could produce significant conformational changes in rotavirus protein. A comparison of the electrophoretic mobilities of simian rotavirus SA-11 protein capsid and RNA extracted from normal and chlorine-treated virus was conducted using polyacrylamide gel electrophoresis. Samples of SA-11, suspended in pH 7.0 phosphate-carbonate buffer, were exposed for 60 s to chlorine concentrations of 1.3 and 2.6 mg/liter. Six to nine major protein bands could be visualized in untreated protein preparations. Chlorine treatment caused increases in the molecular weights of several of the major polypeptide moieties (molecular weights of approximately 35,000, 40,000, and 60,000). The amount of shift was proportional to the chlorine concentration. RNA segments extracted from treated viruses, on the other hand, failed to show any significant alterations.

Chlorine Dioxide

Chlorine dioxide is an oxidizing agent which historically has been used in conjunction with chlorine for the control of colors, tastes, and odors in water. The compound is extremely soluble in water (five times more soluble than chlorine [Grabow, 1982]) and, unlike chlorine, does not react with ammonia or organic compounds to form trihalomethanes (White, 1978; see Nonhalogen Disinfectants, below). Because of these properties, considerable interest has been focused on its use as a primary disinfecting agent, possibly as a replacement for chlorine.

Inactivation process

While chlorine dioxide has been used for color and odor control in water since the mid-19th century, it was nearly 100 years before its bactericidal potential was demonstrated in the laboratory (Ridenour and Ingols, 1947). Bernard et al. (1965), assessing disinfectant effects on *E. coli* suspended in sewage effluent, noted that removal of 90% of the organisms within a 5-min period required five times more chlorine than chlorine dioxide. Numerous studies initiated during the late 1970s have addressed the virucidal properties of chlorine dioxide. Tifft et al. (1977) indicated that chlorine dioxide was twice as efficient as chlorine for the inactivation of poliovirus type 1 and various bacteriophages in wastewater. Support for the superior virucidal capacity of chlorine dioxide over that of chlorine was also presented by Berg et al. (1979), Aieta et al. (1980), Longley et al. (1980), and Roberts et al. (1981).

Several more recent studies have expanded upon the above comparisons to include several additional water disinfectants. Vilagines (1982), investigating enterovirus and bacteriophage removal from treated effluent, found chlorine dioxide to be the most efficient virucide, followed by ozone, chlorine, and combined chlorine-bromine treatment. Taylor and Butler (1982), comparing the virucidal properties of chlorine, chlorine dioxide, bromine chloride, and iodine, indicated chlorine to be most effective at pH 7.0, with chlorine dioxide and iodine being more efficient at pH 9.0 and bromine chloride the best at pH 5.0. They noted that the presence of ammonia profoundly inhibited the action of chlorine and, to a lesser extent, bromine and iodine. No influence on chlorine dioxide was observed, attesting to its potential for the treatment of sewage effluent.

The effects of pH on the efficiency of chlorine dioxide disinfection, described above, have been noted by several authors. Brett and Ridgeway (1981) demonstrated that a 0.04-mg/liter disinfectant residual, which had removed 99% of drinking water-suspended $E.$ $coli$ within 2 min at pH 8.5, required 11 min for the same reduction at pH 6.5. Alvarez and O'Brien (1982b) reported that a 2-log reduction of poliovirus within 10 min at pH 6.0 was increased to 4 log within the same time sequence at pH 10 using the same disinfectant concentration. Enhancement of inactivation with increasing alkalinity was also recently reported by Chen and Vaughn (1990). Mechanistic studies of chlorine dioxide reactivity with viral nucleic acids (Hauchman et al., 1986) also demonstrated enhanced effects with increasing pH.

In spite of the clear advantages offered by the use of chlorine dioxide, its application as a general water disinfectant is not without problems. Its optimum efficiency at alkaline pH compromises its suitability for use with wastewater effluents, which tend to be somewhat acidic. Chlorine dioxide treatment of water may produce potentially toxic chlorite and chlorate ions, as well as carcinogenic epoxides. Its use in drinking water treatment may therefore be suspect. Chlorine dioxide may contain trace amounts of free chlorine which may react with organic compounds to form toxic trihalomethanes, the very reaction which had initiated the search for a chlorine substitute.

Inactivation mechanisms

Prior to 1982, little direct information about the mechanism of virus inactivation by chlorine dioxide was available. White (1978), for example, had speculated that the viral protein coat was the most likely target, based upon absorptive interactions between the disinfectant and peptone solutions. Olivieri et al. first reported the results of their mechanistic studies in 1982. Studies conducted with poliovirus type 1 and the bacterial viruses f2 and φX174 demonstrated a rate relationship between virus inactivation and the inhibition of virus attachment to host cell. The authors also noted an interaction between the disinfectant and viral nucleic acid, specifically the GMP moiety. However, this reaction was too slow to account for virus inactivation rates observed during kinetic studies, and the authors concluded that the primary site of lethal lesion was probably the virus protein coat. These findings were later supported by those of Hauchman et al. (1986) and Noss et al. (1986). In the former, the authors noted a singular lack of relationship between the rates of inactivation of intact f2 phage versus its infectious RNA and concluded that disinfectant action on nucleic acid could not be the principal factor responsible for inactivation of the intact virus particle. In the latter study (Noss et al., 1986), tyrosine residues on the capsid and the A protein moiety of f2 phage were suggested as the likely sites of chlorine dioxide activity. The authors concluded that the interaction produced an altered configuration of coat protein, resulting in a decreased affinity for host cell receptor sites.

Contrary to the above, several studies have indicated the viral nucleic acid to be the principal site of chlorine dioxide activity. Alvarez and O'Brien (1982b) revealed that poliovirus RNA separated from the capsid following treatment, resulting in virion structural conversion from 156S structure to 80S particles. The authors observed no concurrent loss in the ability of chlorine dioxide-treated viruses to adsorb, penetrate, and uncoat normally within their host cell. They concluded that the likely inactivation mechanism involved an interaction between the disinfectant and viral RNA, resulting in the impairment of its normal replication function. Taylor (1982) and Taylor and Butler (1982), working with poliovirus type 1 and f2 phage, similarly observed no effect of chlorine dioxide treatment on virus adsorption to host cell (although very high disinfectant doses did produce capsid damage), concluding that the most likely mechanism involved cleavage and release of the viral genome.

NONHALOGEN DISINFECTANTS

The use of chlorine for the disinfection of groundwater, surface waters, and wastewater has come under increasing scrutiny because of the potential for the formation of toxic chlorinated hydrocarbons in waters containing anthropogenic and naturally occurring organic compounds (Wang et al., 1978). Work by Schnoor et al. (1979) demonstrated the formation of trihalomethanes following the reaction of chlorine with low-molecular-weight compounds (i.e., <3,000 molecular weight). This finding was significant, since many halogenated organic compounds possess a demonstrable toxic (Tardiff and Deinzer, 1973) and mutagenic (Simmon and Tardiff, 1978) potential. In studies reported by Scully et al. (1988), the chlorination of natural waters containing algal proteins contributed significantly to the formation of trihalomethanes. Recently, Krasner et al. (1989) tested for the presence of disinfection by-products at 35 United States water treatment plants. Trihalomethanes, including chloroform, dichlorobromomethane, dibromochloromethane, and bromoform, represented the largest class of halogenated compounds detected, followed by mono-, di-, and trichloroacetic acids, mono- and dibromoacetic acids, formaldehyde, and acetaldehyde. The authors noted a reduction in by-product production with the addition of chloramine compounds, but this process also resulted in the increased production of undesirable cyanogen chloride. (Note: a 60% reduction of trihalomethanes with no apparent cyanogen formation had been reported by Rav-Acha et al. [1985], who reacted water containing trihalomethane precursors with chlorine dioxide prior to the introduction of chlorine.)

Concern over the potential adverse health effects associated with the formation of halogenated by-products, including carcinogens, has prompted a concerted search for alternative methods of disinfection. Among alternates receiving considerable attention during the past decade are chlorine dioxide, discussed above, ozone, and UV light. The efficacy and mechanism of antiviral action of the latter agents will be discussed in the following pages.

Ozone

Inactivation process

Ozone, a powerful oxidizing agent which can be generated by the passage of an electric discharge through a stream of oxygen, was shown to be an effective virucide in early studies conducted by Kessel et al. (1943). These authors reported rapid inactivation of a crude extract of poliovirus type 1 with ozone concentrations ranging from 0.05 to 0.45 mg/liter. Majumdar et al. (1973), using a more purified form of the same virus, identified a dose-response relationship in the form of an ozone threshold level (0.06 mg/liter) below which virus inactivation was drastically slowed. Studies conducted during the 1970s and 1980s established the effectiveness of ozone against a variety of virus types, including polio-, coxsackie-, echo-, and adenoviruses (Snyder and Chang, 1974; Katzenelson et al., 1979; Roy et al., 1982b; Farooq and Akhlaque, 1983; Harakeh and Butler, 1984; Warriner et al., 1985), vesicular stomatitis and encephalomycarditis viruses (Burleson et al., 1975), porcine picornaviruses (Foster et al., 1980), simian rotavirus SA-11 (Harakeh and Butler, 1984; Vaughn et al., 1987a), human rotaviruses (Vaughn et al., 1987a), bacteriophages (Carazzone and Vanini, 1969; Tittlebaum et al., 1971; Greets and Fomichev, 1985; Rogers and Lauer, 1986), hepatitis A virus (Botzenhart and Herbold, 1988; Herbold et al., 1989; Vaughn et al., 1990), and naturally occurring human viruses in sewage-contaminated river water (Joret et al., 1986).

With the advent of studies using partially or highly purified virus stocks tested under well-controlled laboratory conditions, characteristic patterns of virus response to ozone treatment and the influence of physical conditions in the suspending medium (e.g., pH, temperature, turbidity) on virus inactivation rates were elucidated. Snyder and Chang (1974), working with partially purified strains of enteroviruses and a single type of adenovirus suspended in either distilled water or river water, noted considerable variations in strain susceptibility to ozone treatment. While most strains were inactivated to undetectable levels within 10 min of contact, individual strain sensitivities varied significantly. For example, elimination of echovirus type 12 required nearly 10 times the contact time needed for similar reductions of echovirus type 29 (in ozone-demand-free water). Similarly, poliovirus type 2 was nearly 2.5 times more resistant than type 3 and more than 3 times more resistant than type 1. Foster et al. (1980) reported little variation between inactivation rates of purified poliovirus, coxsackievirus, and porcine picornaviruses suspended in phosphate buffer, with all strains completely inactivated within 20 s when exposed to ozone residuals ranging from 0.005 to 0.067 mg/liter. Significant strain variations were observed by Roy et al. (1982b), who exposed a variety of partially purified enterovirus strains to ozone in a mixed, continuous flow reactor. The authors reported the relative resistance of virus types to be as follows: poliovirus type 2 > echovirus 1 > poliovirus type 1 > coxsackievirus B5 > echovirus 5 > coxsackievirus A9. Under the controlled experimental conditions, poliovirus type 2, the least sensitive strain, with 1.5-log removal in 120 s at an ozone residual of 0.15 mg/liter, was some 40 times more resistant than coxsackievirus A9 (most sensitive with 3.5-log removal in 120 s with 0.15 mg of ozone per liter). Harakeh and Butler (1984) also noted sensitivity variations in viruses suspended in activated sludge effluent. The authors reported the relative viral resistances to an ozone concentration of 0.26 mg/liter to be human rotavirus > poliovirus type 1 > simian rotavirus SA-11 > echovirus 1 > coxsackievirus B5 > f2 bacteriophage. The relative sensitivity demonstrated by human rotavirus may have been an artifact related to the use of high-ozone-demand fecal suspensions as the virus source. Support for this hypothesis was provided by Vaughn et al. (1987a), who compared inactivation rates of purified, single-particle suspensions of human rotavirus and SA-11 in defined ozone-demand-free buffer. Both virus types were highly sensitive to ozone, with 5-log reductions in concentration occurring within

10 s at ozone residuals of ≥0.25 mg/liter. The human strain was shown to be considerably more ozone labile than the simian strain.

Three recently published reports addressed the inactivation dynamics of hepatitis A virus by ozone. In studies reported from the same laboratory (Botzenhart and Herbold, 1988; Herbold et al., 1989), unpurified hepatitis A viral stocks (HAV/HFS/GBM), along with poliovirus type 1, were suspended in phosphate-buffered saline and treated with ozone under constant flow conditions. The authors reported significantly differing inactivation patterns between the two virus types, with hepatitis A virus requiring nearly three times the ozone concentration to effect complete inactivation. They indicated a disinfectant residual of 0.38 mg/liter to be sufficient to nearly instantaneously inactivate all input virus. In the most recent report of hepatitis A virus inactivation by ozone (Vaughn et al., 1990), highly purified viral stocks (a large-focus-forming variant of HM-175), consisting primarily of single particles suspended in ionically defined buffer, were exposed to various ozone concentrations. Unlike the above reports, the hepatitis A virus strain used in this study was considerably more tolerant to ozone, with 1.0-mg/liter residuals being required to recreate the rapid infectivity reductions resulting from treatment with 0.1- to 0.2-mg/liter residuals in the earlier work. In their discussion of this apparent disparity in viral tolerance, the authors noted differences in virus strains used, varying detection method sensitivities, and procedural differences. Given the relative ozone tolerance of hepatitis A virus in all studies to date, the authors urged the use of higher ozone residuals than the 0.4-mg/liter concentration previously recommended by Block et al. (1981) for the disinfection of drinking water. The 1-mg/liter threshold concentration originally proposed by Majumdar et al. (1973) may better serve this end.

The roles of pH and temperature in virus inactivation by ozone have been analyzed in relatively few studies, most researchers opting to conduct experiments at the ambient pH and temperature levels of natural waters or wastewater effluents. Roy et al. (1982b), working with partially purified enteroviruses in a continuous-flow system, reported greater viral resistance to ozone at low pH than at neutrality. Survival of poliovirus type 1 exposed to 0.15-mg/liter ozone residuals was four times greater at pH 4.3 than at pH 7.2, and echovirus was more than twice as resistant at the acidic pH. Vaughn et al. (1987a), investigating inactivation in purified single-particle suspensions of human (Wa) and simian (SA-11) rotaviruses, observed significant pH effects with only one strain (SA-11) and only when ozone residuals of ≤0.15 mg/liter were used. Unlike the results of Roy et al. (1982b), SA-11 survival was enhanced with increasing alkalinity. The human strain, however, was so sensitive to ozone inactivation that pH effects were negligible. In a recent study of hepatitis A virus inactivation by ozone, these same authors reported similar marginal pH effects. Roy et al. (1982b) addressed temperature effects on virus inactivation in a continuous-flow system, reporting a twofold decrease in poliovirus resistance to a 0.15-mg/liter ozone residual when the temperature was increased from 5 to 10°C.

Many of the studies referenced above, conducted in ozone-demand-free buffers and using partially or highly purified virus stocks, were specifically designed to delineate basic characteristics of ozone-induced virus inactivation. In using optimum conditions for disinfectant activity (indeed, most demonstrated significant, often complete, virus inactivation within 60 to 120 s of ozone contact), these studies were unable to assess disinfection dynamics in natural waters and wastewater, which may contain considerable ozone demand as well as highly aggregated or particle-bound viruses. These pragmatic considerations have been addressed by several investigations during the past two decades. Foster et al. (1980) compared inactivation rates of three picornaviruses (poliovirus type 1, coxsackievirus A9, and porcine picornavirus type 3) tested as unassociated, feces-associated, and tissue culture-associated suspensions in phosphate buffer. All nonassociated viruses were reduced to undetectable levels within 10 s at initial ozone concentrations of 0.096 to 0.85 mg/liter. Association of virus with fecal particles to a turbidity of 5 nephelometric units appeared to provide sufficient ozone demand to significantly enhance virus survival. Protection of host cell-associated (1 nephelometric unit) poliovirus and coxsackievirus was even more dramatic, with demonstrable survival after 30 s of exposure to initial ozone concentrations of 2.84 to 4.68 mg/liter. In a follow-up to these studies, Emerson et al. (1982) investigated ozone inactivation of cell-associated poliovirus type 1 and coxsackievirus A9 suspensions (adjusted to 5 nephelometric units) in a continuous-flow ozonation system. In this system, unassociated viral controls were completely inactivated within 10 s by the application of 0.081 mg of ozone per liter. Cell-associated poliovirus, however, required a 2-min contact period with an ozone concentration of 6.82 mg/liter (4.70 mg/liter residual) to effect a similar reduction. Host cell-associated coxsackievirus exposed to an initial ozone dose of 4.81 mg/liter (2.18 mg/liter residual) survived for nearly 5 min. These results were later corroborated by Hartemann et al. (1983), who tested the effects of "earth extract" on poliovirus disinfection in a laboratory-scale ozonation treatment plant. A comparison of the resulting linear regression curve slopes clearly demonstrated slope increases which reflected the ozone concentration augmentations required to obtain identical disinfection levels in suspensions containing virus only (slope = -0.015), 1/6,000 earth extract (slope = -0.17), and 1/1,000 earth extract (slope = -0.20).

Studies comparing virus sensitivity to ozone treatment in buffered solutions versus natural waters or wastewater further substantiated the protective effects described above. Farooq and Akhlaque (1983), conducting experiments with poliovirus type 1 suspended in both phosphate buffer and activated sludge effluent, reported that considerably more ozone had to be applied to the effluent to obtain residuals comparable to the clean system. Comparison of inactivation curves indicated that somewhat higher ozone residuals were required in effluent (0.29 to 0.36 mg/liter) to achieve an inactivation rate similar to that occurring in buffer (0.23 to 0.26 mg/liter). Hartemann et al. (1983) noted significant tolerance differences between a laboratory strain of poliovirus type 1 suspended in drinking water (slope = -0.15) and indigenous enteroviruses in wastewater (slope = -0.38). Harakeh and Butler (1984), investigating the survival of enteroviruses and simian rotavirus in ozone-treated activated sludge effluents, reported the need to significantly extend ozone contact times because of effluent quality and the degree of virus aggregation. Warriner et al. (1985) investigated the removal of poliovirus type 1 from the effluent of a bench-scale advanced wastewater treatment (AWT) process (line or alum treatment followed by mixed-medium filtration). The authors indicated that ozone residuals of 2 to 4 mg/liter were required to effect complete virus removal within a 4-min contact period. Joret et al. (1986) successfully used lower ozone residuals for the removal of indigenous viruses from raw river water processed through a full-scale drinking water treatment system (storage, coagulation-filtration, settling, sand filtration) by adding an ozonation step (0.5 to 1 ppm for 1 min) prior to the treatment train, followed by a standard posttreatment ozonation (1.4 to 1.6 ppm for

10 min). The authors reported that use of this progressive ozonation process reduced enterovirus numbers to below detectable limits.

Efforts to assess the general virucidal effects of ozone in absolute terms, on the basis of comparing data such as those presented above, have historically been hampered by the procedural variabilities introduced by individual researchers. Methodological aspects which have consistently lacked uniformity include nature and state of the test virus used (type, strain, purity, physical state [i.e., aggregated, single particle], etc.); the virus detection methods used; the nature of the suspending medium (e.g., distilled water, buffered salt solutions, natural waters [e.g., river water, groundwater], finished drinking water, secondary and tertiary wastewater effluent); the physical properties of the suspending medium (e.g., temperature, pH, ionic strength); the ozone levels used and their specific methods of measurement; and the specific experimental conditions (e.g., use of demand-free vessels, small/large volume static system, bench-scale flow-through system, full-scale treatment facility). A novel procedure for "normalizing" some of the data from different studies was proposed by Greets and Fomichev (1985). The authors demonstrated that inactivation data, plotted on the basis of ozone concentration per virus particle, was described graphically by a curve that was typical for the specific strain under study, suspension densities notwithstanding. Through this operation, the results of different studies using the same test viruses might be directly compared.

Relatively few attempts have been made to mathematically describe the inactivation of viruses by ozone. Majumdar et al. (1973) determined rate equations for poliovirus inactivation with respect to contact time, ozone concentration, and the type of suspending medium (distilled water, primary and secondary wastewater effluents). The authors identified two distinct rate mechanisms, one above and the other below their established threshold concentration of 1 mg/liter. On the basis of their own experimental data, the authors developed the following general equation for poliovirus inactivation in each suspending medium; $\partial C_v/\partial t = -k_1 C v^n C^m$, where v is the concentration of virus particles at any time t (PFU/milliliter); t is contact time (minutes); and C is ozone concentration (milligrams per liter). The values k_1 (reaction constant of inactivation), n, and m (orders of the reaction) varied with C such that when C was >1 mg/liter, k_1, n, and m were 3.06×10^{-1} ml/PFU/mg/min, 1.22, and 0.40, respectively (secondary effluent only). Values of k_1, n, and m when C was <1 mg/liter were calculated to be 7.05×10^{-5} ml/PFU/mg/min, 2.32, and 1.65, respectively, in distilled water and 9.45×10^{-4} ml/PFU/mg/min, 2.10, and 2.08, respectively, in primary and secondary effluent. Although experimental variables such as ozone concentration and the type of suspending medium used prevented the development of a unified inactivation equation, Majumdar et al. noted good agreement when applying appropriate equations to either batch or continuous-flow conditions.

In a more recent paper, Roy et al. (1982a) presented a mathematical model of poliovirus inactivation by ozone based upon kinetic and mechanistic information developed by the authors in an earlier study (Roy et al., 1980). Observing that ozone-induced poliovirus inactivation did not follow single-order kinetics, they speculated that the cause involved inherent properties of the virus rather than virus population heterogeneity. After testing several models, they proposed the equation $S = 1 + A \times t/(1 + B \times t)^2$, where S is virus survival rates obtained from experimental data, t is detention time (reactor volume/flow rate), and A and B are constants (these constants were shown to follow the Arrhenius equation and were thought to be analogous to reaction rate constants). The empirical model was capable of predicting the experimental inactivation data within ±0.5% accuracy. Since constants A and B followed the Arrhenius equation, their value could be predicted at any temperature.

Inactivation mechanisms

Mechanistic studies of the inactivation of microorganisms by ozone have primarily focused on bacteria. Giese and Christensen (1954) indicated the bacterial cell surface to be the primary target of ozone. In the same year, Barron (1954) suggested the primary mechanism to involve the oxidation of enzyme-linked sulfhydryl groups. Bringman (1955) proposed a mechanism in which ozone acted by oxidizing bacterial cytoplasm. Murray et al. (1965) concluded that attack of the bacterial cell wall would alter cell permeability and lead eventually to lysis.

On the basis of the brief description of basic virus characteristics presented above (General Properties of Viruses), none of the foregoing mechanisms could be considered to be relevant to the process of virus inactivation by ozone. The principal viral targets for ozone attack would likely include the capsid protein, the nucleic acid moiety, and, in appropriate species, the lipid envelope. (Since enveloped viruses are not readily transmitted through water, relatively few studies have focused on their inactivation by ozone. They will not be considered in this section.) Interaction between capsid protein and ozone would likely influence the adsorption/penetration sequence between a virus and its specific host cell. The initial step of the virus infection cycle involves a specific reaction between receptor areas, located on both the virus and host cell. The relationship between capsid polypeptide integrity and infectivity has been demonstrated in several studies (Crowell and Philipson, 1971; Longer-Holm and Kornat, 1972; Breindl and Koch, 1972; Cords et al., 1975; Rekosh, 1977). Ozone interactions with viral nucleic acid, on the other hand, might adversely affect the complex virus-specific biosynthetic processes which occur following penetration of the host cell. Studies reviewed by Mudd et al. (1969) demonstrated the oxidizing effects of ozone on purified amino acids including alanine, cysteine, cystine, histidine, methionine, phenylalanine, tryptophan, and tyrosine. Christensen and Giese (1954) reported apparent reductions in nucleic acid integrity following ozone treatment. Riesser et al. (1977), investigating ozone-treated poliovirus, observed damage to the viral capsid. This was interpreted by the authors as evidence for disruption of host cell attachment being the primary mechanism of ozone effect. However, as their data also demonstrated little concurrent reduction of viral penetration, it became clear that other inactivation mechanisms had to be considered.

Roy et al. (1981) made significant contributions to the understanding of the mechanism of ozone-induced virus inactivation. Their experiments involved the treatment of whole poliovirus particles, labeled with either [^{3}H]uridine or ^{14}C-amino acids, with low ozone residuals (0.3 to 0.8 mg/liter). Analytical techniques ranging from sucrose density gradient centrifugation to sodium dodecyl sulfate-polyacrylamide gel electrophoresis were then applied to whole virus particles or to capsid protein and RNA extracted from previously treated virions. Ozone treatment did not appear to cause major disruption of particle integrity, as viruses did not dissociate to subunits or form aggregates. Electrophoresis of capsid proteins revealed significant damage to major polypeptides VP1 and VP2, but VP4, the small polypeptide responsible for host cell attachment, did not appear to be damaged.

This latter finding was supported by the results of companion experiments which demonstrated little alteration in the attachment potential of similarly treated viruses. A comparison of nucleic acid sedimentation profiles from treated and untreated virions delineated apparent damage to the viral genome. The authors concluded that the inactivation resulting from relatively low ozone dosages was a function of RNA damage (Roy et al., 1981). They also noted correlation with earlier kinetic studies which indicated inactivation to be rate limited by the period during which the ozone diffuses through the protein coat to the RNA core.

Recent studies presented by Chen et al. (1987) demonstrated the effects of ozone treatment on the structural integrity of human (Wa) and simian (SA-11) rotaviruses. Here, whole virus particle preparations were exposed for 1 min to ozone concentrations ranging from 0.1 to 0.45 mg/liter. Viral capsid proteins and RNA were then analyzed by sodium dodecyl sulfate-polyacrylamide gel electrophoresis, and the resulting electrophoretic banding patterns were compared with those from untreated virus particles. (Capsid proteins of the double-coated rotaviruses usually separate, upon electrophoresis, into two major polypeptide moieties and seven or eight minor groups. Rotavirus double-stranded RNA normally separates into 11 distinct segments.) Treatment with ozone concentrations as low as 0.1 ml/liter appeared to cause cleavage of capsid proteins to smaller subunits, as evidenced by dramatic losses of material from the major and minor polypeptide groups. Molecular weights of remaining polypeptide moieties were significantly altered. The extent of protein loss and molecular weight shift was directly proportional to ozone concentration for both virus types. When ozone concentrations of ≥ 0.3 mg/liter were used, at least one of the major polypeptide groups split into three distinct protein bands. At the higher ozone concentrations, nearly all of the minor bands were lost. Effects on viral RNA involved proportional losses of material from banded segments with increasing ozone concentration. Treatment with residuals of ≥ 0.3 mg/liter resulted in the loss of all RNA segments. Results of protein and nucleic acid studies were consistent with earlier kinetic data (Vaughn et al. 1987a), which demonstrated extremely rapid (i.e., within a few seconds) inactivation of both virus types with ozone concentrations as low as 0.15 mg/liter. In presently unpublished studies conducted by these authors, electron micrographs of rotavirus particles treated with 0.3 mg of ozone per liter revealed extensive structural damage to both outer and inner capsids, as well as intrusion of stain (phosphotungstic acid) into the area usually occupied by the nucleoprotein core. The observed extent of damage was consistent with the above kinetic and mechanistic studies. Contrary to the findings of Roy et al. (1981) with poliovirus, the rotavirus target for ozone action appeared to include both capsid protein and viral genome. It is possible that capsid may serve as the primary site of interaction, with RNA a secondary target following degradation of sufficient portions of the protein coat. The apparent dichotomy between the studies likely reflects basic structural differences between poliovirus (single capsid, single-stranded RNA genome) and rotavirus (double capsid, double-stranded RNA). Future mechanistic studies of ozone effect may reveal additional structurally related variations.

UV Light

Inactivation process

The many laboratory studies conducted to date have clearly demonstrated the virucidal efficiency of UV light as well as its overall potential as an alternative to chlorine in the disinfection of water and wastewater. Hill et al. (1970) reported the rapid inactivation of eight enterovirus types suspended in estuarine water. Morris (1972), studying UV irradiation of bacterial and viral species, noted that the dose necessary for a 98% reduction of poliovirus was more than twice that required for a similar inactivation of *E. coli*. Von Brodorotti and Mahnel (1982), working with nine different virus species, noted UV susceptibility differences coinciding with structural differences among the test strains. They reported that, in general, viruses with single-stranded genomes were more sensitive than those with double strands, DNA viruses were more resistant than RNA viruses, and within similar groups, UV sensitivity increased with increasing nucleic acid molecular weight. Chang et al. (1985) determined that while inactivation rates for poliovirus and simian rotavirus (SA-11) were similar, these viruses were three times more tolerant to UV treatment than *E. coli*. These authors indicated the average dose required for 3-log removal of viruses to range from 28 to 42 mW · s/cm^2.

In spite of the reliability of UV disinfection demonstrated in each of the above studies, difficulties involving UV dose measurement have prevented direct comparison between different studies and have hindered systematic comparisons between UV efficiencies and those of chlorine treatment. These problems, reviewed by Johnson et al. (1982), include (i) the inability to accurately measure the three-dimensional UV intensity to which a target cell is exposed; (ii) the inability to estimate the average UV dose rate within a reaction vessel on the basis of measurement from a single fixed detector; (iii) the UV absorbence measurement errors influenced by the nonspecific scattering of UV light by particulates; (iv) the use of inappropriate UV dose equations; and (v) the inability to accurately relate exposure times in flow-through systems. With these problems in mind, Qualls and Johnson (1985) developed a mathematical model which predicted bacterial survival in UV-treated flow-through systems. The authors' calculations accounted for variable intensity and flow patterns and nonlinear curves of bacterial survival versus UV dose. Survivors in each fraction of the residence time distribution were calculated separately and then summed to reveal the average survival. Model input data included the system's average UV intensity, the residence time distribution, and a dose-survival curve, which had been experimentally determined. The authors reported that predictions based upon the model correlated well with kinetic data from a UV pilot plant study.

In a earlier study, Severin et al. (1983) applied the results of *E. coli*, *Candida* sp., and bacterial virus f2 inactivation experiments to two kinetic models. One of these, a series-event model, was judged to be superior. In this model, an event is assumed to denote a unit of damage, each event occurring in a stepwise fashion, with each step assumed to be an integer function. Organisms pass from one event level to the next at a rate which is first order with respect to UV intensity. Damage proceeds with continued exposure to a threshold where organisms achieving an event level greater than the threshold are inactivated, while those below this level survive. Thresholds vary according to conditions used, but for a defined set of conditions, the threshold is constant. The resulting equation, which can be used to predict inactivation in a completely mixed, flow-through reaction, was given as $N_s/N_I = 1 - [1 + (1/mk\, I_0\, t)]^{-n}$, where N_s is the density of surviving organisms (number per milliliter); N_I is the initial density of organisms (number per milliliter); m is the intensity factor in an annular reactor; k is a kinetic constant (per centimeter per microwatt per second); I_0 is the intensity at the surface of the quartz tube (micromolar per square centimeter); and t is the theoretical contact time in a completely mixed flow-through reactor (seconds). The authors noted that for general applications, the quantity mI_0 could be replaced by the average intensity within the reaction vessel.

Inactivation mechanisms

Studies by Roizman et al. (1959) and LeBouvier (1959) demonstrated alterations in the antigenic structure of UV-treated poliovirus. Katagiri et al. (1967) presented a correlation between these changes and the loss of viral infectivity. However, DeSena and Jarvis (1981) and Rodgers et al. (1982) published findings indicating that significant modifications to viral structural integrity (as determined by electron microscopy) occurred well after the loss of virus infectivity. A subsequent study of Rodgers et al. (1985) revealed a similar sequence of events following UV irradiation of human rotavirus particles. Results of the latter studies were seen to lend support to the suggestion by DeSena and Jarvis (1981) that the initial target for UV-induced damage was the viral genome, damage of which resulted in the retardation of normal virus replication processes.

REFERENCES

Aieta, E. M., J. D. Berg, P. V. Roberts, and R. C. Cooper. 1980. Comparison of chlorine dioxide and chlorine in wastewater disinfection. *J. Water Pollut. Control Fed.* **52:**810–824.

Alberts, B., D. Bray, J. Lewis, M. Raff, K. Roberts, and J. D. Watson. 1989. *Molecular Biology of the Cell*, 2nd ed. Garland Publishing Inc., New York.

Alvarez, M. E., and R. T. O'Brien. 1982a. Effects of chlorine concentration on the structure of poliovirus. *Appl. Environ. Microbiol.* **43:**237–239.

Alvarez, M. E., and R. T. O'Brien. 1982b. Mechanisms of inactivation of poliovirus by chlorine dioxide and iodine. *Appl. Environ. Microbiol.* **44:**1064–1071.

Barron, E. S. 1954. The role of free radicals of oxygen in reactions produced by ionizing radiations. *Radiat. Res.* **1:**109.

Berg, J. D., E. M. Aieta, P. V. Roberts, and R. C. Cooper. 1979. Effectiveness of chlorine dioxide as a wastewater disinfectant, Sec. 3, p. 61–71. In *Proceedings of the National Symposium on Progress in Wastewater Disinfection Technology.* EPA-600/9-79-018. U.S. Environmental Protection Agency, Cincinnati.

Bernard, M. A., B. M. Israel, V. P. Olivieri, and M. L. Granstrom. 1965. Efficiency of chlorine dioxide as a bactericide. *Appl. Microbiol.* **13:**776–780.

Blacklow, N. R., and G. Cukor. 1982. Viruses and gastrointestinal disease, p. 75–87. In D. A. J. Tyrrell and A. Z. Kapikian (ed.), *Virus Infections of the Gastrointestinal Tract.* Marcel Dekker, Inc., New York.

Block, J. C. 1983. Viruses in environmental waters, p. 117–145. In G. Berg (ed.), *Viral Pollution of the Environment.* CDC Press, Inc., Boca Raton, Fla.

Block, J. C., Y. Richard, P. Hartemann, and J. M. Foliguet. 1981. Disinfection des eaux par l'ozone. *Eau Ind.* **58:**69–76.

Botzenhart, K., and K. Herbold. 1988. Abotung von Hepatitis A Virus im Wasser durch Ozon. *Z. Gesamte Hyg.* **34:**508–510.

Breindl, M., and G. Koch. 1972. Competence of HeLa cells for inactivation by inactivated poliovirus particles and by isolated viral RNA. *Virology* **48:**136–144.

Brett, R. W., and J. W. Ridgeway. 1981. Experience with chlorine dioxide in the Southern Water Authority and Water Research Center. *J. Inst. Water Eng. Sci.* **35:**135–142.

Bringman, G. 1955. Determination of lethal activity of chlorine and ozone on *E. coli. Water Pollut. Abstr.* **28:**12–18.

Burleson, G. R., T. M. Murray, and M. Pollard. 1975. Inactivation of viruses and bacteria by ozone with and without sonication. *Appl. Microbiol.* **29:**340–344.

Carazzone, M. N., and G. C. Vanini. 1969. Experimental studies of the effect of ozone on viruses. 1. Effect of bacteriophage T1. *G. Batteriol. Virol. Immunol.* **62:**828.

Caul, E. O., and S. I. Egglestone. 1982. Coronaviruses in humans, p. 179–193. *In* D. A. J. Tyrrell and A. Z. Kapikian (ed.), *Virus Infections of the Gastrointestinal Tract*. Marcel Dekker, Inc., New York.
Centers for Disease Control. 1990. Viral agents of gastroenteritis: public health importance and outbreak management. *Morbid. Mortal. Weekly Rep.* **39:**1–24.
Chambers, C. W. 1971. Chlorination for control of bacteria and viruses in treatment plant effluents. *J. Water Pollut. Control Fed.* **43:**228–241.
Chang, J. C. H., S. F. Ossoff, D. C. Lobe, M. H. Dorfman, C. M. Dumais, R. G. Qualls, and J. D. Johnson. 1985. UV inactivation of pathogenic and indicator organisms. *Appl. Environ. Microbiol.* **49:**1361–1365.
Chen, Y. S., and J. M. Vaughn. 1990. Inactivation of human and simian rotaviruses by chlorine dioxide. *Appl. Environ. Microbiol.* **56:**1363–1366.
Chen, Y. S., J. M. Vaughn, and R. M. Niles. 1987. Rotavirus RNA and protein alterations resulting from ozone treatment. Abstr. Annu. Meet. Am. Soc. Microbiol. 1987, Q-22, p. 285.
Christensen, E., and A. C. Giese. 1954. Changes in absorption spectra of nucleic acids and their derivatives following exposure to ozone and ultraviolet radiation. *Arch. Biochem. Biophys.* **51:**208–216.
Clark, N. A., R. E. Stevenson, and P. W. Kabler. 1956. The inactivation of purified type 3 adenovirus in water by chlorine. *Am. J. Hyg.* **64:**314–319.
Cords, C. E., C. G. James, and L. C. McLaren. 1975. Alterations of capsid proteins of coxsackievirus A13 by low ionic concentrations. *J. Virol.* **15:**244–252.
Craun, G. F. 1986. *Waterborne Diseases in the United States*. CRC Press, Inc., Boca Raton, Fla.
Crowell, R. L., and L. Philipson. 1971. Specific alterations of coxsackievirus B3 eluted from HeLa cells. *J. Virol.* **8:**509–515.
Davis, B. D., R. Dulbecco, H. N. Eisen, and H. S. Ginsberg. 1990. *Microbiology*, 4th ed. J.B. Lippincott Inc., Philadelphia.
Dennis, W. H., V. P. Olivieri, and C. W. Kruse. 1979a. Mechanism of disinfection: incorporation of Cl-36 into f2 virus. *Water Res.* **13:**363–369.
Dennis, W. H., V. P. Olivieri, and C. W. Kruse. 1979b. The reaction of nucleotides with aqueous hypochlorous acid. *Water Res.* **13:**357–362.
DeSena, J., and D. L. Jarvis. 1981. Modification of the poliovirus capsid by ultraviolet light. *Can. J. Microbiol.* **27:**1185–1193.
Emerson, M. A., O. J. Sproul, and C. Buck. 1982. Ozone inactivation of cell-associated viruses. *Appl. Environ. Microbiol.* **43:**603–608.
Engelbrecht, R. S. 1978. Virus sensitivity to chlorine disinfection of water supplies. Environmental Protection Technology Report (EPA-600/2-78-123). U.S. Environmental Protection Agency, Cincinnati.
Engelbrecht, R. S., M. J. Weber, B. L. Salter, and C. A. Schmidt. 1980. Comparative inactivation of viruses by chlorine. *Appl. Environ. Microbiol.* **40:**249–256.
Farooq, I., and S. Akhlaque. 1983. Comparative response of mixed cultures of bacteria and virus to ozonation. *Water Res.* **17:**809–812.
Farrah, S. R., and S. A. Schaub. 1983. Viruses in wastewater sludges, p. 147–161. *In* G. Berg (ed.), *Viral Pollution of the Environment*. CRC Press, Inc., Boca Raton, Fla.
Foster, D. M., M. A. Emerson, C. E. Buck, D. S. Walsh, and O. J. Sproul. 1980. Ozone inactivation of cell- and fecal-associated viruses and bacteria. *J. Water Pollut. Control Fed.* **52:**2174–2184.
Fraenkel-Conrat, H. 1970. *The Chemistry and Biology of Viruses*. Academic Press, Inc., New York.
Fraenkel-Conrat, H., and B. Singer. 1957. Virus reconstitution. II. Combination of protein and nucleic acid from different strains. *Biochim. Biophys. Acta* **24:**540–548.
Gerba, C. P. 1988. Viral disease transmission by seafoods. *Food Technol.* **42:**99–103.
Gierer, A., and G. Schramm. 1956. Infectivity of ribonucleic acid from tobacco mosaic virus. *Nature* (London) **177:**702–703.
Giese, A. C., and E. Christensen. 1954. Effects of ozone on organisms. *Physiol. Zool.* **27:**101.
Grabow, W. O. K. 1982. Disinfection by halogens, p. 216–260. *In* M. Butler, A. R. Medlin, and

R. Morris (ed.), *Proceedings of the International Symposium on Viruses and Disinfection of Water and Wastewater*. University of Surrey Press, Guildford, Surrey, United Kingdom.
Greets, N. V., and A. Y. Fomichev. 1985. The dependence of ozone antibacterial and antiphage action on the concentration of cells and phage particles in the reaction medium. *Mikrobiologiya* **54**:410–413.
Harakeh, M., and M. Butler. 1984. Inactivation of human rotavirus, SA-11 and other enteric viruses in effluent by disinfectants. *J. Hyg.* **93**:157–163.
Hartemann, P., J. C. Block, J. C. Joret, J. M. Foliguet, and Y. Richard. 1983. Virological study of drinking and wastewater disinfection by ozonation. *Water. Sci. Technol.* **15**:145–154.
Hauchman, F. S., C. I. Noss, and V. P. Olivieri. 1986. Chlorine dioxide reactivity with nucleic acids. *Water Res.* **20**:357–361.
Herbold, K., B. Flehmig, and K. Botzenhart. 1989. Comparison of ozone inactivation, in flowing water, of hepatitis A virus, poliovirus 1, and indicator organisms. *Appl. Environ. Microbiol.* **55**:2949–2953.
Hershey, A. D., and M. Chase. 1952. Independent functions of viral protein and nucleic acid in growth of bacteriophage. *J. Gen. Physiol.* **36**:44–71.
Hill, W. F., F. E. Hamblet, W. H. Benton, and E. W. Akin. 1970. Ultraviolet devitalization of eight selected enteric viruses in estuarine water. *Appl. Microbiol.* **19**:805–812.
Holmes, I. H. 1982. Basic rotavirus virology in humans, p. 111–124. *In* D. A. J. Tyrrell and A. Z. Kapikian (ed.), *Virus Infections of the Gastrointestinal Tract*. Marcel Dekker, Inc., New York.
Iwanowski, D. 1892. Uber die Mosaikkrankheit Tabakspflanze. *Bull. Acad. Imp. Sci. St. Petersbourg* **3**:67–70.
Jensen, H., K. Thomas, and D. G. Sharp. 1980. Inactivation of coxsackievirus B3 and B5 in water by chlorine. *Appl. Environ. Microbiol.* **39**:633–640.
Johnson, J. D., R. G. Qualls, K. H. Aldrich, and M. P. Flynn. 1982. Ultraviolet dose in wastewater disinfection, p. 362–377. *In* M. Butler, A. R. Medlin, and R. Morris (ed.), *Proceedings of the International Symposium on Viruses and Disinfection of Water and Wastewater*. University of Surrey Press, Guildford, Surrey, United Kingdom.
Joret, J.-C., J. Hassen, M. M. Bourbigot, F. Agbalika, P. Hartemann, and J. M. Foliguet. 1986. Virus inactivation during water treatment by a progressive ozonation unit. *Water Res.* **20**:871–876.
Kapikian, A. Z., H. B. Greenberg, R. G. Wyatt, A. R. Kalica, and R. M. Chanock. 1982. The Norwalk group of viruses—agents associated with epidemic viral gastroenteritis, p. 147–177. *In* D. A. J. Tyrrell and A. Z. Kapikian (ed.), *Virus Infections of the Gastrointestinal Tract*. Marcel Dekker, Inc., New York.
Katagiri, S., Y. Hinuma, and N. Ishida. 1967. Biophysical properties of poliovirus particles irradiated with ultraviolet light. *Virology* **32**:337–343.
Katzenelson, E., G. Koerner, N. Biedermann, M. Peleg, and H. I. Shuval. 1979. Measurement of the inactivation kinetics of poliovirus by ozone in a fast-flow mixer. *Appl. Environ. Microbiol.* **37**:715–718.
Kessel, J. F., D. K. Allison, F. J. Moore, and M. Kaime. 1943. Comparison of chlorine and ozone as virucidal agents of poliomyelitis virus. *Proc. Soc. Exp. Biol. Med.* **53**:71–73.
Krasner, S. W., M. J. McGuire, J. G. Jacangelo, N. L. Patania, K. M. Reagan, and E. M. Aieta. 1989. The occurrence of disinfection by-products in US drinking water. *J. Am. Water Works Assoc.* **81**:41–53.
LeBouvier, G. L. 1959. The D → C change in poliovirus particles. *Br. J. Exp. Pathol.* **40**:605–620.
Liu, O. C., H. R. Seraichekas, E. W. Akin, D. A. Brashear, E. L. Katz, and J. Hill, Jr. 1971. Relative resistance of twenty human enteric viruses to free chlorine in Potomac water, p. 171–197. *In* V. Snoeynik (ed.), *Virus and Water Quality: Occurrence and Control*. Proceedings of the 13th Water Quality Conference, vol. 69, no. 1. University Bulletin. Department of Civil Engineering, University of Illinois, Urbana-Champaign.
Loeffler, F., and P. Frosch. 1898. Berichte der Kommission zur Erforschung der Maul- und Klanenseuche bei dem Institut für Infektionskrankheiten in Berlin. *Zentralbl. Bakteriol. Abt. 1 Orig.* **23**:371–391.

Longer-Holm, K., and B. D. Kornat. 1972. Early interaction of rhinovirus with host cells. *J. Virol.* **9**:29–40.
Longley, K. E., B. E. Moore, and C. A. Sorber. 1980. Comparison of chlorine and chlorine dioxide as disinfectants. *J. Water Pollut. Control Fed.* **52**:2098–2105.
Lothrop, T. L., and O. J. Sproul. 1969. High-level inactivation of viruses in wastewater by chlorination. *J. Water Pollut. Control Fed.* **41**:567–575.
Luria, S. E. 1958. The multiplication of viruses. *Handb. Protoplasmaforsch.* **4**:1–58.
Lwoff, A. 1957. The concepts of virus. *J. Gen. Microbiol.* **17**:239–253.
Majumdar, S., W. H. Ceckler, and O. J. Sproul. 1973. Inactivation of poliovirus in water by ozonation. *J. Water Pollut. Control Fed.* **45**:2433–2443.
Moore, B. E., B. P. Sagik, and J. F. Malina. 1975. Viral association with suspended solids. *Water Res.* **9**:197–203.
Morris, E. J. 1972. The practical use of ultraviolet radiation for disinfection purposes. *Med. Lab. Technol.* **29**:41–47.
Mudd, J. B., R. Leavitt, A. Ongun, and T. T. McManus. 1969. Reaction of ozone with amino acids and proteins. *Atmos. Environ.* **3**:669–682.
Murray, R. G. E., S. Pamela, and H. E. Elson. 1965. Location of mucopeptide sections of the cell wall of *E. coli* and other gram-negative bacteria. *Can. J. Microbiol.* **11**:547–553.
Noss, C. I., F. S. Hauchman, and V. P. Olivieri. 1986. Chlorine dioxide reactivity with proteins. *Water Res.* **20**:351–356.
O'Brien, R. T., and J. Newman. 1979. Structural and compositional changes associated with chlorine inactivation of polioviruses. *Appl. Environ. Microbiol.* **38**:1034–1039.
Olivieri, V. P., F. S. Hauchman, C. I. Noss, R. Vasl, M. P. Neeper, and D. O. Cliver. 1982. Mode of action of chlorine dioxide on selected bacterial and enteric viruses, p. 261–288. *In* M. Butler, A. R. Medlin, and R. Morris (ed.), *Proceedings of the International Symposium on Viruses and Disinfection of Water and Wastewater*. University of Surrey Press, Guildford, Surrey, United Kingdom.
Olivieri, V. P., C. W. Kruse, Y. C. Hsu, A. C. Griffiths, and K. Katawa. 1975. The comparative mode of action of chlorine, bromine, and iodine on f2 bacteriophage, p. 145–162. *In* J. D. Johnson (ed.), *Disinfection of Waters and Waste Water*. Ann Arbor Sciences, Ann Arbor, Mich.
Qualls, R. G., and J. D. Johnson. 1985. Modeling and efficiency of ultraviolet disinfection systems. *Water Res.* **19**:1039–1046.
Rav-Acha, C., E. Choshen, A. Serri, and B. Limoni. 1985. The role of formation and reduction of THM and chlorite concentrations in the disinfection of water with Cl_2 and ClO_2. *Environ. Pollut.* (Ser. B) **10**:47–60.
Rekosh, D. M. K. 1977. The molecular biology of picornaviruses, p. 63–110. *In* D. P. Nayak (ed.), *Molecular Biology of Animal Viruses*, vol. 1. Marcel Dekker, Inc., New York.
Richards, G. 1985. Outbreaks of shellfish-associated enteric virus illness in the United States: requisite for development of viral guidelines. *J. Food Protect.* **48**:815–823.
Ridenour, G. M., and R. S. Ingols. 1947. Bactericidal properties of chlorine dioxide. *J. Am. Water Works Assoc.* **39**:561–567.
Riesser, V. M., J. R. Perrich, B. B. Silver, and J. R. McCammon. 1977. Possible mechanisms of poliovirus inactivation by ozone, p. 186–192. *In* Forum on ozone disinfection, Proceedings. International Ozone Institute, Syracuse, N.Y.
Roberts, P. V., E. M. Aieta, J. D. Berg, and B. M. Chow. 1981. *Chlorine Dioxide for Wastewater Disinfection. A Feasibility Evaluation*. U.S. National Technical Information Service, Springfield, Va.
Rodgers, F. G., R. Gurrle, and M. Bulmore. 1982. Effect of faecal material on the inactivation of poliovirus type 1 by ultraviolet irradiation, p. 378–384. *In* M. Butler, A. R. Medlin, and R. Morris (ed.), *Proceedings of the International Symposium on Viruses and Disinfection of Water and Wastewater*. University of Surrey Press, Guildford, Surrey, United Kingdom.
Rodgers, F. G., P. Hufton, E. Kurzawska, C. Molloy, and S. Morgan. 1985. Morphological response of human rotavirus to ultra-violet radiation, heat and disinfectants. *J. Med. Microbiol.* **20**:123–130.

Rogers, S. E., and W. C. Lauer. 1986. Disinfection for potable reuse. *J. Water Pollut. Control Fed.* **58**:193–198.
Roizman, B., M. M. Mayer, and P. R. Roane. 1959. Immuno-chemical studies of poliovirus. IV. Alteration of the immunological specificity of purified poliomyelitis virus by heat and ultraviolet light. *J. Immunol.* **82**:19–25.
Roy, D., E. S. K. Chian, and R. S. Engelbrecht. 1982a. Mathematical model for enterovirus inactivation by ozone. *Water Res.* **16**:667–673.
Roy, D., R. S. Engelbrecht, and E. S. K. Chian. 1982b. Comparative inactivation of six enteroviruses by ozone. *J. Am. Water Works Assoc.* **74**:660–664.
Roy, D., R. S. Engelbrecht, P. K. Y. Wong, and E. S. K. Chian. 1980. Inactivation of enteroviruses by ozone. *Prog. Water Technol.* **12**:819–823.
Roy, D., P. K. Y. Wong, R. S. Engelbrecht, and E. S. K. Chian. 1981. Mechanism of enteroviral inactivation by ozone. *Appl. Environ. Microbiol.* **41**:718–723.
Sawyer, C. N., and P. L. McCarty. 1978. *Chemistry for Environmental Engineering*, p. 391. McGraw-Hill, Inc., New York.
Scarpino, P. V., G. Berg, S. L. Chang, D. Darling, and M. Lucas. 1972. A comparative study of the inactivation of viruses in water by chlorine. *Water Res.* **6**:959–965.
Scarpino, P. V., F. A. O. Brigano, S. Cronier, and M. L. Zink. 1979. Effect of particulates on disinfection of enteroviruses in water by chlorine dioxide, p. 38. Environmental Protection Technology Report (EPA-600/2-79-054). U.S. Environmental Protection Agency, Cincinnati.
Schnoor, J. L., J. L. Nitzschke, R. D. Lucas, and J. N. Veenstra. 1979. Trihalomethane yields as a function of precursor molecular weight. *Environ. Sci. Technol.* **13**:1134–1138.
Scully, F. E., G. D. Howell, R. Kravitz, and J. T. Jewell. 1988. Proteins in natural waters and their relation to the formation of chlorinated organics during water disinfection. *Environ. Sci. Technol.* **22**:537–542.
Severin, B. F., M. T. Suidan, and R. S. Englebrecht. 1983. Kinetic modeling of UV disinfection of water. *Water Res.* **17**:1669–1678.
Sharp, D. G. 1976. Virus particle aggregation and halogen disinfection of water supplies. (EPA-600/2-76-287.) U.S. Environmental Protection Agency, Cincinnati.
Sharp, D. G., R. Floyd, and J. D. Johnson. 1975. Nature of the surviving plaque-forming unit of reovirus in water containing bromine. *Appl. Microbiol.* **29**:94–101.
Sharp, D. G., and J. Leong. 1980. Inactivation of poliovirus I (Brunhilde) single particles by chlorine in water. *Appl. Environ. Microbiol.* **40**:381–385.
Simmon, V. F., and R. G. Tardiff. 1978. The mutagenic activity of halogenated compounds found in chlorinated drinking water, p. 127–146. *In* R. L. Jolley, H. Gorchev, and D. H. Hamilton (ed.), *Water Chlorination: Environmental Impact and Health Effects*, vol. 2. Ann Arbor Science, Ann Arbor, Mich.
Slade, J. S., and B. J. Ford. 1983. Discharge to the environment of viruses in wastewater: sludges and aerosols, p. 4–15. *In* G. Berg (ed.), *Viral Pollution of the Environment*. CRC Press, Inc., Boca Raton, Fla.
Snyder, J. E., and P. W. Chang. 1974. Relative resistance of eight human enteric viruses to ozonation in Saugatucket River water, p. 82–99. Aquatic Applications of Ozone Workshop, Boston, Mass., 19–20 September 1974. International Ozone Institute, Boston, Mass.
Sung, R. D. 1974. Effects of organic constituents in wastewater on the chlorination process. Ph.D. dissertation. University of California, Davis.
Tardiff, R. G., and M. Deinzer. 1973. Toxicity of organic compounds in drinking water. Internal report. U.S. Environmental Protection Agency, Cincinnati.
Taylor, G. R. 1982. The effect of disinfectants on picornavirus structure and infectivity, p. 289–297. *In* M. Butler, A. R. Medlin, and R. Morris (ed.), *Proceedings of the International Symposium on Viruses and Disinfection of Water and Wastewater*. University of Surrey Press, Guildford, Surrey, United Kingdom.
Taylor, G. R., and M. Butler. 1982. A comparison of the virucidal properties of chlorine, chlorine dioxide, bromine chloride and iodine. *J. Hyg.* **89**:321–328.
Tifft, E. C., P. E. Moffa, S. L. Richardson, and R. I. Field. 1977. Enhancement of high rate disinfection by sequential addition of chlorine and chlorine dioxide. *J. Water Pollut. Control Fed.* **49**:1652–1658.

Tittlebaum, M. E., J. L. Pavoni, H. T. Spencer, and M. Fleischman. 1971. *Ozone Disinfection of Viruses.* Institute on Ozonation in Sewage Treatment, University of Wisconsin, Milwaukee.
Tyrrell, D. A. J. 1982. Some aspects of the classification and basic biology of viruses of the gastrointestinal tract, p. 1–12. *In* D. A. J. Tyrrell and A. Z. Kapikian (ed.), *Virus Infections of the Gastrointestinal Tract.* Marcel Dekker, Inc., New York.
Vaughn, J. M., Y. S. Chen, K. Lindburg, and D. Morales. 1987a. Inactivation of human and simian rotaviruses by ozone. *Appl. Environ. Microbiol.* **53**:2218–2221.
Vaughn, J. M., Y. S. Chen, and D. Morales. 1987b. Inactivation of hepatitis A virus by chlorine. Abstr. Annu. Meet. Am. Soc. Microbiol. 1987, Q-21, p. 285.
Vaughn, J. M., Y. S. Chen, J. F. Novotny, and D. Strout. 1990. Effects of ozone treatment of the infectivity of hepatitis A virus. *Can. J. Microbiol.* **36**:557–560.
Vaughn, J. M., Y. S. Chen, and M. Z. Thomas. 1986. Inactivation of human and simian rotaviruses by chlorine. *Appl. Environ. Microbiol.* **51**:391–394.
Vaughn, J. M., and E. F. Landry. 1983. Viruses in soils and groundwater, p. 163–210. *In* G. Berg (ed.), *Viral Pollution of the Environment.* CRC Press, Inc., Boca Raton, Fla.
Vilagines, R. 1982. Inactivation of viruses in effluent by oxides and other oxidizing agents, p. 306–323. *In* M. Butler, A. R. Medlin, and R. Morris (ed.), *Proceedings of the International Symposium on Viruses and Disinfection of Water and Wastewater.* University of Surrey Press, Guildford, Surrey, United Kingdom.
von Brodorotti, H. S., and H. Mahnel. 1982. Comparative studies on susceptibility of viruses to ultraviolet rays. *Zentralbl. Veterinaermed. Reihe B* **29**:129–136.
Wallis, C., and J. L. Melnick. 1967. Virus aggregation as the cause of the non-neutralizable fraction. *Virology* **1**:478–488.
Wang, M. H., L. K. Wang, G. G. Peery, and R. C. M. Cheung. 1978. General theories of chemical disinfection and sterilization of sludge: part 3. *Water Sewage Works* **125**:99–104.
Warriner, R., K. D. Kostenbader, D. O. Cliver, and W.-C. Ku. 1985. Disinfection of advanced wastewater effluent by chlorine, chlorine dioxide and ozone. *Water Res.* **19**:1515–1526.
Weidenkopf, S. J. 1958. Inactivation of type 1 poliomyelitis virus with chlorine. *Virology* **5**:56–67.
White, G. C. 1978. *Disinfection of Wastewater and Water for Reuse.* Van Nostrand Reinhold Co., New York.

Chapter 11

Model of *Giardia lamblia* Inactivation by Free Chlorine

Robert M. Clark

Introduction 242
Background 243
　Problem 243
　Objective 244
　Significance 244
Disinfection Theory 245
Data and Modeling Approach 246
　Infectivity versus In Vitro Findings 247
　Data Assessment 248
　Development of Model 249
Summary and Conclusions 251
References 253

The 1986 amendments to the Safe Drinking Water Act require the U.S. Environmental Protection Agency (EPA) to promulgate primary drinking water regulations (i) specifying criteria under which filtration would be required, (ii) requiring disinfection as a treatment technique for all public water systems, and (iii) establishing maximum contaminant levels or treatment requirements for control of *Giardia lamblia*, viruses, *Legionella* spp., heterotrophic plate count bacteria, and turbidity. EPA has promulgated treatment technique requirements to fulfill the Safe Drinking Water Act requirement for systems using surface waters and groundwaters under the direct influence of surface water (Federal Register, 1989). Additional regulations specifying disinfection requirements for systems using groundwater sources not under the direct influence of surface water will be proposed and promulgated at a later date. This paper presents a model that relates pH, temperature, chlorine concentration, and inactivation level to *Giardia* inactivation

---

Robert M. Clark • Drinking Water Research Division, Risk Reduction Engineering Laboratory, Cincinnati, Ohio 45268.

by free chlorine. Because *G. lamblia* is known to be one of the most resistant organisms to disinfection by chlorine found in water, much interest and effort has been devoted to determination of $C \cdot t$ (the product of disinfectant concentration [milligrams per liter] and disinfectant contact [minutes]) values for this organism.

BACKGROUND

Under the Surface Water Treatment Rule (SWTR) all community and noncommunity public water systems using surface water, or groundwater under the direct influence of surface water, are required to provide minimum disinfection to control *G. lamblia*, enteric viruses, and bacteria (Federal Register, 1989). In addition, unless the source water is well protected and meets certain water quality criteria (total or fecal coliforms and turbidity limits), treatment must also include filtration. The treatment provided, in any case, is required to achieve at least 99.9% removal or inactivation of *G. lamblia* cysts and at least 99.99% removal or inactivation of viruses (i.e., virus of fecal origin and infectious to humans). For unfiltered systems, it must be demonstrated that disinfection alone achieves the minimum performance requirements; this is determined by monitoring disinfectant residual(s), disinfectant contact time(s), pH (if chlorine is used), and water temperature. These data must be applied to determine whether their "$C \cdot t$" value equals or exceeds the $C \cdot t$ values for *G. lamblia* specified in the SWTR (Federal Register, 1989). With the exception of chloramines, when ammonia is added prior to chlorine, the disinfectant $C \cdot t$ values set for *G. lamblia* also are adequate to achieve greater than 99.99% inactivation of viruses. For filtered systems, states are required to specify the level of disinfection for each system to ensure that the overall treatment in the system achieves at least 99.9% and 99.99% removal or inactivation of *G. lamblia* cysts and viruses, respectively (Federal Register, 1989).

In the Guidance Manual to the SWTR, the EPA recommends $C \cdot t$ values for different disinfectants to achieve levels of inactivation for unfiltered systems (EPA, 1989). The percent inactivation that filtered systems should achieve as a function of the filtration technology in place and source water quality conditions is also recommended (EPA, 1989).

Problem

The destruction of pathogens by chlorination is dependent on a number of factors including water temperature, pH, disinfectant contact time, degree of mixing, turbidity, presence of interfering substances, and concentration of chlorine available. The pH has a significant effect on inactivation efficiency because it determines the species of chlorine found in solution, and each species has a different inactivation efficiency.

The impact of temperature on disinfection efficiency is shown by the example that, for Clarke's work in virus destruction by chlorine, contact time had to be increased two to three times when the temperature was lowered 10°C (Clarke et al., 1962). Disinfection by chlorination can inactivate *Giardia* cysts, but only under rigorous conditions. More recently, Hoff et al. (1984) concluded that (i) these cysts are among the most resistant pathogens known, (ii) disinfection at low temperatures is especially difficult, and (iii) treatment processes prior to disinfection are important.

Table 1. $C \cdot t$ values for 99% inactivation of *G. lamblia* cysts by free chlorine[a]

Temp (°C)	pH	Disinfectant concn (mg/liter)	Time (min)	$C \cdot t$	Mean $C \cdot t$	No. of expts
5	6	1.0–8.0	6–47	47–84	65	4
	7	2.0–8.0	7–42	56–152	97	3
	8	2.0–8.0	72–164	72–164	110	3
15	6	2.5–3.0	7	18–21	20	2
	7	2.5–3.0	6–18	18–45	32	2
	8	2.5–3.0	7–21	21–52	37	2
25	6	1.5	<6	<9	<9	1
	7	1.5	<7	<10	<10	1
	8	1.5	<8	<12	<12	1

[a] Data from Jarroll et al. (1981).

Typical $C \cdot t$ values for 99% inactivation of *G. lamblia* by free chlorine at different temperatures and pH values are shown in Table 1. It can be seen that inactivation rates decrease at lower temperatures and at higher pH values, as indicated by the higher $C \cdot t$ values. Jarroll et al. (1981), using in vitro excystation to determine cyst viability, showed that greater than 99.8% of *G. lamblia* cysts could be killed by exposure to 2.5 mg of chlorine per liter for 10 min at 15°C and pH 6, or after 60 min at pH 7 or 8. At 5°C, exposure to 2 mg of chlorine per liter killed at least 99.8% of all cysts at pH 6 and 7 after 60 min (Jarroll et al., 1981). While it required 8 mg/liter to kill the same percentage of cysts at pH 6 and 7 after 10 min, 8 mg/liter was required to inactivate cysts to the same level at pH 8 after 30 min.

Because of the obvious interactions among such water characteristics as concentration of disinfectant, temperature, pH, and time, when evaluating drinking water disinfection processes it is helpful that a model be developed for predicting $C \cdot t$ values under the various conditions that may exist in drinking water systems.

Objective

As indicated, many factors influence *G. lamblia* reaction kinetics. The objective of the study described in this paper is to develop an equation that will relate $C \cdot t$ values for *Giardia* inactivation by chlorine to such factors as pH, temperature, level of inactivation, and chlorine concentration.

Significance

The significance of these efforts relates to the fact that EPA's Office of Drinking Water has adopted the $C \cdot t$ concept to quantify the inactivation of *G. lamblia* by disinfection with free chlorine. Whether or not a utility is forced to install surface water treatment will depend on its ability to meet the $C \cdot t$ values specified by the SWTR. Even if the utility is not required to install filtration, it may have to make significant investments in holding basins and disinfection capacity to meet these

Model of *Giardia* Inactivation by Free Chlorine   245

requirements. Therefore, $C \cdot t$ values established under the SWTR will be extremely important to the drinking water industry, and it is important that the industry understand the basis for the procedures used to estimate these values.

DISINFECTION THEORY

Current disinfection theory is based on the Chick or Chick-Watson model. Chick's Law expresses the rate of destruction of microorganisms based on a first-order chemical reaction (Chick, 1908):

$$\partial N_t/\partial t = -kt \quad (1)$$

which when integrated yields

$$\ln N_t/N_0 = -t \quad (2)$$

where N_t is the number of organisms present at time t in minutes; N_0 is the number of organisms present at time 0; k is a rate constant characteristic of type of disinfectant, microorganism, and water quality aspects of the system; and t is the time in minutes.

Watson, using Chick's data, refined this equation to produce an empirical relation that included changes in the disinfectant concentration (Watson, 1908):

$$\ln (N_t/N_0) = r\, C^n t \quad (3)$$

where C is the concentration of disinfectant in milligrams per liter, r is the coefficient of specific lethality, and n is the coefficient of dilution, or

$$(1/r) \ln (N_t/N_0) = C^n t \quad (4)$$

For a given level of survival such as $N/N_0 = 0.001$ (3-log reduction) the left-hand side of equation 4 is a constant K, or

$$K = C^n t \quad (5)$$

The value K will vary depending on the level of inactivation. This concept can be combined with the previous analysis as follows (Hom, 1972):

$$\log_{10} N_t/N_0 = R\, C^a t^b \mathrm{pH}^d \mathrm{temp}^e \quad (6)$$

in which N_t is the number of organisms at time t in minutes; N_0 is the initial number of organisms at time 0; C is the concentration of disinfectant (chlorine) in milligrams per liter; pH is the pH at which the experiment was conducted; t is time in minutes; temp is the temperature in degrees Celsius; and R, a, b, d, and e are constants.
Letting

$$N_t/N_0 = I \quad (7)$$

Table 2. Characterization of G. lamblia free chlorine inactivation studies used in predictive model[a]

Reference	Cyst source	Viability assay	Comments
Jarroll et al. (1981)	Symptomatic human	Excystation	Conventional survival curves based on multiple samples. Endpoint, ~0.1% survival.
Hibler et al. (1987)	Gerbils; adapted from infected humans (CDS isolate)	Gerbil infectivity (five animals per sample)	No survival curves. Endpoint sought, ~0.01% survival.
Rice et al. (1982)	Symptomatic and nonsymptomatic humans	Excystation	Conventional survival curves based on multiple samples. Endpoint, ~0.1% survival.
Rubin et al. (1989)	Gerbils; adapted from infected humans (several isolates used)	Excystation	Conventional survival curves based on multiple samples. Endpoint, ~0.1% survival.

[a] Data provided by John Hoff, formerly of EPA.

we have

$$\log_{10} I = R\ C^a t^b \text{pH}^d \text{temp}^e \quad (8)$$

or

$$\log_{10}(\log_{10} I) = \log R + a \log C + b \log t + d \log \text{pH} + e \log \text{temp} \quad (9)$$

Equation 9 will be used to estimate parameters for the model to be used in this analysis.

DATA AND MODELING APPROACH

Several data sets are available to be considered as a data base for estimating the parameters in equation 9. These data sets are summarized in Table 2.

The Jarroll et al., Rice et al., and Rubin et al. data are based on excystation techniques. The Hibler et al. data are based on animal infectivity data. Hoff et al. (1984) compared mouse infectivity and excystation for determining the viability of *Giardia muris* cysts exposed to chlorine and reported that both methods yielded similar results. Both the excystation and animal infectivity techniques are described below (Hoff, 1986).

Most *G. lamblia* inactivation experiments have been performed using excystation techniques. Although the approach varies slightly among investigators, the techniques described by Jarroll et al. (1981) are typical of those used by most researchers. Jarroll et al. concentrated *G. lamblia* cysts from human fecal species and stored them in deionized water at 3°C. A hemacytometer was used to determine cyst numbers. A demand-free buffer was prepared and mixed in proportions that yielded solutions of pH 6, 7, and 8 diluted to a 0.01 M concentration. This held the pH constant but was not detrimental to the cysts.

Chlorine stock solutions were made and added to the diluted demand-free buffer. A 250-ml preparation of 0.01 M demand-free chlorinated buffer was added to each of several 800-ml glass beakers with covers. Each beaker was maintained at either 25, 15, or 5°C. A cyst suspension was added to each beaker in an attempt to maintain a final concentration of approximately 650 cysts per ml. Chlorine demand for each cyst preparation was predetermined and additional chlorine was added to compensate for cyst demand. After contact times of 10, 30, or 60 min the chlorine residual was verified for the solution in each beaker, and simultaneously a 47-ml sample of each solution was removed and the chlorine action was stopped by the addition of a thiosulfate compound.

These samples were then centrifuged, and the resulting cyst pellets were exposed to an excystation procedure. Controls were maintained. Experimental protocols involved simultaneously conducting buffer and thiosulfate controls for the specified temperature, pH, and contact time. Disinfection experiments were performed in duplicate or triplicate. At least 1,000 cysts were counted at each trial, and the excystation percentage was determined. Results were presented as percent cyst survival.

The use of in vitro excystation to determine cyst viability has shown that greater than 99.8% of *Giardia* cysts can be killed by exposure to 2.5 mg of chlorine per liter for 10 min at 15°C and pH 6, or after 60 min at pH 7 or 8. At 5°C, exposure to 2 mg of chlorine per liter for 60 min killed at least 99.8% of all cysts at pH 6 and 7 (Jarroll et al., 1981). The same percentage of cysts were killed by 8 mg/liter at pH 6 and 7 after 10 min as were killed by 8 mg/liter at pH 8 after 30 min. $C \cdot t$ values for 99% inactivation of *G. lamblia* by free chlorine at different temperatures and pH values are shown in Table 1. The higher $C \cdot t$ values indicated that inactivation rates decreased at lower temperatures and at higher pH values.

Infectivity versus In Vitro Findings

Most *Giardia* inactivation data have been based on excystation techniques because few *Giardia* cyst inactivation data are available based on the use of animal infectivity as a measure of cyst viability (Hoff, 1986). Some investigators have compared mouse infectivity versus excystation for determining the viability of *G. muris* cysts exposed to chlorine and have reported that both methods gave similar results (Hoff et al., 1984). A recent experiment used Mongolian gerbils to determine the effects of chlorine on *G. lamblia* cysts (Hibler et al., 1987). In a series of experiments, cysts were exposed for various time periods to free chlorine concentrations ranging from 0.4 to 4.2 mg/liter at 0.5, 2.5, and 5.0°C and pH 7, 8, and 9. Each of five gerbils was fed 5×10^4 of the chlorine-exposed cysts and subsequently examined for evidence of infection. Since the test animals had each received a dose of 5×10^4 cysts and since infectivity studies with unchlorinated cysts showed that approximately 5 cysts usually constituted an infective dose, the following assumptions were made depending on the infectivity patterns occurring in the animals receiving chlorine-exposed cysts. If all five animals were infected, then it was assumed that the $C \cdot t$ had produced less than 99.99% inactivation; if no animals were infected, the $C \cdot t$ had produced greater than 99.99% inactivation. It is impossible to determine the exact level of inactivation for these results. If, however, one to four animals were infected, it was assumed that the level of viable cysts was five per animal and that 99.99% of the original cyst population had been inactivated.

Data Assessment

In an earlier study a $C \cdot t$ model was developed based on the Hibler et al. animal infectivity data (Clark et al., 1989). Unfortunately the Hibler et al. data yield $C \cdot t$ values at only one inactivation level.

Although the Hibler et al. data were considered as being essential for inclusion in determining the $C \cdot t$ values under the SWTR, there are some problems associated with these data. The Hibler et al. inactivation data are based on quantal (binary) data for only five animals per dose. This is something of a problem, although some measurement error was accounted for by using interval estimates, i.e., confidence intervals. That is, there may be some fluctuation in $C \cdot t$ levels due to the fact that it is difficult to determine an exact level of inactivation. By looking at a confidence interval for the mean $C \cdot t$ required, however, these fluctuations are largely accounted for. More animals per dose would have narrowed these intervals.

In the previous analysis, results in which 0/5 or 5/5 animals were infected were not included (Clark et al., 1989). This approach is similar to the multiple-tube dilution coliform enumeration method because these extreme values provide inexact results. An estimate of five cysts per animal as a minimum infective dose was used. This is crude, but is accepted within the field.

For development of $C \cdot t$ values for the SWTR, it was decided that data from at least one of the three excystation studies available at the time of rule development should be included in the SWTR. It was also decided that the Hibler data set should be included in all combinations considered because it was the largest data set, it was based on animal infectivity, and it reflected a higher percent inactivation than required under the SWTR. Since the data based on excystation, with the exception of a few data points, only reflected percent inactivation up to 1 log, or less than that required under the SWTR, inclusion of the Hibler data was considered essential for developing a model that could predict disinfection conditions for achieving 99.9% inactivation with minimum uncertainty. However, as mentioned earlier, it was determined that the data from at least one of the excystation studies were essential.

A fundamental question that needed to be addressed was the statistical compatibility of the data sets. An indicator random variable procedure was used to test the compatibility of the data sets using the following equation:

$$t = R \, I^f \, C^g \, \text{pH}^h \, \text{temp}^i \, 10^{jz} \quad (10)$$

in which t is time in minutes; I is the inactivation level; C is the concentration in milligrams per liter; pH is pH units; temp is temperature in degrees Celsius; z is an indicator random variable; and R, f, g, h, i, and j are constants to be determined from regression.

The indicator random variable was defined as follows:

$$\begin{aligned} z &= 0 \text{ if Hibler data} \\ &= 1 \text{ if other data} \end{aligned} \quad (11)$$

Equation 10 can be transformed as follows:

$$\log t = \log R + f \log I + g \log C + h \log \text{pH} + i \log \text{temp} + jz \quad (12)$$

Model of *Giardia* Inactivation by Free Chlorine   249

In equation 12, when $z = 0$, it is defined over the Hibler data set, and

$$t = R\, I^f\, C^g\, \mathrm{pH}^c\, \mathrm{temp}^i \qquad (13)$$

where $z = 1$; then

$$t = (R \cdot 10^j)\, I^f\, C^g\, \mathrm{pH}^h\, \mathrm{temp}^i \qquad (14)$$

Table 3 displays the data set combinations and regression diagnostics. Note that z is the indicator random variable.

As indicated in column 4 of Table 3, residual plots were used to determine constant variance and normality. Fortunately, a strict assumption of normality is not required. As stated by Neter et al. (1985), "Small departures from normality do not create any serious problems. Major departures, on the other hand, should be of concern." Further, Neter et al. write, "Unless the departures from normality are serious, particularly with respect to skewness, the actual confidence coefficients and risks of error will be close to the levels for exact normality." In addition, because of the large sample size, one would expect the Central Limit Theorem to apply.

It was found that 90% of the Hibler-Jarroll data fell within ± 1.64 standard deviations of the mean. In addition, 75% of the data fell within ± 1 standard deviation, which gives support for the normality assumption. For a permanent normal distribution we would expect 68% of the data to lie within ± 1 standard deviation. Similarly, we would expect 90% to lie within ± 1.64 standard deviation of the mean.

The indicator random variable for the intercept variable using the Hibler-Jarroll data base was not significant ($P = 0.3372$). All other data bases considered had a significant indicator random variable at the 0.005 level of significance. A formal test for differences of intercept and/or slope between the Hibler and Jarroll data sets was conducted, and no difference was detected. Thus statistical analysis supports the use of the Hibler-Jarroll data base for extending the model development, and the parameters in equation 13 were estimated using these data, as shown in Table 4 in the log-transformed form (Draper and Smith, 1981).

In Table 4, column 7, entitled "Variance inflation factor," is defined as $1/(1 - R_k^2)$, where R_k^2 is the coefficient of multiple determination when X_k is regressed on the other independent variables in the model. The minimum value of the variance inflation factor is 1 if there is no multicolinearity. As shown in column 7, all of the variance inflation factors are close to 1.

Development of Model

As an extension of the model (equation 13) developed for the SWTR, the Chick-Watson model discussed earlier in equation 6 will be applied to the Hibler-Jarroll data set.

Equation 6 can be reformulated as follows:

$$-1/R \log_{10} (N/N_0)^{1/b}\, C^{-a/b}\, \mathrm{pH}^{-d/b}\, \mathrm{temp}^{e/b} = t \qquad (15)$$

Table 3. Diagnostic results from data set combination analysis

Data sets considered[a]	r^2	Variables	Plots
Hibler, Rice, Jarroll, Rubin	0.6801	Intercept and temp not significant	Nonnormal data, nonconstant variable
Hibler, Rice, Jarroll, Rubin, z	0.7316	Intercept and temp not significant	Nonnormal data, nonconstant variable
Hibler, Rice, Rubin	0.6649	Intercept and temp not significant	Nonnormal data, nonconstant variable
Hibler, Rice, Rubin, z	0.7899	Intercept not significant	Nonnormal data, nonconstant variable
Hibler, Jarroll, Rubin	0.6424	Intercept and temp not significant	Nonnormal data, nonconstant variable
Hibler, Jarroll, Rubin, z	0.6879	Intercept and temp not significant	Nonnormal data, nonconstant variable
Hibler, Rice, Jarroll	0.8619	All variables significant	Nonnormal data, nonconstant variable
Hibler, Rice, Jarroll, z	0.865	All variables significant	Nonnormal data, nonconstant variable
Hibler, Rubin	0.6483	Temp not significant	Nonnormal data, nonconstant variable
Hibler, Rubin, z	0.7593	Intercept not significant	Nonnormal data, constant variable
Hibler, Rice	0.8548	All variables significant	Nonnormal data, constant variable
Hibler, Rice, z	0.8678	All variables significant	Nonnormal data, constant variable
Hibler, Jarroll	0.8452	All variables significant	Nonnormal data, constant variable
Hibler, Jarroll, z	0.8459	z not significant	Nonnormal data, constant variable

[a] Sources: Hibler, Hibler et al. (1987); Rice, Rice et al. (1982); Jarroll, Jarroll et al. (1981); Rubin, Rubin et al. (1989); z, indicator random variable (see the text and equation 10).

Letting $N/N_0 = I$ yields

$$-1/R \, [\log_{10} I]^{1/b} \, C^{-a/b} \, \text{pH}^{-d/b} \, \text{temp}^{-e/b} = t \qquad (16)$$

The parameters in equation 16 can be estimated by taking a log transformation and using linear regression. Parameter estimates for this equation are given in the equation below:

$$t = 0.32(-\log_{10} I)^{1.18} \, C^{-0.86} \, \text{pH}^{2.49} \, (\text{temp})^{-0.17} \qquad (17)$$

Table 4. Parameter estimates for equation 14

Variable	Value	Statistical analysis			Variance inflation factor
		SE	t for HO[a]	$P > 0$[b]	
Log R · 10^j	−0.902	0.200	−4.518	0.0001	0.000
f	−0.268	0.014	−19.420	0.0001	1.183
g	−0.812	0.042	−19.136	0.0001	1.033
h	2.544	0.221	11.535	0.0001	1.032
i	−0.146	0.028	−5.117	0.0001	1.179

[a] Parameter = 0. If the absolute value of this number is 2 or greater, then this variable would be considered significant in terms of its contribution to the overall regression model.
[b] If this value is 0.05 or greater, then the variable would not be considered significant in terms of its contribution to the overall regression model.

The r^2 for equation 17 is 0.87. The corresponding $C \cdot t$ equation is as follows:

$$C \cdot t = 0.32(-\log_{10} I)^{1.18} C^{0.14} pH^{2.49} pH^{-0.17} \qquad (18)$$

and 99% confidence intervals for the parameter estimates for equation 18 based on the Bonferroni method (Belsley et al., 1980) are: R, (0.10, 1.00); $1/b$, (0.03, 1.34); $1 + a/b$, (0.03, 0.25); d/b, (1.93, 3.05); d/c, (−0.24, −0.10).

The effectiveness of equation 18 is illustrated by Fig. 1 which is a plot of actual versus predicted levels of $C \cdot t$.

The partial derivatives for equation 18 with respect to each independent variable are as follows:

$$\partial(C \cdot t)/\partial(-\log_{10} I) = 0.18 \, (-\log_{10} I)^{0.18} C^{0.14} pH^{2.49} temp^{-0.17} \qquad (19)$$

$$\partial(C \cdot t)/\partial C = 0.15 \, (-\log_{10} I)^{1.18} C^{-0.86} pH^{2.49} temp^{-0.17} \qquad (20)$$

$$\partial(C \cdot t)/\partial(pH) = 0.18 \, (-\log_{10} I)^{1.18} C^{0.14} pH^{1.49} temp^{-0.17} \qquad (21)$$

$$\partial(C \cdot t)/\partial(temp) = 0.06 \, (-\log_{10} I)^{1.18} C^{0.14} pH^{2.49} temp^{-1.17} \qquad (22)$$

Table 5 summarizes the derivatives for the log transform of equation 18 for a given independent variable with respect to all the other variables, assuming the derivative of the $C \cdot t$ equation is zero and that all the other variables are constant.

Equation 18 predicts mean $C \cdot t$ values which are below the regulatory targets chosen for use in the SWTR. To derive equivalent values, a 99% confidence interval should be estimated at 4 logs of inactivation and projected to lower levels of inactivation using first-order kinetics.

SUMMARY AND CONCLUSIONS

Amendments to the Safe Drinking Water Act clearly require that all surface water suppliers in the United States filter and/or disinfect to protect the health of

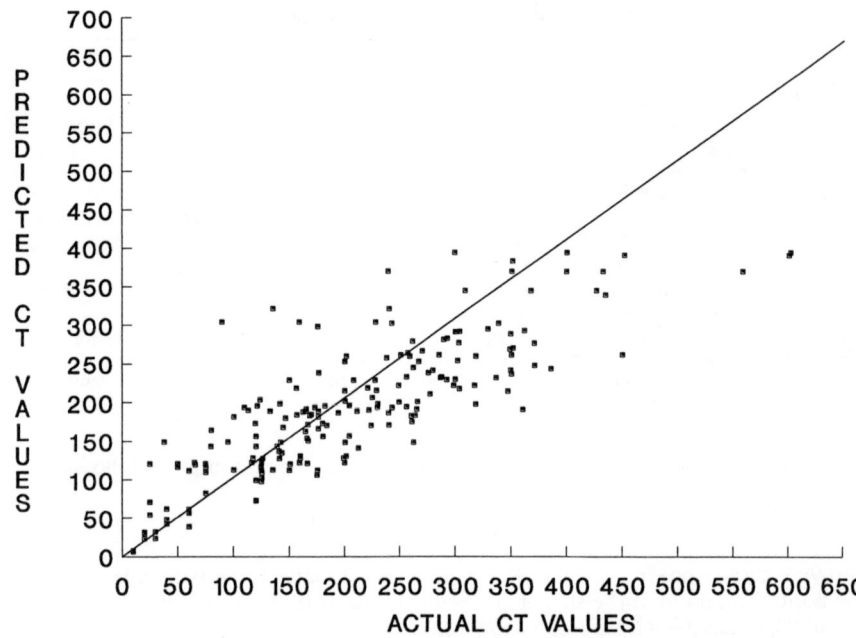

Figure 1. Comparison of actual and predicted $C \cdot t$ values using equation 18.

their customers. *G. lamblia* has been identified as one of the leading causes of waterborne disease outbreaks in the United States. *G. lamblia* cysts are also one of the most resistant organisms to disinfection by free chlorine. EPA's Office of Drinking Water has adopted the $C \cdot t$ concept to quantify microbial inactivation. If a utility can assure that a large enough $C \cdot t$ can be maintained to ensure adequate disinfection, then, depending upon site-specific factors, it may not be required to install filtration. Similarly, the $C \cdot t$ concept can be applied to filtered systems for determining appropriate levels of protection.

In this paper, an equation has been developed that can be used to predict $C \cdot t$ values for the inactivation of *G. lamblia* by free chlorine based on the interaction of disinfectant concentration, temperature, pH, and inactivation level. The parameters for this equation have been derived from a set of animal infectivity data and excystation data. The equation can be used to predict $C \cdot t$ values for achieving 0.5

Table 5. Derivatives of log transform for equation 18

$y =$	$x = dx/dy$			
	$\text{Log}_{10}(-\text{log}_{10}I)$	$\text{Log}_{10} C$	Log_{10} pH	Log_{10} temp
$\text{Log}_{10}(-\text{log}_{10}I)$		-8.383	-0.475	6.793
$\text{Log}_{10} C$	-0.119		-0.475	0.810
Log_{10} pH	-2.107	-17.600		14.310
Log_{10} temp	-0.147	1.234	0.070	

to 4 logs of inactivation within temperature ranges of 0.5 to 5°C, chlorine concentration ranges up to 4 mg/liter, and pH levels of 6 to 8. Although the equation was not based on pH values above 8, the model is still considered applicable to pH levels of 9 for reasons discussed elsewhere. The equation shows the effect of disproportionate increases of $C \cdot t$ versus inactivation levels. With the use of 99% confidence intervals at the 4-log inactivation level and by applying first-order kinetics to these endpoints, a conservative regulatory strategy for defining $C \cdot t$ at various levels of inactivation has been proposed.

I acknowledge Patricia Pierson and Jean Lillie of the Drinking Water Research Division for assistance in preparing this manuscript and Deanna Wild of the Computer Sciences Corporation for helpful technical suggestions.

REFERENCES

Belsley, D. A., E. Kuh, and R. E. Welsch. 1980. *Regression Diagnostics*. John Wiley & Sons, Inc., New York.
Chick, H. 1908. An investigation of the laws of disinfection. *J. Hyg.* **8**:92–158.
Clark, R. M., E. J. Read, and J. C. Hoff. 1989. Analysis of inactivation of *Giardia lamblia* by chlorine. *J. Environ. Eng.* **115**:80–90.
Clarke, N. A., C. Berg, P. W. Kabler, and S. L. Chang. 1962. Human enteric viruses in water: source, survival, and removability. International Conference on Water Pollutions Research, London.
Draper, N., and H. Smith. 1981. *Applied Regression Analysis*, 2nd ed. John Wiley & Sons, Inc., New York.
Federal Register. 1989. National primary drinking water regulations: filtration, disinfection, turbidity, *Giardia lamblia*, viruses, *Legionella*, and heterotrophic bacteria. Final rule, 40 CFR parts 141 and 142. *Fed. Regist.* **54**:27486–27541.
Hibler, C. P., C. M. Hancock, L. M. Perger, J. G. Vegrzyn, and K. D. Swabby. 1987. Inactivation of *Giardia* cysts with chlorine at 0.5°C to 5.0°C. American Water Works Association Research Foundation, Denver, Colo.
Hoff, J. C. 1986. Inactivation of microbial agents by chemical disinfectants. Publication no. EPA/600-2-86-067. U.S. Environmental Protection Agency, Washington, D.C.
Hoff, J. C., E. W. Rice, and F. W. Schaefer III. 1984. Disinfection and the control of waterborne giardiasis. In *Proceedings of the 1984 Specialty Conference*. Environmental Engineering Division, American Society of Chemical Engineers, Los Angeles.
Hom, L. W. 1972. Kinetics of chlorine disinfection in an ecosystem. *J. Sanitary Eng. Div. ASCE* **98**:183–194.
Jarroll, E. L., A. K. Bingham, and E. A. Meyer. 1981. Effect of chlorine on *Giardia lamblia* cyst viability. *Appl. Environ. Microbiol.* **41**:483–487.
Neter, J., W. Wasserman, and M. Kutner. 1985. *Applied Linear Statistical Models*, 2nd ed. Irwin, Homewood, Ill.
Rice, E. W., J. C. Hoff, and F. W. Schaefer III. 1982. Inactivation of *Giardia* cysts by chlorine. *Appl. Environ. Microbiol.* **43**:250–251.
Rubin, A. J., D. P. Evers, C. M. Eyman, and E. L. Jarroll. 1989. Inactivation of gerbil-cultured *Giardia lamblia* cysts by free chlorine. *Appl. Environ. Microbiol.* **55**:2592–2594.
U.S. Environmental Protection Agency. 1989. Draft Guidance Manual for Compliance with the Surface Water Treatment Requirements for Public Water Systems, March 21, 1989. Criteria and Standards Division, Office of Drinking Water, U. S. Environmental Protection Agency, Washington, D.C.
Watson, H. E. 1908. A note on the variation of the rate of disinfection with change in the concentration of the disinfectant. *J. Hyg.* **8**:536–542.

V. BIOFILMS

Chapter 12

Background and Models for Bacterial Biofilm Formation and Function in Water Distribution Systems

Betty H. Olson, Richard McCleary, and James Meeker

Introduction 255
Biofilm Formation 256
  Physicochemical Factors 256
  Biological Factors 270
  Modeling Biofilms...................................... 275
    Difference of Means Tests............................ 278
    Regressions 280
  References... 283

Bacterial biofilm formation on solid surfaces has been studied for over 65 years. The focal points of these studies are extremely diverse because investigators from many scientific disciplines have contributed to the biofilm research literature. For example, those studying energy loss processes (chemical and civil engineers) and those studying cellular adhesion to surfaces (bacterial physiologists and ecologists) have only rarely collaborated. There is a considerable gap in background between these groups of researchers, and consequently, energy losses due to biofilm fouling have not been adequately described in microbiological terms. Likewise, microbial observations of biofilms that appear in the literature are often difficult to interpret quantitatively in terms of actual energy losses.
  The most recent trend in biofilm research has been, therefore, to bridge the gaps in that which has been learned about the physical, chemical, and biological phenomena of biofilm formation and fouling. An overall review of biofilms by engineers and scientists (Characklis and Werner, 1989) indicates that the consolidating opinions of Marshall (1976) and Baier et al. (1968) show that the most important areas of research still involve the understanding and control of microbial

---

Betty H. Olson and Richard McCleary • Environmental Analysis and Design, Program in Social Ecology, University of California, Irvine, California 92717. *James Meeker* • Criminology, Law and Society, Program in Social Ecology, University of California, Irvine, California 92717.

fouling, specifically (i) the effect of different surfaces on macromolecular adsorption; (ii) the reaction between microbial polymers and adsorbed macromolecular films; (iii) the nature of microbial exopolymers (adhesions) that anchor cells to surfaces; and (iv) the properties of colonizing bacteria.

Information gained on biofilm fouling of heat exchangers and other materials can be used to improve our understanding of biofilm formation, the effect of biofilms on water quality, and control of biofilms in water distribution systems. This review covers the physicochemical and biological factors that influence biofilm formation and release and the models that could be used to study biofilms in water distribution systems. The processes that contribute to biofilm formation are, of course, linked to the problematic effects of biofilms. Ways in which biofilms could affect water quality in distribution systems and models to describe potential interactions are discussed.

BIOFILM FORMATION

It has been generally conceded that if bacteria are present in an aqueous environment they invariably become the primary (or initial) colonizers of any wetted solid surface in that environment. Virtually all materials are susceptible to bacterial biofilm formation. Even very complex biofilms, in which a variety of microbial biota may exist, invariably begin as bacterial biofilms. Thus, bacterial biofilm formation is usually the appropriate phenomenon with which to begin any study of biological accumulations onto surfaces; this is especially true for water distribution systems.

The events that occur at solid-liquid interfaces in aqueous systems prior to bacterial colonization determine the in situ characteristics of the surface to which bacteria adsorb. A clear understanding of these events is necessary in order to describe the formation of biofilms.

For the present discussion we will use the following terms: absorbance, to describe adhesion to surfaces; adsorbent, to describe the surface upon which the adhesion occurs; and adsorbate, to mean the molecular or colloidal material that becomes adhered (Daniels, 1972).

Physicochemical Factors

Adsorption of organic molecules onto surfaces

The surface characteristics of an adsorbent are instantaneously modified after exposure to natural aqueous environments. In aqueous systems containing even very low concentrations of organic matter, this modification is caused by the adsorption of an organic film onto surfaces (Baier, 1970; Characklis, 1981). This reorientation of suspended organic matter is really the concentration of molecules at solid-liquid, liquid-liquid, or air-liquid interfaces, and it results in the decrease of the interfacial tension that exists at these interfaces (Israelachvili and McGuiggin, 1988). Thus, there is a loss of free energy (and an increase in entropy) that occurs when organic molecules adsorb to surfaces.

Many natural aqueous environments, e.g., lakes, rivers, and oceans, have very low concentrations of dissolved organic material (usually in the order of grams per liter). Macromolecular adsorption can be observed and measured even in such dilute solutions as these. For example, the formation of an adsorbed organic film

Figure 1. Contact angle (R) of methylene iodide with (A) clean platinum surface (R < 10); (B) organic surface-conditioning film on platinum after immersion in oceanic or estuarine water (R = 30); (C) cell walls of E. coli air dried at 80°C on platinum (R = 64). (Data from Loeb and Neihof, 1975; Baier, 1970; and Dexter, 1979.)

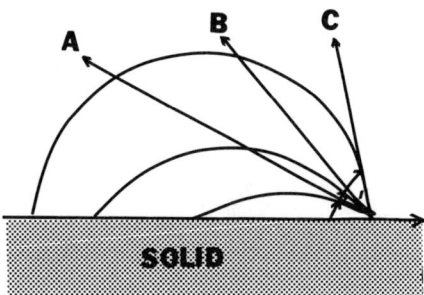

onto platinum plates immersed in Atlantic Ocean and Chesapeake Bay water was observed and measured using ellipsometry, contact angle measurements, and fluorescence spectroscopy (Loeb and Neihof, 1975; Neihof and Loeb, 1973). A film accumulated on these surfaces within a few minutes after immersion; the thickness of the film 1 h after immersion was roughly estimated to be less than about 250 Å (25 nm). Changes in the surface "wettability" were observed by measuring liquid contact angles before and after immersion in oceanic or estuarine waters. Figure 1 shows that the tendency of a test liquid to spread on a surface (wettability) can be described by measuring the angle (R) that a drop makes with the surface. The less wettable the surface, the greater the contact angle. Comparison of contact angles between drops of methylene iodine or water on clean platinum, on platinum plates rinsed in organic nutrient medium, and platinum plates after immersion in oceanic or estuarine waters indicated that hydrophobic film rapidly accumulated on the surfaces. Fluorescence spectra of aqueous solutions and adsorbed films in contact with aqueous media have been used to study such adsorbed films (Loeb and Neihof, 1975; Baier et al., 1968). The organic materials that accumulated on surfaces immersed in oceanic and estuarine waters were very similar to humic substances (Loeb and Neihof, 1975). Humus is the product of complex and poorly understood processes that involve biological compounds excreted from living organisms or decomposed from dead organisms. The complex nature of humic substances makes structural analysis difficult, to say the least. There is evidence that the composition of humus varies greatly depending on the environment from which it has been obtained (Loeb and Neihof, 1975). Generally, humus is anionic, with many carboxyl groups (it is weakly acidic) and 1 to 5% nitrogen (50% of the nitrogen in heterocyclic molecules; the rest of the nitrogen in amino sugars, ammonia, or amino nitrogen; up to 5% of the total nitrogen is in peptide linkages). Fluorescence spectra of adsorbed organic films indicate that proteinaceous materials adhere to solid surfaces exposed to natural waters (Loeb and Neihof, 1975; Baier et al., 1968).

The rapid formation of organic macromolecular films onto a submerged surface (germanium prism) was observed by multiple attenuated internal reflection infrared spectroscopy, and in situ films were qualitatively described (Baier, 1973). The infrared spectra of films that accumulated on surfaces immersed in natural waters indicated that glycoproteinaceous, polysaccharidous, and hydrocarbon molecules spontaneously adsorb. Surface-conditioning films precede biological fouling of inorganic surfaces (Dempsy, 1981; Characklis, 1981). Cellular adhesion depends on the adsorption of macromolecular films (Characklis, 1981). The kinetics of adsorp-

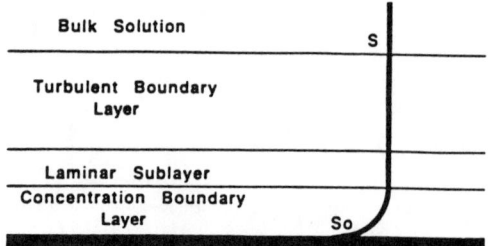

Figure 2. Description of layers that are adjacent to a surface exposed to flow. Adapted from O'Melia (1980) and Characklis and Werner (1989).

tion of macromolecules such as proteins onto hydrophobic surfaces immersed in natural waters has not been adequately studied, but there is evidence that adsorption isotherms probably have the Langmuir-type form (Zisman, 1957; Loeb and Neihof, 1975; Rosenberg and Kjellerberg, 1986).

In nature, then, the attachment of bacterial cells is never to a clean surface, but rather to an adsorbed organic film; the organic film is a surface-conditioning layer, and its formation is actually the first step in the development of bacterial biofilms. Water distribution systems share several of the above measured characteristics. The water is likely to contain biologically available carbon in microgram quantities, and the surface material of the pipe will certainly be preconditioned with a variety of organic and inorganic constituents. The surfaces, except in the case of polyvinylchloride pipe materials, are highly wettable, resulting in a small contact angle.

Transport of microorganisms to surfaces

There are three physical processes that contribute to small particle transport in aquatic environments (Characklis, 1981; O'Melia, 1980). They are the results of thermal effects (Brownian motion; molecular diffusion), flow effects (turbulent or laminar fluid shear), and gravity effects (differential settling; sedimentation).

In moving fluids the effects of fluid shear are usually dominant. For example, in turbulent flow regimes, suspended particles including bacterial cells are transported to solid surfaces primarily by turbulent eddy transport (Stolzenbach, 1989). Velocity differences or gradients occur in all real flowing fluids. Particles following the motion of the suspending fluid will travel at different velocities depending upon these velocity gradients. Particle transport in this case depends on the mean velocity gradient (Spielman, 1977). The flux of particles at the walls of the fluid container increases with particle concentration.

Sedimentation of planktonic (unattached) bacterial cells is probably negligible in the absence of fluid "downsweeps." The buoyant density of bacterial cells is not great enough so that single cells sink in still water. Bacterial cells do sink to lower depths in aquatic environments if they are already associated with particles like phytoplankton, fecal pellets, detritus, or inorganic material (Rheinheimer, 1980). The biological processes that occur because of bacterial attachment and metabolism on particles are dealt with in another section of this review.

The net effect of Brownian diffusion on the movement of bacteria in aquatic systems is usually nil. In water distribution systems (Fig. 2), movement within the boundary layer is by diffusion (O'Melia, 1980). Thermal gradients may, however, contribute to the transport of microbial cells to or away from surfaces (Spielman,

Table 1. Adsorption of bacterial cells to solid surfaces immersed in dilute organic solutions

Phase of adsorption	Mechanism	Net effect	Physicochemical factors
Reversible	Balance between electrostatic double-layer repulsion and London-van der Waals attraction	Loosely adherent cells are a finite distance (0.20 Å) from the adsorbent; cells are easily removed by rinsing	pH electrolyte concentration, organic concentration, fluid motion
Irreversible	Adhesion due to biologically mediated adhesions (e.g., polyanionic carbohydrates)	Tenaciously adherent cells; biologically mediated adhesions bridge the gap between cells and surfaces; cells can be removed by alternation of adhesions or by harsh physical methods (e.g., brushing or scraping)	Time, temperature, surface free energy (q); critical surface tension (q_c), intraction parameters (d); free energy of adhesion (WF)

1977). Molecular diffusion is, of course, much more dependent upon thermal effects than is the movement of bacterial cells. The constant bombardment of ions, molecules, and macromolecules by water molecules is the process by which diffusion occurs. But the chemical environment can also affect bacterial transport to surfaces.

The chemical environment in which a bacterium exists influences the direction of taxis (Chet et al., 1975; Young and Mitchell, 1973a; Bell and Mitchell, 1972). The movement of flagellated bacterial cells is not merely undirected motion. Biofilm formation is influenced by directional taxis (Doetsch and Seymour, 1970; Young and Mitchell, 1973a). Chemicals (e.g., carbohydrates and amino sugars) that elicit positive (attractive) chemotactic responses were shown to enhance the rate of bacterial attachment to artificial surfaces. Conversely, chemicals (e.g., $CHCl_3$ and $CuSO_4$) that cause negative (repulsive) chemotactic responses lead to active avoidance of regions with high concentrations of the chemical. Sublethal concentrations of a wide variety of bacterial toxins including heavy metals and hydrocarbons are known to cause negative chemotaxis (Young and Mitchell, 1973b). The negative chemotactic response of certain bacteria to sublethal concentrations of toxins takes precedence even when higher concentrations of nutrients or other chemicals that usually elicit a positive chemotactic response are present. Negative chemotaxis to sublethal concentrations of toxins was suggested as a means to control marine fouling (Young and Mitchell, 1973b). Mathematical models to describe positive and negative microbial chemotaxis in flowing water have been devised.

Adsorption of bacteria to solid surfaces

The scenario that we endeavor to describe is bacterial adsorption to inert solid surfaces immersed in very dilute organic solutions. Table 1 is offered as the framework upon which we have organized this discussion of physicochemical factors that contribute to a bacterial adsorption. Of course, any model or description of such diverse phenomena as bacterial adsorption must admit certain oversimplifications. For example, we have purposely excluded bacterial adsorption to surfaces other than inert solids. These omissions include bacterial adsorption to charged resins (ion-exchange), water-soluble polyelectrolytes (flocculation), and whole-cell immobilization by chemical sorption (covalent bonding). Although the first two constituents may be found in water distribution systems from filter failure or breakthrough, these topics are beyond the scope of this review. Daniels (1972) has reviewed the literature in these areas.

The process of bacterial adsorption to solid surfaces was first described by ZoBell (1943), and his conclusions are today largely unrefuted. The process of bacterial adsorption to such surfaces is a two-stage mechanism (Table 1). The first step is reversible and is controlled by electrostatic interactions between the cell surface and the adsorbent. The second step forms a much more tenacious bond between adsorbent and adsorbate and is therefore generally referred to as irreversible. This process is the result of the biological activity of the adsorbate (bacterium).

Reversibly adsorbed bacterial cells are easily displaced by gentle rinsing or by fluid shear (Rittman, 1989). Marshall et al. (1971b) observed that reversible adsorption of bacterial cells (*Achromobacter* sp. and *Pseudomonas* sp.) to glass coverslips could be explained in terms of the Derjaguin-Landau and Verwey-Overbeek (DLVO) theory because of the effects of different concentrations of mono- and divalent electrolytes. Simply, the DLVO theory states that the distance of separation between colloidal adsorbents and adsorbates in an electrolyte is that distance at which the repulsive (V_R) and attractive (V_A) forces are balanced. Curtis (1973, 1978) suggested that the repulsive forces (electrostatic double layer) and the attractive forces (principally London dispersion force) are significant in cellular adhesion processes.

The energy of interaction (V_T) between a bacterium and a solid surface is

$$V_T = V_R + V_A \tag{1}$$

The repulsive electrostatic force between lyophobic colloids (used as models for living cells) was described by Derjaguin and Landau in 1941 and Verwey and Overbeek in 1948. The repulsive force is the result of the electrical double layer that exists on any charged surface in an electrolyte solution such that the ionizable groups on that surface tend to attract charged groups of the opposite sign. Bacterial cell surfaces are generally polyanionic, i.e., they are negatively charged (Marshall, 1976). The electrostatic double layer that forms around bacteria in an electrolyte consists of positively charged ions on the outside (the electrolyte) and negatively charged organic compounds (e.g., extracellular carbohydrates) on the inside. Consequently, there is a difference in electrical potential across the electrostatic double layer that is called the surface potential (c). The average c is calculable from electrokinetic measurements because the mobility of a charged particle in an electric field (the electrophoretic mobility) is related to the zeta potential, z. The outer portion of the double layer will shear in an electric field at some radius, leaving the surface with a net charge, z, which is proportional to the electrophoretic mobility, V:

Table 2. Reversibly bound *Achromobacter* sp. strain R8 (Marshall et al., 1971)

Electrolyte concn (M)	Bacterial concn in electrolyte:			
	Univalent (NaCl)		Divalent (MgSO$_4$)	
	$1/K$ (Å)	Bacteria ($\times 10^3$)/mm^2	$1/K$ (Å)	Bacteria ($\times 10^3$)/mm^3
10^{-1}	10	3,500	5	4,100
10^{-2}	31	2,280	15	3,050
10^{-3}	100	950	50	1,920
10^{-4}	310	0	150	810
10^{-5}	1,000	0	500	0

$$z = \frac{4V!n}{e} \quad (2)$$

where e is the dielectric constant and n is the viscosity of the liquid. Under many circumstances the electrokinetic plane of shear should coincide with the surface of the cell so that z and c are equal. However, c may not be equal to z under such conditions as low ionic concentrations or irregular cell surfaces like highly filamented bacteria (Daniels, 1972). Both of these can occur in water distribution systems (Fig. 3).

The surface potential decreases with increasing electrolyte concentration; trivalent ions lower c more than divalent ions, and divalent ions lower c more than monovalent ions. There is experimental evidence relating surface potential and cellular adsorption to surfaces. These data generally show that there is an inverse correlation between adsorption and cell surface potential.

The reason that adsorption and c are related can be explained by the thickness of the electrostatic double layer ($1/K$) across which the potential c exists in aqueous solutions. This is calculable for symmetrical electrolytes at 25°C:

$$1/K = \frac{3.06 \times 10^{-18}}{z\ c} \quad (3)$$

where K is the inverse Debye-Huckel parameter for the diffuse electrical double layer, z is the valency of electrolyte, and c is the molar concentration of electrolyte (Fowler and Mackay, 1980; Marshall, 1984). Table 2 shows the thickness of the double layer ($1/K$ in Ångstroms) at different concentrations of electrolyte and the number of reversibly bound *Achromobacter* sp. strain R8 (Marshall et al., 1971a). The thickness of the electrostatic double layer increases with decreasing electrolyte concentration (as does c). Thus, in high electrolyte concentration, $1/K$ and c are small; the most reversible bacterial adsorption occurs under these circumstances. Marshall et al. (1971b) published the solutions for the following calculations of the energy of interaction between glass and *Pseudomonas* sp. strain R3:

$$V_R = \frac{e}{x_1}\left(\frac{a_1\ a_2}{a_1 + a_2}\right)[(c_1 + c_2)^2 \ln(1 + e^{-KH}) + (c - c_2)^2 \ln(1 - e^{KH})] \quad (4)$$

and

$$V_A = -\frac{A}{x_2} \cdot \frac{a_1 a_2}{a_1 + a_2} \cdot \frac{1}{H} \qquad (5)$$

where a_1 and a_2 are the radii of curvature for the adsorbent and adsorbate, respectively; x_1 and x_2 are the surface potentials of the adsorbent and adsorbate, respectively; H is the distance of closest approach of the adsorbent and adsorbate; c is the dielectric constant of water; A is Hamaker's constant; and K is the inverse Debye-Huckel length. The attractive energy of interaction between bacterium-size colloids and surfaces, V_A, is the London-van der Waals force (Baier, 1973). If the electrostatic values of the dielectric constants of the two surfaces differ from that of the suspending liquid, then an attractive force will exist between the two surfaces and will be inversely proportional to the distance between them.

In conditions of high electrolyte, an attraction also exists, though at some finite distance (about 10 Å) from the adsorbent. The point at which V_R and V_A are balanced (the abscissa equals 0) is the distance at which reversibly adsorbed bacteria are held. The repulsive hump that exists closer to the adsorbent prevents a closer approach by the bacterium. The energy required to exceed this repulsion hump is greater than the motive force of *Pseudomonas* sp. strain R3. It is this gap that must be bridged by extracellular materials in order for bacteria to become irreversibly adsorbed.

In conditions of low electrolyte concentration, no net attraction trough can be demonstrated; there were no reversibly adsorbed bacteria observed experimentally. Figure 4 illustrates reversible bacterial adsorption in a solution of high electrolyte concentration ($\geq 10^{-3}$ M monovalent or $\geq 10^{-4}$ M divalent electrolyte). As can be seen in Table 2, the majority of monovalent and divalent cations are below this level, indicating that once contact occurs between a pipe wall surface and a bacterium, it is likely to be irreversible.

For cells to become firmly (or irreversibly) adherent to surfaces, they must bridge the separation depicted in Fig. 5 (usually on the order of 10 Å or so). Bacterial cells achieve this by producing extracellular adhesions. The surface physics of adhesion are difficult to model. Currently, the predicted strengths of adhesive joints are a good deal greater than the actual adhesive strengths, and apparently little correlation exists between the two (Daniels, 1972), indicating the need to improve models. Nevertheless, refinement of existing models should improve predictions of adhesive strengths when better estimates of inherent flaws that exist in any adhesive joint are considered. These adhesive flaws are the primary source of weakness in macroscopic adhesive bonds. The adhesive joint between a bacterium and a solid surface is probably nearly free of flaws because the bacterium itself is the source of the adhesive and because the area of adsorption (per cell) is very small (Daniels, 1972). Thus the predicted adhesion strengths may very closely approximate the actual adhesive strength of the bond between bacterium and substrate. The process of adhesion is thermodynamically described in terms of surface free energy.

Figure 3. Filamentous biofilm from a freshwater aqueduct. Magnification, ×670 (left); ×6,900 (right).

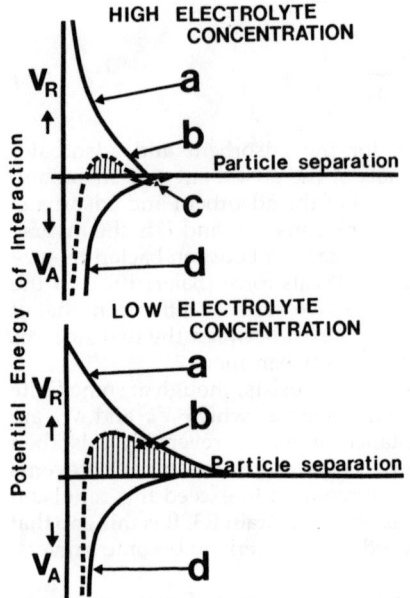

Figure 4. The interaction energy between a particle and substrate in electrolyte solution (Dexter, 1975).

The surface free energy is that energy which is associated with the location of a boundary of a substance where there is a very sharp change in the concentration or in the properties of that substance. This location of the substance may be called its surface. The surface free energy is defined as the energy required to form a new unit area of surface:

$$\frac{\partial G}{\partial A} = q \qquad (6)$$

where G is Gibbs free energy, A is area, and q is surface free energy per unit volume.

An empirical way of looking at q is as follows. Consider an imaginary line, l, drawn through some liquid a distance W_s by a force F. By the definition of q,

$$q = \frac{FW_s}{lW_s} \; (FL^{-1}) \qquad (7)$$

because FW_s is a measure of energy (e.g., force distance in dynes per centimeter equals energy in ergs) and lW_s is the area (i.e., lW_s describes the area after the imaginary line, l, has been drawn through the liquid for the distance W_s). Thus, for liquids, the quantity q is really force per unit length (FL^{-1}), which is called the coefficient of surface tension and can be empirically described as the force required to part a unit length of surface (Daniels, 1972).

The coefficient of surface tension for liquids can be rather easily measured. A common standard method employs the du Nory torsion balance, in which a wettable solid (usually platinum) is drawn through the liquid-air interface and the force required to part a unit length of the interface is measured. The coefficient of surface tension for liquids can also be determined by observing the rise or decline of the liquid in a capillary tube. In the capillary method, the value q is determined from the equilibrium between the upward force of surface tension exerted on the circular ring at the liquid-air interface (equal to q dynes for every unit length of capillary tube; $q2!r$) and the downward force on the cylinder of fluid in the capillary (equal to its volume times its weight per unit volume; $!r^2hdg$). Taken together, this equals

$$q = 1/2\ r\ \cdot\ h\ \cdot\ d\ \cdot\ g (FL^{-1}) \qquad (8)$$

where r is radius of the capillary tube, h is the height that the liquid rises (h is negative if the level of the fluid declines), d is the density of the fluid, and g is the gravitational constant. The dimensions of surface tension are FL^{-1}.

Unfortunately, the surface tension and the surface free energy of a solid are not necessarily equal. Anyway, the sorts of tests used to determine the surface tension of liquids are obviously inappropriate for solids (e.g., the breaking strength of solids involves too many practical considerations to be accurate [Daniels, 1972]). The most successful method of measuring the surface free energy of solids is that of Zisman (1967, 1972). In this method an empirical parameter of wettability called the critical surface tension (q_c) is determined by measuring the contact angle (R; see Fig. 1) between each of a series of droplets, each with known and different surface tensions, and the solid surface. The plot of cosine R for each droplet versus the known surface tension of the droplet is fitted with the best straight line. The intercept of the line with the cosine $R = 1$ axis is at the value for critical surface tension of the substrate because q_c is the coefficient of surface tension at which the solid is completely wetted; i.e., $R = 0$ and cosine $0 = 1$. The value q_c is only an

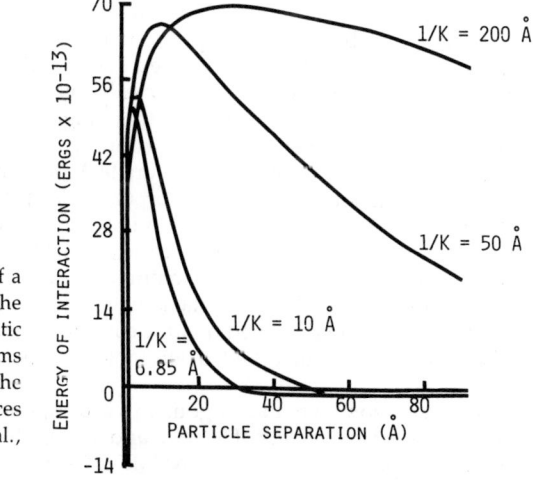

Figure 5. Reversible adsorption of a bacterium to a solid surface. The bacterium is held by electrostatic forces a finite distance in Ångstroms from the solid surface because of the net attraction between the interfaces (Daniels, 1972; Marshall et al., 1971b).

approximation of the solid-surface free energy and may not be equal to it. Usually, q_c is somewhat less than the actual surface free energy of the solid surface because of error from a variety of interactions between the solid and the particular liquid droplets used in the determination (Daniels, 1972). Young's (1805) equation describes the balance of forces that exist for a liquid drop on a surface:

$$q_{sv} = q_{sl} + q_{lv} \cos R \qquad (9)$$

where the subscripts sv, sl, and lv refer to the surface free energies between the solid-vapor, solid-liquid, and liquid-vapor interfaces, respectively. The interfacial tension between a liquid and a solid is represented as

$$q_{sl} = q_{sv}^{1/2} + q_{lv}^{(-2d_{sl})(q_{sv}q_{lv})} \qquad (10)$$

when d is the interaction parameter between the solid and the liquid, and remembering that the interfacial tension and free energy of liquids (q_{lv}) are equivalent. The parameter d_{sl} is dependent upon the molecular properties of the two substances. If the intermolecular forces across the interface of the two phases are the same as they are within both the phases, then $d = 1$. If the intermolecular forces are different between the phases, then $d < 1$; $d > 1$ if there is interaction between the phases (e.g., solubility). Thus the surface free energy of a solid surface (q_{sv}) can be approximated from the wettability (q_c) because $q_{lv} = q_c$ when cosine d = 1 (combine equations 9 and 10):

$$q_{sv} = \frac{q_c}{d_{sl}^2} \qquad (11)$$

The estimate of q_{sv} is therefore dependent upon a good estimate of d_{sl}, which may be obtained from the value of dipole moment, ionization energies, and polarizability of both the liquid and the solid phases (Daniels, 1972).

Investigators have reported that there exists a range in q_c (20 to 30 dyn/cm) for which minimum bacterial adsorption occurs (Baier et al., 1968). This "biocompatibility range" of q_c can be demonstrated in situ (Dexter et al., 1975). Bioincompatible surfaces (i.e., surfaces which are less easily fouled) were colonized by only 1 to 10% as many bacteria as surfaces outside this range for q_c (Fig. 4 in Dexter et al., 1975). These observations were made by counting the attached bacteria (microscopic direct cell counting or biomass estimates) per unit area of solid surface after immersion in laboratory medium or natural aquatic ecosystems. Remembering that, in situ, the first event in biofilm formation is always the spontaneous adsorption of an organic film, we conclude from these experiments that q_c must strongly influence the organic macromolecular conditioning films because it is to this conditioning film that bacteria actually adsorb. The translation of solid surface characteristics to the adsorbed organic conditioning film is probably in terms of the configuration of the adsorbed macromolecules (Caldwell et al., 1981; Costerton 1980).

However, somewhat analogous in vitro experiments have yielded what seem to be contradictory results (Fletcher and Loeb, 1976). Attachment of the marine bacterium *Pseudomonas* sp. strain NCMB-2021 to various surfaces in assays during which the cells were suspended in autoclaved and filtered (0.2-μm porosity) seawater was as follows. The most densely colonized surfaces were hydrophobic (i.e., low wettability, low q_c, and high R); attachment decreased continuously with increasing hydrophilicity. Incidentally, these results support what has been previously stated concerning the inverse relationship between c and bacterial adsorption because hydrophobic surfaces had lower c than hydrophilic, highly charged surfaces. There was a direct relationship between increasing contact angle (i.e., decreasing wettability) and irreversible bacterial adsorption. No biocompatibility range was reported. The contradictory data are shown schematically in Fig. 1 in Dexter (1974).

Dexter et al. (1975) discussed the disparity in attachment versus wettability data and developed a model that appears to unify the results. The data that have been generated (see Fig. 1 in Dexter, 1979) may be summarized as follows. (i) The location of the biocompatibility range of q_c varies according to the liquid medium. In seawater the range is 20 to 25 dyn/cm; in blood the range is 25 to 30 dyn/cm. (ii) The minimum in the attachment curve occurs either in the approximate range 20 to 30 dyn/cm or at one end of the wettability scale. (iii) When solid surfaces are contaminated with water vapor or organic material, the q_c shifts towards the range 20 to 30 dyn/cm from either higher or lower values of q_c (Baier et al., 1968). The unification of these facts has been attempted by considering the interaction parameters (d) and then predicting the shape of the bacterial adsorption attachment-versus-q_c curve.

To understand adsorption as a function of surface wettability, q_c, we must begin by looking at the thermodynamic work of adhesion, w_a. The free energy of adhesion is the difference between the surface free energies of newly formed interfaces and the surface free energies of the interface prior to separation.

$$w_a = q_{sv} + q_{lv} - q_{sl} \tag{12}$$

Or, for the case when an organic conditioning film adsorbs to a surface, the change in free energy for the process at constant temperature, pressure, and volume, neglecting entropy changes, is written in terms of the Helmholtz free energy (WF):

$$WF = q_{so} + q_{ow} - q_{sw} \tag{13}$$

where the subscripts so, ow, and sw refer to the solid-organic, the organic-water, and the solid-water interfaces, respectively (Baier et al., 1968). The value of q_{sw} will be greater than the sum of $q_{so} + q_{ow}$ (i.e., WF will be negative) when the film adsorbs to the surface spontaneously.

It is not possible to calculate the exact value of q_{sv} and q_{lv} in equation 12 because not enough information from experiments with organic surface conditioning films is available at the present time. However, the following assumptions regarding each q term in equation 13 were put forward by Dexter (1979) in order to approximate WF as a function of q_c.

First, the q_{ow} term is probably small because the interface is diffuse and the wettability of the organic film is probably not very different from the "wettability" of water. q_{ow} should not change regardless of the solid's surface wettability because the organic species adsorbed are the same regardless of the solid surface q_c; only the kinetics of adsorption vary with q_c:

$$q_c{}^w = 22 \tag{14}$$

This assumption may not be valid, however, because q_c may affect the configuration of the adsorbed macromolecules (Daniels, 1972). The effect of molecular configuration changes on q_{ow} has not yet been considered.

Second, the q_{so} term is much larger than q_{ow}, less than q_{sw} (because adsorption of organics is spontaneous), and variable in a regular way with q_c. q_{so} probably varies with q_c similarly to the way q_{sw} varies with q_c. This reasoning follows because the interaction between the solid and water is probably not unlike the interaction between the solid and the organic film, since conditioning films are up to 98% water.

Third, the q_{sw} can be closely approximated because

$$q_{sw} = q_{sv} + q_{wl} - 2d\,(q_{sv}\,q_{wl})^{1/2} \tag{15}$$

The terms q_{sv} and q_{wl} can be calculated as shown:

$$q_{sv} = \frac{q_c}{d_{sl}{}^2} \tag{16}$$

$$q_{wl} = \frac{q_c{}^w}{d_{wl}{}^2} \tag{17}$$

so that

$$q_{sw} = \frac{q_c}{d_{sl}{}^2} + \frac{q_c{}^w}{d_{wl}{}^2} - 2d_{sw} \frac{q_c}{d_{sl}{}^2} \frac{q_c{}^w}{d_{wl}{}^2}\,1/2 \tag{18}$$

when the interaction parameters between the solid-liquid and water-liquid interfaces respectively are known. The wettability of pure water against nonpolar organic liquids is known ($d_{sl}{}^2$; $d_{wl}{}^2$) (Daniels, 1972; Dexter, 1974). Thus the interaction ($q_c{}^w$) parameter w_{wl} is equal to the square root of the quotient, which is the wettability of water divided by its surface tension ($q_c{}^w/q_{wl}$), or

$$\frac{22}{72} = 0.55 \tag{19}$$

When the q_{sw}-versus-wettability curves for various values of $d_{wl}{}^2$ and $d_{sl}{}^2$ are plotted they closely resemble the different types of bacterial adsorption-versus-wettability curves that exist in the literature (Dexter, 1979).

If the assumptions apply as outlined above, then the WF versus wettability must be the mirror image of the q_{sw}-versus-wettability curve as shown in Fig. 1 of Dexter (1979). The experimental evidence of bacterial adsorption requires the conclusion that the organic conditioning film adsorbs to the surface first. Therefore, bacterial adsorption versus q_c probably resembles the q_{sw}-versus-q_c curve.

On the basis of this model, the three summarizing facts about bacterial adsorption versus q_c data are unified as follows.

(i) The location of the minimum in the bacterial adsorption-versus-q_c curve is at the same place as the minimum in the q_{sw}-versus-q_c curve; this minimum is clearly a function of d_{sl} and d_{wl}. The value for d_{wl} is not known for seawater or blood, but there is no reason to expect that they are equal to 0.55 (d_{wl} for pure water). Accordingly, the minimum point in q_{sw} versus q_c will vary slightly with medium type.

(ii) Changes in the d_{sl} term affect the shape of the q_{sw}-versus-q_c curve when d_{sw} and d_{wl} are constant such that at high values of d_{sl} (>0.75) there is no biocompatibility range (minimum value of q_{sw} is at one end of the wettability scale). Values of d_{sl} less than about 0.6 yield q_{sw}-versus-q_c curves that have minimum q_{sw} values in or near to the biocompatibility range.

(iii) The q_c of surfaces contaminated with water vapor or organics approaches the biocompatibility range because q_c^w is 22 dyn/cm; presumably the interfacial tension of organic surface conditioning films is similar since they are up to 98% water.

From this discussion it is patently obvious that much more information about macromolecular adsorption to surfaces and also the interaction of microbial adhesions with the macromolecular surface-conditioning films is required in order to successfully model the physicochemical phenomena of bacterial adsorption to inert surfaces.

A significant improvement in the determination of solid surface energies from contact angle measurements and liquid surface energies uses the empirically derived relationship between d and q_{sl}:

$$d = -0.0075\, q_{sl} + 1.00 \qquad (20)$$

so that

$$q_{sl} = \frac{(q_{sv} - q_{lv})^2}{1 - 0.015\, q_{sv}\, q_{lv}} \qquad (21)$$

and with Young's equation,

$$\cos d = \frac{(0.015\, q_{sv} - 2.00)\, q_{sv}\, q_{lv} + q_{lv}}{q_{lv}\,(0.015\, q_{sv}\, q_{lv} - 1)} \qquad (22)$$

It is possible to calculate all of the terms in the free energy balance

$$WF^{adh} = q_{sl} = q_{Bl} - q_{sB} \qquad (23)$$

where the subscripts sl, Bl, and sB are solid-liquid, bacterial-liquid, and solid-bacterial, respectively. The prediction of bacterial adsorption as a function of substrate surface energy is dependent on the relationship of the interfacial tension of the liquid (q_{lv}) and the surface free energy of the bacterial cells (q_{Bv}). It is interesting to note the similarity of angles B and C in Fig. 1. This similarity indicates

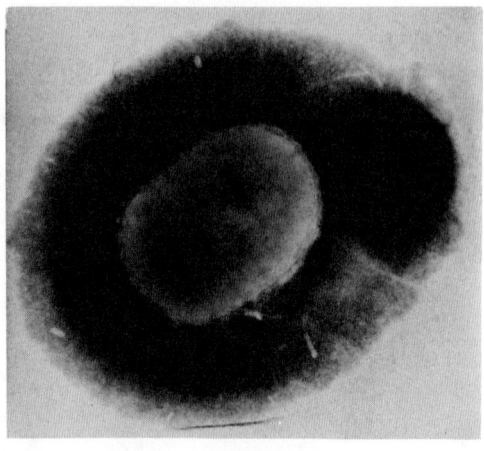

Figure 6. Scanning electron micrograph of appendages used by bacteria to establish contact with pipe surfaces in water distribution systems.

that the biocompatibility range which is observed in in situ bioadhesion studies occurs under the condition where $q_{lv} < q_{Bv}$.

Biological Factors

Bacteria overcome the electrostatic double-layer forces of repulsion that prevent contact with substrates by bridging the distance with polymers or appendages or both (Fig. 6). It is this process that accounts for the tenacious adhesion of sessile bacteria, and it is therefore the biological factors of biofilm formation that contribute most significantly to what often results in biofilm fouling. Recent advances in the field of microbial ultrastructure have marvelously demonstrated the exquisite form and function of the bacterial cell surface polymers and appendages that mediate biofilm formation.

The bacterial glycocalyx

Traditional concepts of bacterial cell surface structure have been formulated largely from observations made in the quite artificial circumstances of liquid pure culture. However, these concepts have now been revolutionized by eliminating the natural selective pressures in pure culture and by considering instead the ultrastructure of natural in situ bacteria. The bacterial exopolymers of adhesion have been demonstrated in an overwhelming majority of microscopic observations of natural sessile populations (Costerton et al., 1981; Costerton et al., 1985; Characklis and Werner, 1989). Figure 7 illustrates the exopolymers observed to be associated with natural sessile bacterial populations; these exopolymers are involved with the adhesion of bacteria to solid surfaces. Hence, the environmental studies of microbial ultrastructure have largely determined the direction that has been taken towards a more accurate portrayal of the outermost regions of the bacterial cell surface.

It is now most apparent that the predominant morphological cell surface feature in bacterial adsorption to inert surfaces is the glycocalyx (Allison and Sutherland, 1987). By analogy to the fuzzy carbohydrate-rich glycocalyx that

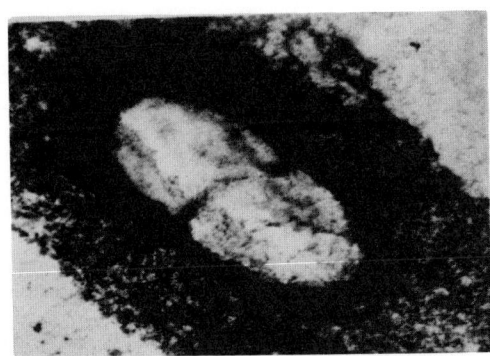

Figure 7. Example of exopolymers produced by bacteria in water distribution systems.

surrounds certain animal cells, Costerton et al. (1981) stated that the bacterial glycocalyx is any polysaccharide-containing bacterial surface structure that is distal to the surface of the outer membrane of gram-negative bacteria (the lipopolysaccharide-containing layer) or to the surface of the peptidoglycan layer of gram-positive bacteria (Costerton et al., 1981; Costerton et al., 1985). The glycocalyces of bacteria may be categorized by subdivision into two main types: the S layer and the capsule. The S-layer glycocalyx has been observed in a few species (e.g., *Spirillum serpens* and *Bacteroides ruminicoa*) and has been described as a regular two-dimensional array of globular glycoprotein units. Most by far, however, of the adherent bacterial species that have been observed in nature possess a fibrous matrix of polysaccharide which is up to 98% water and has by tradition been called the capsule. Capsules have been further categorized by certain nonexclusive properties: rigid, flexible, integral, and peripheral. Rigid capsules are sufficiently organized so as to resist penetration of particulate dyes like nigrosin or India ink, while flexible capsules are not so endowed. Integral capsules are normally intimately associated with the cell surface such that centrifugation is usually unable to effect dispersion of the exopolysaccharide, while peripheral capsules are not so endowed. This terminology provides a generally useful open-ended series of descriptors that is applicable for most of the known forms of bacterial exopolysaccharides. Accordingly, for example, Costerton et al. have described the exopolysaccharides of *Klebsiella aerogenes* and *Pseudomonas aeruginosa* as rigid integral and flexible peripheral capsular glycocalyces, respectively (Costerton et al., 1981; Costerton et al., 1985).

Some sort of capsular glycocalyx can be observed around nearly all of the cells in natural biofilms (Characklis and Werner, 1989). As has been stated, many glycocalyces are entirely permeable to particulate dyes, and therefore they are usually not visible in the light microscope. Light microscopy has been used, however, in concert with capsule-specific immunoglobulin proteins, to render the very hydrated flexible glycocalyces more refractile; this occurs because of the additional density of the surface structure due to the antibody-polysaccharide complex. This antibody stabilization technique is known as the quellung reaction and was used as early as 1896 (Costerton et al., 1981). The elaborate organization and extent of capsular glycocalyces were not realized before the application of specific electron-dense stains for transmission electron microscopy of anionic polymers like mucins (Costerton et al., 1981). Antibody stabilization with capsule-specific immunoglobulins or lectins, followed by treatment with ruthenium red (an electron-dense, anionic-specific stain), is usually required to preserve and visualize the capsular glycocalyces of bacteria because fixation and dehydration procedures for transmission electron microscopy drastically collapse the very hydrated surface organelle (Costerton et al., 1978).

The capsules of gram-positive bacteria like *Bacillus subtilis* are composed of teichoic acids; the teichoic acids consist of a medley of polymers including polyol phosphates (glycerol phosphate or ribitol phosphate) and carbohydrates like glucose, galactose, N-acetylglucosamine, and N-acetylgalactosamine. The capsules of gram-negative bacteria may be composed of a wide variety of homo- and heteropolysaccharides, and some strains produce more than one polymer at the same time.

The function of the bacterial glycocalyx with respect to biofilm formation is as an adhesive or cement. It allows the bacterium to become firmly adherent or "irreversibly adsorbed" on an inert surface and then to multiply if available nutrients are present. Bacterial glycocalyces provide the organism with a myriad of environmental advantages beyond the ability to firmly adsorb to surfaces, including resistance to extremely low nutrient conditions, bacteriophage attack, and the action of antibiotics (Bitton, 1979). The structure and function of the bacterial glycocalyx have been recently reviewed by Costerton and co-workers (Costerton et al., 1978; Costerton et al., 1981; Costerton et al., 1985).

Bacterial adsorption via attachment organelles

It has been suggested that irreversible bacterial adsorption to solid surfaces can occur as the result of the adhesive properties of cell surface organelles or structures. There is actually scant experimental evidence linking biofilm formation on inert solid surfaces to cell surface processes not involving the glycocalyx. Nevertheless, bacterial adhesion to inert solid surfaces via certain other cell processes probably does occur. For example, flagellar adsorption to glass slides may mediate bacterial adsorption. The role of attachment organelles in bacterial adsorption to inert solid surfaces is really not clear with respect to mechanism or extent, however.

Bacterial fimbriae (also commonly referred to as pili) have probably been mistakenly identified in scanning electron micrographs of biofilm bacteria (Duguid, 1959; Dempsy, 1981) because the resolution of scanning electron microscopy is not sufficient to visualize these structures. However, the transmission electron micrographs published by Weiss (1973) demonstrate that bacteria of the genus *Sulfolobus* (chemolithotrophic, acidophilic, thermophilic, gram-negative, lobed cells that use elemental sulfur as an energy source) do indeed adhere to sulfur crystals in hot springs by nonflagellar appendages. The sulfur crystals to which *Sulfolobus* adhere become the substrate for metabolism, and therefore sulfur is most certainly not an inert adsorbent in this case. The specific adhesion of bacteria to many noninert surfaces (e.g., plant and animal tissues) is known to occur via nonflagellar surface appendages.

In 1950 Houwink and van Iterson described surface structures on *Escherichia coli* that are morphologically distinct from flagella; the suggestion that these structures act as adhesins dates to the original observation. Duguid (1959) named the structures fimbriae (= fibers) and described several types according to their morphological characteristics. The most commonly observed fimbria was designated type 1 and is known by that name today. The name pili (= hairs) was suggested by Brinton and Lauffer (1959) to describe several other surface structures, some of which are susceptible to bacteriophage. Ever since these names were given, the distinction between fimbriae and pili has been blurred. The terms are generally considered synonymous, but a distinction according to function is now being adopted (Orskov and Orskov, 1983). Accordingly, there are presently two recognized groups of nonflagellar surface structures on *E. coli*. Group 1 consists of the adhesive factors, and Orskov and Orskov (1983) have proposed that these structures should be called fimbriae; group 2 structures include all of the surface organelles that mediate conjugal chromosomal transfer (i.e., the sex pili) and are typed according to phage susceptibility. Generally, *E. coli* has only one or two sex pili per cell, in contrast to 100 to 400 fimbriae per cell.

There are other bacterial cell surface components that are involved in adsorption to surfaces but have only recently been described. Corpe (1979) has reviewed some of the diverse literature in this area. For example, goblet-shaped proteinaceous subunits from the cell wall of the gliding bacterium *Flexibacter polymorphus* probably act as cell surface "pores" through which some sort of adhesin (perhaps a glycoprotein [H. F. Ridgway, personal communication]) is excreted, thus mediating gliding motility on surfaces. The adhesin is extruded through a duct located in the stem of the goblet-shaped subunit; both of these structures are associated with the outer lipopolysaccharide component of the cell membrane.

Bacteria in certain genera are able to adhere to solid surfaces by means of stalks or holdfast surface processes (Poindexter, 1964). For example, the genus *Caulobacter* is ubiquitous in nature, and particularly in freshwater environments, surfaces are frequently colonized by this organism. In fact, an easy way to observe *Caulobacter* in freshwater habitats is simply to suspend glass microscope slides in the water for several days and then observe the adherent biofilm. *Caulobacter* organisms possess stalks with which they are able to adhere to surfaces, often in regular groups called rosettes. Apparently a polysaccharidous adhesin is secreted from the distal end of the stalk; the cells in rosettes are separated by a common mass of this adhesin, which is located at the center of the rosette (Corpe, 1979). Bacteria with more complex cell cycles and surface structures, like *Caulobacter* and *Hyphomicrobium*, are oligotrophic (i.e., grow slowly at very low nutrient concentrations); they also may require rather complex growth factors. Hence, Corpe and co-workers (Corpe et al., 1975; Corpe, 1979) reasoned that the prosthecate bacteria in biofilms are probably not the initial or primary colonizers of the surface, but rather they are "secondary periphytes"; i.e., they are genera that adhere after a primary biofilm has been established. There is experimental evidence to support this hypothesis (Corpe, 1979). Thus, we have generally assumed that there is a regular (i.e., in some ways predictable) succession of bacterial genera on surfaces during the development of biofilms in nature.

A commonly reported observation is that filamentous bacteria predominate in more fully developed biofilms (Costerton et al., 1985). This phenomenon has been usually explained in terms of a succession of bacterial genera. There is evidence, however, that normally bacillary bacteria that are common initial surface colonizers (e.g., *Pseudomonas* spp) may continue to grow without dividing in biofilms and in some cases may produce very long filaments up to 15 to 20 μm in length. Therefore, the morphological changes in biofilms may be due, in part, to the unique physiological factors that affect adherent bacteria.

Metabolism in bacterial biofilms

The advantages that are realized when a bacterium is immobilized in a biofilm generally result in greater heterotrophic activity relative to the cell's unattached counterparts (Zeikus, 1983; Hamilton and Maxwell, 1987; Characklis and Werner, 1989). Although this fact can be demonstrated in laboratory and in environmental studies, a dispute has continued with regard to its significance in nature, particularly in the marine environment. Azam and Hodson (1977) showed that >90% of the total heterotrophic activity in seawater is apparently due to planktonic (unattached) bacteria; others have similarly come to this conclusion with regard to seawater (Bell and Albright, 1981). However, there is considerable evidence that controverts this conclusion. More recent studies that utilize in situ sediment traps show that the bacteria associated with large sedimenting particles (>35 µm) may contribute as much as 100 times the biomass carbon that is associated with the suspended (i.e., unattached) fraction of seawater. The sessile bacteria on these sedimenting particles were probably not accurately represented in certain earlier studies because water samples taken with Niskin bottles are undoubtedly biased in favor of the suspended fraction, thereby underestimating the number of larger aggregates in the environment. Thus, in view of the overwhelming evidence that biofilm bacteria are more active than their planktonic counterparts, the contribution to carbon flux by unattached bacteria in the marine environment has probably been greatly exaggerated. The situation in freshwater environments is less confusing because most researchers have demonstrated a strong correlation between heterotrophy and attached bacteria in fresh water (Bell and Albright, 1982); there are those who claim that most of the heterotrophy in fresh water is due to the filterable fraction (<1.0-µm pore size).

Disinfectants, reduced sensitivities of bacteria, and biofilms in water distribution systems

The importance of biofilm formation in water distributions is fourfold: (i) the enhanced growth of opportunistic pathogens such as *Legionella* spp. (Stout et al., 1985); (ii) the development of taste and odor problems; (iii) the reduction of flow; and (iv) the regrowth of coliform organisms which can result in violation of the Coliform Rule (Hall and Moyer, 1989). All of these factors can be dramatically influenced according to how biofilms within distribution systems are managed.

The use of disinfectants for biofilm removal in water distribution systems influences both the nature of the biofilm and the bacteria present in the film. The use of chloramines has been shown to be more effective at reducing biofilms in a model pipe system than chlorine (LeChevallier et al., 1988). The primary reason for this is the mode of action; chlorine mainly destroys through oxidation, while chloramines appear to have a mechanism that penetrates into the biofilm and also the cell. Further, different bacteria have different sensitivities to disinfectants. There appear to be several mechanisms of adaption by bacteria to repeated exposure to a disinfectant (LeChevallier et al., 1989), including unidirectional development to resistance, bimodal resistance, and no ability to develop resistance. These characteristics vary among species in a genus as well as between genera. Importantly, they speak to the diversity of bacterial communities and their response to environmental manipulations.

MODELING BIOFILMS

Traditionally, two types of modeling are applied to biofilms. One is built on a mathematical approach based on a deterministic model, and the other is statistical and based on a probabilistic model. In the former, equations are usually developed to describe existing data and a more general model is derived from extrapolations based upon the existing data. In the latter, it is difficult to obtain enough verifiable data for the situation, so a statistical approach is utilized to predict what is happening. Each type of model is used to answer very different questions. The mathematical approaches to both population dynamics and spatial gradients were recently described in a Dahlem Conference Proceedings (Wanner, 1989; Wimpenny et al., 1989).

Statistical models of biofilm phenomena are restricted by the form of the typical data and the nature of the typical problem (Tukey, 1977). For the moment, however, we ignore the nature of the problem and concentrate on the nature of the data. When organism counts are to be analyzed, the dependent variable data will consist of nonnegative integers. That is,

$$Y_i = 0, 1, 2, \ldots, \qquad (24)$$

where Y_i is the number of organisms in the ith sample. Due to the nature of these data, models must be based on discrete probability density functions. When organism concentrations are analyzed, on the other hand, the dependent variable data will consist of nonnegative rational numbers:

$$0 \leq Y_i < \infty \qquad (25)$$

where Y_i is the number of organisms in the ith sample per area or volume. In this case, models can be approximately based on continuous density functions. Left-hand truncation of Y_i complicates the analysis, however, so translating counts into concentrations does not simplify the problem.

In light of the restrictions posed by the dependent variable, a class of discrete models whose parameters are interpreted as organism concentrations is recommended. To develop these models, suppose that data collection is represented as a Bernoulli process where, in any of n independent samples, an organism is discovered with probability p or is not discovered with probability $1 - p$. The probability of discovering exactly k organisms in n samples is given by the binomial density function (B):

$$B(k; n, p) = \frac{n!}{k!(n-k)!} p^k (1-p)^{n-k} \qquad (26)$$

This process is akin to a sequence of n coin-flip experiments where p is the probability of "heads" on any trial. Instead of heads, of course, the event of interest is the discovery of an organism on a surface sample.

Unfortunately, the binomial model has two practical limitations: (i) it assumes that the probability of discovering an organism is the same for each and every sample, and (ii) it assumes that the size (usually area) of a sample is so small that no more than one organism can be discovered in the ith trial. In principle, the first limitation is insurmountable. If samples are drawn from different surfaces, or if they are drawn from a finite single surface but "without replacement," each of the n samples is likely to have a different value of p. In fact, statistical modeling is ordinarily aimed at explaining why p varies from sample to sample.

Ignoring the implausibility of the homogeneity assumption (i), however, assumption (ii) can be satisfied by a simple redefinition of the model. For small p, the binomial density function is approximated by

$$B(k; n, p) = \lambda^{k_e - \lambda}/k! \qquad (27)$$

where $\lambda = np$. [See Feller (1968) for a derivation of the approximation and also a proof that Poisson is the limiting distribution of $B(k; n, p)$.] As p grows smaller and n grows larger—which is to say, as the sample size grows smaller—the approximation improves; the binomial density on the left-hand side approaches the right-hand side as a limit. Readers may recognize the right-hand side as the Poisson density function. Though mathematically similar to the binomial model, this Poisson model is conceptually distinct. We now write the model explicitly as

$$P(k; \lambda) = \lambda^{k_e - \lambda}/k! \qquad (28)$$

The utility of this model is not limited to approximations of the binomial model. On the contrary, there is reason to believe that bacterial formation processes obey Poisson laws (see Pielou, 1977, and Edelstein-Keshet, 1989).

One immediate use of the Poisson model is derivation of a sampling distribution. For that purpose, the parameter λ is interpreted as the concentration of organisms on a surface. If a total of k organisms are counted in n samples, then the concentration of organisms on the source surface is estimated as

$$\hat{\lambda} = k/n \text{ organisms per sample} \qquad (29)$$

This assumes that each of the n samples has the same arbitrary total area. If the n samples vary in area, then the concentration of organisms on the surface is estimated as

$$\hat{\lambda} = k/\sum_{i=1}^{n} tA_i \text{ organisms per unit area} \qquad (30)$$

where A_i is the area of the ith sample expressed in arbitrary units of t, say, $t = 1$ μm^2. The Poisson density is then generalized to

$$P(k; \lambda t) = (\lambda t)^{k_e - \lambda t}/k! \qquad (31)$$

To illustrate, suppose now that the concentration of organisms on a surface is estimated as

$$\hat{\lambda} = 0.001 \text{ organism per } \mu m^2 \qquad (32)$$

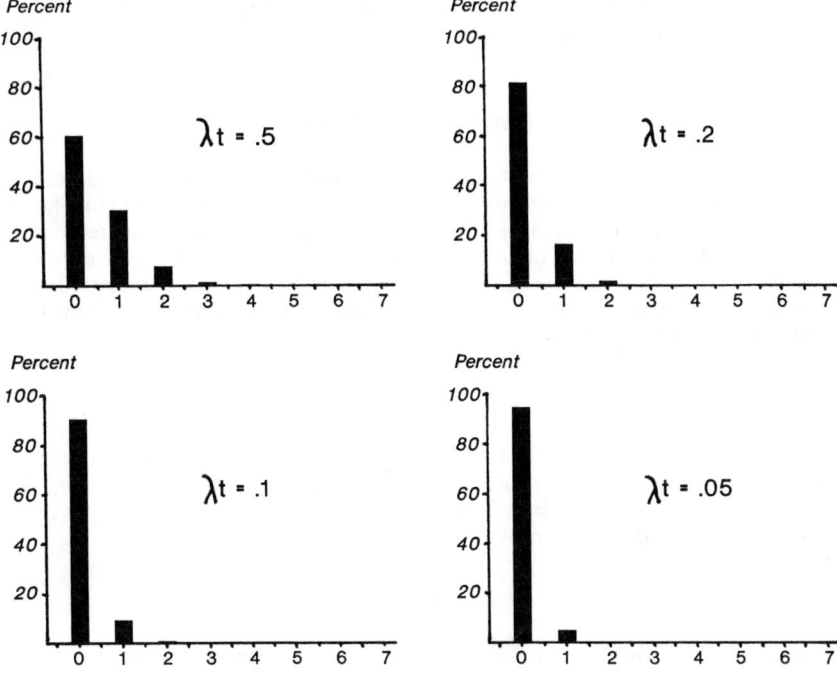

Figure 8. Organism frequency counts for Poisson distributions with various mean parameters.

If n-cm^2 (i.e., $t = 100$) samples are drawn "with replacement" from the surface, by the Poisson density function approximately 90.5% of the samples are expected to have no organisms:

$$P(0;\ \lambda = 0.001,\ t = 100) = [(0.001)(100)]^{0_e - [(0.001)(100)]}/0! \quad (33)$$
$$= [0.1]^{0_e - [0.1]}/0! \simeq 0.905$$

9.05% of the samples are expected to have exactly one organism:

$$P(1;\ \lambda = 0.001,\ t = 100) = [(0.001)(100)]^{1_e - [(0.001)(100)]}/1! \quad (34)$$
$$= [0.1]^{1_e - [0.1]}/1! \simeq 0.0905$$

and 0.46% of the samples are expected to have exactly two organisms:

$$P(2;\ \lambda = 0.001,\ t = 100) = [(0.001)(100)]^{2_e - [(0.001)(100)]}/2! \quad (35)$$
$$= [0.1]^{2_e - [0.1]}/2! \simeq 0.0046$$

Poisson distributions for $\lambda t = 0.5, 0.2, 0.1$, and 0.05 are shown in Fig. 8. In each case, most of the n samples are expected to have no organisms, which gives the Poisson model the alternative name "rare event model."

Difference of Means Tests

Simple analyses of bacterial counts often center on hypothetical differences between two sets of samples corresponding to qualitatively different circumstances. Before addressing these analyses directly, however, we must ensure that the samples are drawn from a Poisson process. Toward that end, we now define the Poisson process in terms of sample counts. Let Y_i and Y_j be the numbers of organisms discovered in the ith and j^{th} surface samples. Then if Y_i and Y_j are homogeneous and independent, that is, if

$$P(Y_i = k|Y_j) = P(Y_i = k) \tag{36}$$

[in this standard notation, $P(Y)$ is the probability of Y; $P(Y|X)$ is the conditional probability of Y, in effect the probability of Y when X is some specific value], then organisms are Poisson distributed and the probability of finding exactly k organisms in the ith sample is given by

$$P(Y_i = k) = \lambda t^{k_e - \lambda t}/k! \tag{37}$$

In light of its homogeneity property and also to distinguish this simplest case from its more sophisticated relatives, we call this process "homogeneous" Poisson.

One consequence of homogeneity is that the parameter λ is both the mean and variance of the Poisson process, and this property suggests a straightforward "goodness-of-fit" test. If n samples are drawn from a homogeneous Poisson surface, the sample mean and variance,

$$\overline{Y} = 1/n \sum_{i=1}^{n} Y_i \text{ and } s^2 = 1/n \sum_{i=1}^{n} (Y_i - \overline{Y})^2 \tag{38}$$

have the same expected value, namely,

$$E(\overline{Y}) = E(s^2) = \lambda \tag{39}$$

The ratio of sample variance to sample mean, furthermore, is distributed as χ^2 with $n - 1$ degrees of freedom (Fisher et al., 1922). That is, under the Poisson null hypothesis,

$$H_0: Y_i \sim P(k; \lambda t); s^2/\overline{Y} \sim \chi^2 \text{ with } df = n - 1 \tag{40}$$

An examination of the formulae for the sample mean and variance illuminates the basis of this test. In fact, the variance-mean ratio is identical to the Pearson (1900) χ^2 statistic:

$$\chi^2 = \left[1/n \sum_{i=1}^{n} (Y_i - \overline{Y})^2 \right] / \left[1/n \sum_{i=1}^{n} Y_i \right] = \sum_{i=1}^{n} \left[(Y_i - \overline{Y})^2 / \sum_{i=1}^{n} Y_i \right] \quad (41)$$

Readers familiar with the χ^2 statistic will recognize that the variance-mean ratio test is a straightforward goodness-of-fit test for the observed and Poisson-expected frequencies over n samples. Due to its notoriously low power, tests based on χ^2 are not generally recommended. Since the variance-mean ratio is identical to χ^2, furthermore, criticisms of χ^2 tests apply to variance-mean ratio tests as well. As a consequence, we recommend the "likelihood ratio" statistic,

$$L^2 = 2 \sum_{i=1}^{n} Y_i \ln(Y_i/\overline{Y}) \quad (42)$$

Despite the apparent differences in their formulae, χ^2 and L^2 give identical results for large n. (L^2 is attributed to Fisher [1924]. See Fisher [1950] on the problem with χ^2 and on the relationship of the two statistics.) Readers who are familiar with χ^2 but not L^2 may be comforted to learn that the two statistics otherwise have identical uses and interpretations. Under the null hypothesis of "fit,"

$$H_0: E(Y_i) = \lambda \quad (43)$$

L^2 is distributed as χ^2 with $n - 1$ degrees of freedom.

If the value of the L^2 statistic is small, say, $P(L^2) < 0.95$, the Poisson null hypothesis is not rejected; we assume accordingly that the n samples were drawn from a homogeneous Poisson source. If L^2 is large, on the other hand, it is possible that the samples were generated by two Poisson processes. Suppose that we can separate the n samples into two sets of n_1 and n_2:

$$n_1 + n_2 = n \quad (44)$$

This assumes auxiliary knowledge, of course. The n_1 and n_2 of samples may have been from the same surface, for example, but at different times, or if at the same time, the n_1 and n_2 of samples may have been measured by different counting methods or with different instruments. In either case, the question of whether the samples come from the same source is addressed by testing the difference of the two sample means for statistical significance.

Readers may be surprised to learn that this test relies on the same theory and procedures used for testing the difference of two normal variates. The difference of two Poisson variates is not a Poisson variate, of course. (While any sum of Poisson variates [e.g., $Z_i + Z_j$] is itself a Poisson variate, the difference of two Poisson variates [e.g., $Z_i - Z_j$] has a more complicated distribution. See Kendall et al. [1983] for derivations.) Under a null hypothesis that two samples come from the same source, however, the difference of two sample means is distributed as a normal variate. The null hypothesis of no difference in means,

$$H_0: \overline{Y_{n_1}} - \overline{Y_{n_2}} = 0 \quad (45)$$

is thus tested by comparing the difference to a normal distribution with mean and variance of λ. Means and variances of the n_1 and n_2 samples are estimated as usual by

$$\overline{Y_{n_1}} = \sum_{i=1}^{n_1} Y_i \text{ and } s_{n_1}^2 = 1/(n_1 - 1) \sum_{i=1}^{n_1} (Y_i - \overline{Y_{n_1}})^2 \tag{46}$$

$$\overline{Y_{n_2}} = \sum_{j=1}^{n_2} Y_j \text{ and } s_{n_2}^2 = 1/(n_1 - 1) \sum_{j=1}^{n_2} (Y_j - \overline{Y_{n_2}}) \tag{47}$$

The difference $(\overline{Y_{n_1}} - \overline{Y_{n_2}})$ is distributed as a normal variate with zero mean and variance:

$$s^2 = (n_1/n)s_{n_1}^2 + (n_2/n)s_{n_2}^2 \tag{48}$$

So the null hypothesis is thus rejected if the test statistic

$$D = [\sqrt{n}/s][\overline{Y_{n_1}} - \overline{Y_{n_2}}] \tag{49}$$

exceeds the critical region.

Regressions

Until now, we concentrated on model restrictions posed by the form of the typical data and ignored restrictions posed by the nature of the typical problem. The typical problem concerns a hypothetical relationship between an organism concentration and one or more independent variables. We have already encountered a special case of the typical problem in the difference-of-means test. The homogeneity assumption implies that all n samples result from only one process. Rejecting a difference-of-means null hypothesis amounts to hypothesizing two Poisson processes, of course. While homogeneity is implausible in that case, strictly speaking, the analysis can proceed with two distinct homogeneous models. When many processes are involved, however, a fundamentally different model is required.

To accommodate heterogeneity in the Poisson model, we now assume that the ith and j^{th} samples come from different sources and, as a result, that Y_i and Y_j have different expected values. That is,

$$E(Y_i) \neq E(Y_j) \neq \lambda \tag{50}$$

This violates the homogeneity assumption. Let Z_i be some environmental feature that distinguishes the ith and j^{th} samples, however, and let λ_i be the expected value of Y_i conditional on Z_i. That is,

$$E(Y_i|Z_i) = \lambda_i \tag{51}$$

Substituting λ_i for λ in the Poisson density function, the probability of discovering exactly k organisms in the ith sample is now given by

$$P(Y_i = k|Z) = (t\lambda_i)^k e^{-t\lambda_i}/k! \tag{52}$$

Since λ varies with Z, each areal unit is the domain of a different Poisson process and the assumption of homogeneity is relaxed.

To illustrate the simplest case of this model, let Z_i be the binary variable that distinguishes two surface types, type A and type B. In other words,

$$Z_i = 1 \text{ if the } i^{th} \text{ sample is a type A surface} \tag{53}$$
$$= 0 \text{ if the } i^{th} \text{ sample is a type B surface}$$

One could simply test whether the n_1 of type A and n_2 of type B have different means, of course. For reasons that will soon be clear, however, we require a more general approach to the difference-of-means null hypothesis. Toward that end, let λ_i be an exponential function of Z:

$$\lambda_i = e^{\alpha + \beta Z_i} \tag{54}$$

Taking natural logarithms,

$$\ln(\lambda_i) = \alpha + \beta Z_i \tag{55}$$

This is why our model is sometimes called log-linear. In either form, the model's interpretation is straightforward. If Y_i is sampled from a type B surface, $Z_i = 0$ and

$$\lambda_i = e^\alpha \text{ or } \ln(\lambda_i) = \alpha \tag{56}$$

If Y_i is sampled from a type A surface, on the other hand, $Z_i = 1$ and

$$\lambda_i = e^{\alpha + \beta} \text{ or } \ln(\lambda_i) = \alpha + \beta \tag{57}$$

Parameters of the Poisson regression model give the mean concentrations for type A and type B surfaces. A simple algorithm for estimating α and β is not obvious, of course. But for one complication, α and β could be estimated as the mean log-transformed concentrations

$$\hat{\alpha} = a/n_1 \sum_{i=1}^{n_1} \ln(Y_i) \text{ and } \hat{\beta} = 1/n_2 \sum_{j=1}^{n_2} \ln(Y_j) - \hat{\alpha} \tag{58}$$

Since bacterial counts are often zero, however, and since $\ln(0)$ is undefined, these most "obvious" estimates are problematic. Although we cannot give the derivation here (see Lawless [1987] for that), maximum likelihood estimates of α and β are found by the solution of

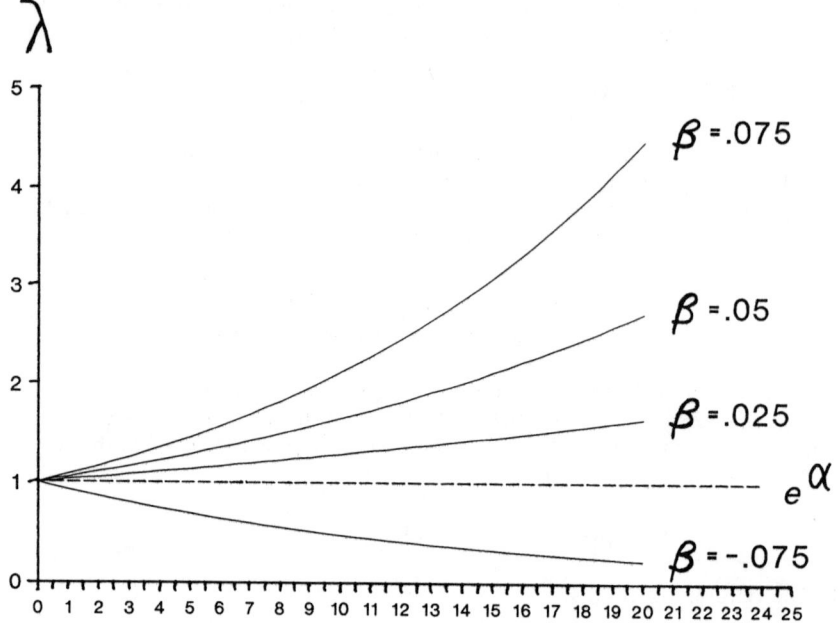

Figure 9. Relationships between expected concentrations and a continuous independent variable for several parameter values.

$$0 = \sum_{i=1}^{n} (Y_i - e^{\alpha + \beta Z_i}) \text{ and } 0 = \sum_{i=1}^{n} (Y_i Z_i - Z_i e^{\alpha + \beta Z_i}) \tag{59}$$

These formulae may be somewhat more complicated than the estimation formulae the reader is used to; the estimates have no closed-form analytic solution, for example. On the other hand, these formulae estimate α and β no matter how the independent variable Z is defined, and this introduces a more realistic, nontrivial problem.

Suppose now that Z_i is not a binary variable but rather has ordinal, interval, or ratio scale properties. For example, Z_i might be the age, temperature, or any other characteristic of the ith sample measured in basic units. In that general case, the Poisson regression model

$$\lambda_i = e^{\alpha + \beta Z_i} \text{ or } \ln(\lambda_i) = \alpha + \beta Z_i \tag{60}$$

gives the expected effect of a unit increase in Z on the expected bacterial concentration. Relationships between Z_i and $E(\lambda_i)$ for several values of β are shown in Fig. 9. The relationships have the same intercept for all values of β. This is because, whenever $Z = 0$,

$$E(\lambda_i) = e^{\alpha} \tag{61}$$

Otherwise, for $\beta > 0$, $E(\lambda_i)$ increases with Z; for $\beta < 0$, $E(\lambda_i)$ decreases with Z. More important, negative bacterial concentrations are not admitted. For $\lambda < 0$, $E(\lambda_i)$ approaches zero as Z approaches infinity. This is an essential property of any bacterial formation model and one that accrues in this case because of the multiplicative form of this Poisson regression model.

REFERENCES

Allison, D. G., and I. W. Sutherland. 1987. The role of exopolysaccharides in adhesion of freshwater bacteria. *J. Gen. Microbiol.* **133**:1319–1327.
Azam, F., and R. E. Hodson. 1977. Size distribution and activity of marine microheterotrophs. *Limnol. Oceanogr.* **22**:492–501.
Baier, R. E. 1970. Surface factors in bioadhesion, p. 15-48. *In* R. S. Manly (ed.), *Adhesion in Biological Systems*. Academic Press, Inc., New York.
Baier, R. E. 1973. Influence of the initial surface condition of materials on bioadhesion, p. 633-639. *In Proceedings of the 3rd International Congress on Marine Corrosion and Fouling*. Northwestern Press, Evanston, Ill.
Baier, R. E., E. G. Sharfrin, and W. A. Zisman. 1968. Adhesion: mechanisms that assist or impede it. *Science* **162**:1360–1368.
Bell, C. R., and L. J. Albright. 1981. Attached and free-floating bacteria in the Fraser River Estuary, British Columbia, Canada. *Mar. Ecol. Prog. Ser.* **6**:317–327.
Bell, C. R., and L. J. Albright. 1982. Attached and free-floating bacteria in a diverse selection of water bodies. *Appl. Environ. Microbiol.* **43**:1227–1237.
Bell, W., and R. Mitchell. 1972. Chemotactic and growth responses of marine bacteria to algal extracellular products. *Biol. Bull.* **143**:265–277.
Bitton, G. 1979. Adsorption of viruses to surfaces: technological and ecological implications, p. 331-373. *In* G. Bitton and K. C. Marshall (ed.), *Adsorption of Microorganisms to Surfaces*. John Wiley & Sons, Inc., New York.
Brinton, C. C., Jr., and M. A. Lauffer. 1959. The electrophoresis of virus, bacteria and cells and the microscopic method of electrophoresis, p. 427-429. *In* M. Bier (ed.), *Electrophoresis—Theory, Methods and Application*. Academic Press, Inc., New York.
Caldwell, D. E., D. K. Brannan, M. E. Morris, and M. R. Betlach. 1981. Quantitation of microbial growth on surfaces. *Microb. Ecol.* **7**:1–11.
Characklis, W. G. 1981. Fouling biofilm development: a process analysis. *Biotechnol. Bioeng.* **23**:1923–1960.
Characklis, W. G., and P. A. Werner. 1989. *Structure and Function of Biofilms*. John Wiley & Sons, Inc., New York.
Chet, I., P. Asketh, and R. Mitchell. 1975. Repulsion of bacteria from marine surfaces. *Appl. Microbiol.* **30**:1043–1045.
Corpe, W. A. 1979. Microbial surface components involved in adsorption of microorganisms onto surfaces, p. 105-144. *In* G. Bitton and K. C. Marshall (ed.), *Adsorption of Microorganisms to Surfaces*. John Wiley & Sons, Inc., New York.
Corpe, W. A., L. Matsuuchi, and B. Ambruster. 1975. Secretion of adhesive polymers and attachment of marine bacteria to surfaces, p. 433-442. *In* J. M. Sharpely and A. M. Kaplan (ed.), *Proceedings of the 3rd International Biodegradation Symposium*. Applied Science Publishers, London.
Costerton, J. W. 1980. Some techniques involved in study of adsorption of microorganisms to surfaces. *In* G. Bitton and K. C. Marshall (ed.), *Adsorption of Microorganisms to Surfaces*. John Willey & Sons, Inc., New York.
Costerton, J. W., R. T. Irwin, and K. J. Cheng. 1981. The bacterial gycocalyx in nature and disease. *Annu. Rev. Microbiol.* **35**:299 324.
Costerton, J. W., T. J. Marrie, and K. J. Cheng. 1985. Phenomena of bacterial adhesion, p. 3-43. *In* D. C. Savage and M. Fletcher (ed.), *Bacterial Adhesion: Mechanisms and Physiological Significance*. Plenum Publishing Corp., New York.
Costerton, W., G. G. Geesey, and K. J. King. 1978. How bacteria stick. *Sci. Am.* **238**:86–97.
Curtis, A. S. G. 1973. Cell adhesion. *Prog. Biophys.* **27**:315–386.

Curtis, A. S. G. 1978. Cell adhesion, p. 1-21. *In Adhesion*, vol. 2. Applied Science Publishers, Essex, England.

Daniels, S. L. 1972. The adsorption of microorganisms onto solid surfaces: a review. *Dev. Ind. Microbiol.* **13**:211-253.

Dempsy, M. J. 1981. Marine bacterial fouling: a scanning electron microscopy study. **61**:305-315.

Derjaguin, B. V., and L. Landau. 1941. Theory of the stability of strongly charged lyophobic soils and of the adhesion of strongly charged particles in solutions of electrolytes. *Acta Physicochim. URSS* **14**:633-662.

Dexter, S. C. 1976. Influence of substrate wettability on the formation of bacterial slime films on solid surfaces immersed in natural sea water, p. 137-144. *In Proceedings of the 4th International Congress on Marine Corrosion and Fouling*. Northwestern Press, Evanston, Ill.

Dexter, S. C. 1979. Influence of substratum contact surface tension on bacterial adhesion in *in situ* studies. *J. Colloid Interface Sci.* **70**:346-353.

Dexter, S. C., J. D. Sullivan, Jr., J. Williams III, and S. W. Watson. 1975. Influence of substrate wettability on the attachment of marine bacteria to various surfaces. *Appl. Microbiol.* **30**:298-308.

Doetsch, R. N., and W. F. K. Seymour. 1970. Negative chemotaxis in bacteria. *Life Sci.* **9**:1029-1037.

Duguid, J. P. 1959. Fimbriae and adhesive properties in Klebsiella strains. *J. Gen. Microbiol.* **21**:271-286.

Edelstein-Keshet, L. 1989. *Mathematical Models in Biology*, p. 80-81. Random House, New York.

Feller, W. 1968. *An Introduction to Probability Theory and Its Applications*, 3rd ed., vol. 1, p. 146-164. John Wiley & Sons, Inc., New York.

Fisher, R. A. 1924. The conditions under which chi square measures the discrepancy between observation and hypothesis. *J. R. Statist. Soc.* **87**:442-450.

Fisher, R. A. 1950. The significance of deviations from expectation in a Poisson series. *Biometrics* **6**:17-24.

Fisher, R. A., H. G. Thornton, and W. A. Mackenzie. 1922. The accuracy of the plating method of estimating the density of bacterial populations with particular reference to the use of Thornton's agar medium with soil samples. *Ann. Appl. Bot.* **9**:325-359.

Fletcher, M., and G. I. Loeb. 1976. The influence of substratum surface properties on the attachment of a marine bacterium. *J. Colloid Interface Sci.* **3**:459-469.

Fowler, H. W., and A. J. Mackay. 1980. The measurement of microbial adhesion, p. 143. *In* R. W. Berkeley, J. M. Lynch, J. Malling, P. R. Rutter, and B. Vincent (ed.), *Microbial Adhesion to Surfaces*. Ellis-Horwood, Chichester, England.

Hall, N. H., and N. P. Moyer. 1989. Evaluation of multiple tube fermentation test and the autoanalysis colilert test for the enumeration of coliforms and Escherichia coli in private well water samples, p. 479-496. *In Technology Conference Proceedings: Advances in Water Analysis and Treatment*.

Hamilton, W. A., and S. Maxwell. 1987. Biofilms: microbial interactions and metabolic activities. *Symp. Soc. Gen. Microbiol.* **41**:361-385.

Houwink, A. L., and W. van Iterson. 1950. Electron microscopical observations on bacterial cytology. II. A study of flagellation. *Biochim. Biophys. Acta* **5**:10-15.

Israelachvili, J. N., and P. M. McGuiggin. 1988. Forces between surfaces in liquids. *Science* **241**:795-800.

Kendall, M. G., A. Stuart, and K. Ord. 1983. *Advanced Theory of Statistics*, 5th ed., vol. 1, p. 127, 366. Griffin, London.

Lawless, J. F. 1987. Regression methods for Poisson process data. *J. Am. Statist. Assoc.* **82**:808-815.

LeChevallier, M. W., C. D. Cawthon, and R. G. Lee. 1988. Inactivation of biofilm bacteria. *Appl. Environ. Microbiol.* **54**:2492-2499.

LeChevallier, M. W., B. H. Olson, and G. A. McFeters. 1989. *Control of Biofilms within Water Distribution Systems*, p. 203. American Water Works Association, Denver, Colo.

Loeb, G. I., and R. A. Neihof. 1975. Marine conditioning films. *Adv. Chem. Ser.* **145**:319-335.

London, F. 1937. The general theory of molecular forces. *Faraday Soc. Trans.* **33**:8-26.

Marshall, K. 1976. *Interfaces in Microbial Ecology*. Harvard University Press, Cambridge, Mass.

Marshall, K., R. Stout, and R. Mitchell. 1971a. Selective sorption of bacteria from seawater. *Can. J. Microbiol.* **17**:1413–1416.
Marshall, K., R. Stout, and R. Mitchell. 1971b. Mechanism of the initial events in the sorption of marine bacteria to surfaces. *J. Gen. Microbiol.* **68**:337–348.
Marshall, K. C. (ed.). 1984. *Microbial Adhesion and Aggregation.* Dahlem Konferenzen. Springer, Berlin.
Neihof, R. A., and G. I. Loeb. 1973. Molecular fouling of surfaces in seawater, p. 710-715. *In* Proceedings of the 3rd International Congress on Marine Corrosion and Fouling. Northwestern Press, Evanston, Ill.
O'Melia, C. R. 1980. Aquasols: the behavior of small particles in aquatic systems. *Environ. Sci. Technol.* **14**:1052–1060.
Orskov, I., and F. Orskov. 1983. Serology of *E. coli* fimbriae. *Prog. Allergy* **33**:80–105.
Pearson, K. 1900. On the criterion that a given system of deviations from the probable in the case of a correlated system of variables is such that it can be reasonably supposed to have arisen from random sampling. *Philosophy Series 5* **50**:157–172.
Pielou, E. C. 1977. *Mathematical Ecology*, p. 135-147. John Wiley & Sons, Inc., New York.
Poindexter, J. S. 1964. Biological properties and classifications of the *Caulobacter* group. *Bacteriol. Rev.* **28**:231–295.
Rheinheimer, G. 1980. *Aquatic Microbiology*, 2nd ed. Wiley Interscience, New York.
Rittman, B. E. 1989. Detachment from biofilms, p. 49-58. *In* W. G. Characklis and P. A. Werner (ed.), *Structure and Function of Biofilms.* John Wiley & Sons, Inc., New York.
Rosenberg, M., and S. Kjellerberg. 1986. Hydrophobic interactions: role of bacterial adhesion. *Adv. Microb. Ecol.* **9**:353–393.
Spielman, L. A. 1977. Particle capture from low speed laminar flows. *Annu. Rev. Fluid Mech.* **9**:297–319.
Stolzenbach, K. D. 1989. Particle transport and attachment, p. 33-48. *In* W. G. Characklis and P. A. Werner (ed.), *Structure and Function of Biofilms.* John Wiley & Sons, Inc., New York.
Stout, J. E., V. L. Yu, and M. G. Best. 1985. Ecology of *Legionella pneumophila* within water distribution systems. *Appl. Environ. Microbiol.* **49**:221–228.
Tukey, J. W. 1977. *Exploratory Data Analysis.* Addison-Wesley, Reading, Mass.
Verwey, E. J. W., and J. T. G. Overbeek. 1948. *Theory of the Stability of Lyophobic Colloids.* Elsevier, Amsterdam.
Wanner, O. 1989. Modeling population dynamics, p. 91-110. *In* W. G. Characklis and P. A. Werner (ed.), *Structure and Function of Biofilms.* John Wiley & Sons, Inc., New York.
Weiss, R. L. 1973. Attachment of bacteria to sulfur in extreme environments. *J. Gen. Microbiol.* **77**:501–507.
Wimpenny, J. W. T., A. Peters, and M. Scourfield. 1989. Modeling spatial gradients, p. 111-128. *In* W. G. Characklis and P. A. Werner (ed.), *Structure and Function of Biofilms.* John Wiley & Sons, Inc., New York.
Young, L. Y., and R. Mitchell. 1973a. The role of chemotactic responses in primary microbial film formation, p. 617-623. *In* Proceedings of the 3rd International Congress on Marine Corrosion and Fouling.
Young, L. Y., and R. Mitchell. 1973b. Negative chemotaxis of marine bacteria to toxic chemicals. *Appl. Microbiol.* **25**:972–975.
Young, T. 1805. An essay on the cohesion of fluids. *Phil. Trans. R. Soc. London* **95**:65–73.
Zeikus, J. G. 1983. Metabolic communication between biodegradation populations in nature. *Symp. Soc. Gen. Microbiol.* **34**:434–462.
Zisman, W. A. 1957. *A Decade of Basic and Applied Science in the Navy*, p. 30. U.S. Government Printing Office, Washington, D.C.
Zisman, W. A. 1967. Contact angle, wettability and adhesion. The Kendall award symposium honoring W. A. Zisman. *Adv. Chem Ser.* **43**.
Zisman, W. A. 1972. Surface energetics of wetting, spreading and adhesion. *J. Paint Technol.* **44**:41–57.
ZoBell, C. E. 1943. The effect of solid surfaces upon bacterial activity. *J. Bacteriol.* **46**:39–56.

Index

Adenovirus, 221, 223
  type 3, 222
Adsorption, 55, 58, 78, 81, 118, 148
  constants, 83
  macromolecular, 269
  of bacteria, 259
Advection, 52, 74, 78
  and dispersion, 84
Advection-dispersion equation, 28, 66, 70, 120, 123
Advection-dispersion model, 58
Aerobiology, 160
Aerodynamic diameter, 162
Aerosol stability, 166
Aerosols
  biological, 161
  infectious, 161
  rehydration, 161
Air entry pressure, 28
Algae, introduction, 3
Antagonism, 10
Appropriate flux, 68
Aquifers
  configuration parameters, 69
  confined, 23
  hydraulic conductivity, 23
  hydraulic parameters, 40
  storage coefficient, 23, 24
  transmissivity, 23
  water table, 23
Arrhenius equation, 233
Astrovirus, 221
Autecology, 5

Bacillus anthracis, 167
Bacillus cereus, 13
Bacillus popilliae, 195
Bacillus sp. strain BCI-INS, 125, 129 131
Bacillus species
Bacillus stearothermophilus, 35
Bacillus subtilis, 125, 167
Bacillus thuringiensis, 13, 192, 195, 196, 198, 203
Bacteria
  adsorption, 260
  aerosols, 167
  attachment, 101
  glycocalyx, 270
  growth, 123
  introduction, 3
Bacteriophage f2, 33, 83, 227, 228
Bacteriophage MS-2, 65, 79, 83, 84, 145, 146, 175
Bacteriophage ϕX174, 174, 227
Bacteriophage S13, 174
Bacteriophage T series, 175
Bacteriophage T2, 145, 146, 222
Bacteriophages
  as tracers, 33, 35
  survival in aerosols, 173–176
Bacteriostasis, 194
Bernoulli process, 275
Bessel function, 72
Binomial model, 275
Biochemical methods, 7
Biocompatibility range, 266
Biodeterioration, 16
Biofilms
  accumulation, 123
  decay rate, 123
  formation, 255
  fouling, 256
  production rate, 123
Biogeochemical activities of bacteria, 2
Biogeochemical cycling, 13–16
Biological aerosols, 161
Biological pesticides, 13
Biomass yield coefficient, 123
Bioremediation, 49
Blackman bilinear model, 199
Boundary condition, 74
Breakthrough curve, 74
Bromine chloride, 227
Brownian diffusion, 162, 258
Brownian motion, 96, 101, 203, 258
Bubble pressure, 28

Calibration of predictive models, 38
Calicivirus, 220
Carbon cycle, 14
Carbon dioxide in biogeochemical cycling, 14
Carrying capacity, 209
Catastrophe model, 181
Catastrophe theory, 181
Caulobacter species, 273
Cell cycle, 202
Central limit theorem, 249

287

Chaos theory in ecology, 211
Charge interactions, 162
Chemical disinfectants, 219–221
Chemotactic movement, 96
Chemotaxis coefficient, 96
Chloramines, 222
Chlorine as disinfectant, 221
Chlorine dioxide as disinfectant, 226
Chromobacterium species, 193
Clostridium acetobutylicum, 18
Coefficient of aquifer storage, 23, 24
Coefficient of biomass yield, 123
Coefficient of chemotaxis, 96
Coefficient of contaminant distribution, 39
Coefficient of dispersion, 120
Coefficient of dispersivity, 39
Coefficient of filtration, 81
Coefficient of retardation, 39, 84
Coefficient of storage, 60
Coefficient of virus adsorption, 60
Colloidal filtration theory, 131
Colloidal transport, 30, 43
Cometabolism, 19
Commensalism, 8
Compatibility of data sets, 140, 141, 248
Competition, 10
Conditioning film, surface, 258, 267
Conductivity in viral inactivation, 150
Confined aquifer, storativity, 23
Contact angle, 258
Contaminant decay, 30
Contaminant distribution coefficient, 39
Continuous transport, 93
Continuum approach, 24
Convection, 74
Covalent bonding, 260
Coxsackievirus, 171, 223
  A9, 229, 231
  B3, 149
  B4, 83
  B5, 224
Critical surface tension, 265
Crop plants, bacterial survival on, 191
Cubic model, 157

Darcy flux, 120
Darcy's Law, 24
Darcy-Richards equation, 28
Debye-Huckel parameter, 261
Decay, 74
Decay or die-off, microbial, 78
Design of models
  experimental, 138
  statistical, 139
Desulfovibrio desulfuricans, 125, 127
Dielectric constant, 261

Die-off rate constants, 80
Diffuse double layer, bacterial, 104
Diffusion, molecular, 29, 84, 258
Diffusional properties, 162
Directional taxis, 259
Discharge flow, 25
Discharge rate, 25
Discontinuous transport, 93
Discrete models, 275
Disinfectants
  chemical, 219–221
  halogen based, 221–228
  non-halogen based, 228–236
Dispersion, 52, 74, 78
  coefficient, 120
  hydraulic, 74
  hydrodynamic, 29, 84
  mechanical, 74
  models, 182
Dispersivity, 86
  coefficient, 39
  distance dependent, 70
  longitudinal, 29
  transverse, 29
Distance-dependent dispersivity, 70
Dormancy, bacterial, 202, 203, 205
Drinking water treatment, 18

Echovirus, 223
  type 1, 83, 224
  type 7, 149
  type 12, 229
Effective porosity, 23, 39
Electrical double layer, 104
Electrical precipitation, 162
Electromagnetic radiation, 164
Electrophoretic mobility, 260
Electrostatic double layer, 260
Encephalomyocarditis virus, 172
Enterobacter aerogenes, 125, 126
Enterobacter cloacae, 151, 153
Enterovirus, 220
Enumeration of microorganisms, 6
Epiphytic bacteria, 193–194
Error function, 69
Erwinia amylovora, 196
Erwinia species, 193
Escherichia coli, 33, 34, 35, 80, 84, 103, 126, 128, 130, 157, 165, 167, 223
  motility mutants, 126
  strain B, 165
Eulerian approach, 24
Exchangeable aluminum, 148
Experimental design for modeling, 138
Expert systems, 201
Exponential decay, 198

Index   289

model, 179
Exponential growth, 198

Fecal coliforms, 80
Fecal streptococci, 80
Fick's Law, 30, 68
Filtration, 58, 78
Filtration coefficient, 81
Flexibacter polymorphus, 273
Flocculation, 260
Flow effects, 258
Foot and mouth disease virus, 172
Francisella tularensis, 168, 195
Freundlich isotherm, 42
Fungi, introduction, 3

Gas hydrate formation, 165
Geochemistry, 44
Geomicrobiology, 2
Germicidal activity of chlorine, 221
Giardia lamblia, inactivation modeling, 242–253
Glycocalyx, bacterial, 270–272
Gravitational constant, 101
Gravitational deposition, 163
Gravitational settling, 162
Gravity effects, 258
Gremmeniella abietina, 200
Groundwater pore velocity, 29
Growth of bacteria, 150
Growth rates, 107
Growth response, 203

Halogen-based disinfectants, 221–228
Hardness, water, 150, 153, 156
Hepatitis A virus (human enterovirus 72), 224, 230
Heterogeneity in the subsurface medium, 26, 56
Heterogeneous (modeling term), 74
Homogeneity of porous medium, 26
Homogeneous (modeling term), 74–75
Human enterovirus 72 (hepatitis A virus), 224, 230
Hydraulic conductivity, 75
Hydraulic dispersion, 74
Hydraulic gradient, 22, 23
Hydraulic head, 23
Hydrodynamic dispersion, 29, 84
Hydrodynamics, 25
Hydrogen cycle, 14
Hyphomicrobium species, 273
Hypobiosis, 194
Hypobiotic, 203

Identification of microorganisms, 5

Immigration, 209
Immobilization, 96
Impaction, 163
  effects, 162
Inactivation rate values, 150
Incompatibility of data sets, 141
Incubation temperature, 156
Inertial properties, 162
Infection threshold, 196
Infectious aerosols, 161
Infective dose, *G. lamblia*, 247
INHIBSIM model, 200
Initial condition, of contaminant, 75
Interactions
  long-range, 103
  of microorganisms, 8
  short-range, 103
Interception
  bacterial, in groundwater, 101
  in aerosols, 162
Interfacial tension, 266
Iodine, as water disinfectant, 227
Ionizing radiation, 164
Iron cycle, 16
Irreversible sorption, 105, 106
Isolation of microorganisms, 5
Isotropy, 26

Kinetic model, 180
Klebsiella oxytoca, 151, 153, 154
Klebsiella pneumoniae, 125, 126, 129, 168
Kriging, 39

Lag period, 203
Lag phase, 205
Lagrangian approach, 24
Landscape ecology, 210
Langmuir-type isotherms, 258
Leaching of metals, 18
Likelihood ratio, 279
Linear regression, 139
  multiple regression, 143
  two-step procedure, 142
Log linear model, 281
Log normality, 210
Log phase, 205
Logistic model, 198
Longitudinal dispersivity, 29
Long-range interactions, bacterial, 103

Macromolecular adsorption, 256, 269
Maillard reactions, 166
Matric pressure, 28
Mechanical dispersion, 74
Mengovirus, 171
Metabolism in biofilms, 273

290   Index

Meteorological conditions, 163
Methods, biochemical, 7
MICROBE-SCREEN model, 200
Microbial activity, 51
Microbial aerosol models, 176
Microbial decay (die-off), 78
Microbial decay rates, 56
Microbial ecology, 1, 5, 202
Microbial enhanced oil recovery, 18
Microbial inactivation, 32
Microbial kinetics, 31
Microbial penetration, 124
Microbial tracer studies, 42
Microbial transport, 78
Microorganisms
  as tracers, 34
  enumeration, 6
  interactions, 8
  isolation and identification, 5
Microspheres, 102
Migration, bacterial, 93
  tactic, 93, 95
  transverse, 95
Mineralization, 14
Minimum effective dose for bacterial infection, 196
Models
  advection-dispersion, 58
  binomial, 275
  Blackman bilinear, 199
  catastrophe, 181
  cubic, 157
  discrete, 275
  dispersion, 182
  exponential decay, 179
  INHIBSIM, 200
  kinetic, 180
  log linear, 281
  logistic, 198
  MICROBE-SCREEN, 200
  microbial aerosols, 176
  Monod equation, 199
  Poisson, 276, 277
  polynomial, 156
  population growth, 198
  potential flow, 61
  Powell diffusional, 199
  quadratic, 156, 157
  SESOIL, 35
  simulation, 199
  stochastic, 43
  Sumatra-1, 36, 43
  virus transport, 58
  Vishnu/Siva, 201
  WORM, 36, 37
Moisture content, soils, 51

Molecular diffusion, 29, 84, 258
Monochloramine, 157
Monod equation model, 199
Monod kinetics, 107, 123
Monte Carlo technique, 43, 204
Mutualism, 9

Neutralism, 8
Nitrogen cycle, 15
Non-halogen disinfectants, 228–236
Normal distribution, 279
Norwalk-like viruses, 220
Nutrient pool, 205

Oil recovery, microbial enhanced, 18
On-site disposal systems, 33
Open air factor, 164
Oxygen cycle, 14–15
Ozone, as disinfectant, 229–234

Parameter estimation, 39
Parasitism, 10
Pathogens
  animal, 12
  human, 11
  plant, 13
Pearson chi-square, 278
Penetration rate, bacterial, 125–128
Permeability, subsurface, 129
Phosphorus cycle, 16
Poisson model, 276–277, 278
Poliovirus, 170, 223, 229, 231, 233, 235
  type 1, 44, 83, 145, 146, 149, 222, 223, 224, 225, 227, 228, 229, 230, 231
  type 2, 229
Polynomial model, 156
Polynomial regression, 156
Population curves, 205
Population growth model, 198
Porcine picornavirus 3, 231
Pore size exclusion, 57
Porosity, 120
  time dependent, 43
Potential flow model, 61
Potentiometric energy, 23
Potentiometric surface, 24
Powell diffusional model, 199
Predation, 11, 40
Prehumidification, of aerosol samples, 161
Pressure
  air entry, 28
  bubble, 28
  matric, 28
Probit function, 197
Protozoa, introduction, 4
Pseudomonas aeruginosa, 98, 194

Index 291

Pseudomonas cepacia, 195, 196, 198, 200, 205
Pseudomonas fluorescens, 151, 153, 191, 196, 198, 200, 205
Pseudomonas putida, 125
Pseudomonas sp. strain NCMB-2021, 266
Pseudomonas sp. strain R3, 261, 263
Pseudomonas species, 193, 273
Pseudomonas syringae, 191, 194, 196, 197, 209, 210

Quadratic model, viral survival, 156–157
Quadratic temperature response curve, 204

Recalcitrant chemical pollution, 18
Regrowth, microbial, effect on studies, 40
Rehydration of aerosols, 161
Relative humidity, 163
Reovirus, 171, 223
Repulsive electrostatic force, 260
Resin-extractable phosphorus, 148
Retardation, 75
Retardation coefficient, 39, 84
Retarded transport, 95
Reversible sorption, 106
Reynolds number, 124
Rotavirus, 171, 220, 224
  SA-11, 83, 224, 226, 230, 234, 235
  type 2 strain Wa, 224, 230, 234

Salmonella species, 80
Salmonella typhimurium, 80
Salt, species and concentration effects, 51
Sampling, 5
Sarcina lutea, 195
Saturated medium, 75
Saturation pH, 148
Sediment traps, 274
Semliki Forest virus, 172
Serratia marcescens, 169, 195
SESOIL model, 35
Settling, 101
Shigella species, 80
Short-range interactions, 103
Simian virus 40, 173
Simulation models, 199
Sink, 75
Soil hydraulic properties, 60
Solar radiation, 205
Solute transport, 22
Sorption, 98
  irreversible, 105, 106
  reversible, 106
Source, 75
Specific yield, 23
Spores, 119, 192, 198

Staphylococcus aureus, 169
Statistical design, 139
Stochastic modeling, 43
Storage coefficient, 60
Storativity, 23
Straining, 97, 118
Streptococcus faecalis, 80
Sulfolobus species, 272
Sulfur cycle, 15
Sumatra-1 model, 36, 43
Surface-conditioning film, 258, 269
Survival curves, 195
Survival, viruses
  in soil, 141–143, 145–149
  in wastewater sludge, 139–141
  in water, 149–156
Survival slope values, viral, 142, 146
Suspended solids, 150
Synecology, 5
Synergism, 9

Tactic migration, 93, 95
Tylor series expansion, 68
Temperature, 155
  aerosols, microbial survival, 164
  viral survival, 155–156
Temperature response curve, 204
Thermal gradients, 258
Thiobacillus denitrificans, 125
Thiobacillus species, 9, 15, 18
Thiobacillus thiooxidans, 6
Time-dependent porosity, 43
Tracer bacteriophage, 33, 35
Tracer microorganisms, 34
Transmissivity of aquifer, 23
Transport
  characteristics, 60
  continuous, 93
  discontinuous, 93
  in porous media, 56
  retarded, 95
Transverse dispersivity, 29
Transverse migration, 95
Trihalomethanes, 228
Turbidity, 150, 154, 155

Ultramicrobacteria, 119
Unsaturated medium, 75
UV light, as disinfectant, 235–236

Vadose zone, 23
Validation, 38
Vapor pressure, 164
Variance inflation factor, 249
Venezuelan equine encephalitis virus, 173
Verification, of predictive models, 38

Viability models, aerosolized microorganisms, 179
Vibrio proteolytica, 98
Viral inactivation rates, 143
Virus adsorption coefficient, 60
Virus association with soil, 51
Virus inactivation (decay) rate, 60
Virus transport models, 58
Viruses
  general properties, 218–219
  introduction, 4
  survival in aerosols, 170–173
Vishnu/Siva model, 201

Waste treatment, 17
Water hardness, 150, 153, 156
Water lattice modification, 165
Water table aquifer, 23
WORM model, 36, 37

Xanthomonas campestris, 196, 197
Xanthomonas species, 193
Xenobiotic compounds, 19

Yellow fever virus, 172
Yersinia pestis, 195